Managing Resource Abundance and Wealth

Managing Resource Abundance and Wealth

The Norwegian Experience

Jonathon W. Moses and Bjørn Letnes

OXFORD
UNIVERSITY PRESS

OXFORD
UNIVERSITY PRESS

Great Clarendon Street, Oxford, OX2 6DP,
United Kingdom

Oxford University Press is a department of the University of Oxford.
It furthers the University's objective of excellence in research, scholarship,
and education by publishing worldwide. Oxford is a registered trade mark of
Oxford University Press in the UK and in certain other countries

First Edition published in 2017

Published in the United States of America by Oxford University Press
198 Madison Avenue, New York, NY 10016, United States of America

British Library Cataloguing in Publication Data
Data available

Library of Congress Control Number: 2016958735

ISBN 978–0–19–878717–4

To our children

Acknowledgments

We are fortunate to work at one of Europe's premiere technical universities (the Norwegian University of Science and Technology, or NTNU), where petroleum engineers are trained to address the enormous technical challenges associated with bringing Norway's offshore oil and gas to market. Several years ago, we began a collaboration with NTNU's Institute for Petroleum Technology to teach engineering students about the social, political, economic, and ethical challenges of petroleum management. We jumped at the opportunity to develop a new course, and the text before you is just one tangible result of our efforts.

Over the years, we have become indebted to our students—both graduate and undergraduate, engineers and social scientists. We are proud to see how the course has functioned as a bridge between the social science and engineering fields, allowing students and staff alike to pass back and forth on fruitful journeys of discovery and growth. Among those exchanges are our annual sojourns to the Department of Chemical and Mining Engineering at the University of Dar es Salaam in Tanzania, where we have enjoyed getting to know our African students in a more intensive (and hence necessarily more intimate) setting. We are grateful to all these students, in both Trondheim and Dar es Salaam, for their very useful input over the years. We have learned a great deal from the exchange.

We are also grateful to the responsible academics in these programs, who have opened up their programs to us, and forgiven our trespasses. More than that, our engineering colleagues have made us feel very welcome indeed. We would have never continued with this effort had it not been for the support and encouragement of Ambrose Itika, Jon Kleppe, and Egil Tjåland. We wish to thank these colleagues for helping us build and maintain this bridge.

As we learned more about the industry, we became indebted to several friendly informants along the way: in the field, within the industry, and among consulting colleagues. We have learned much from those who willingly shared their knowledge with us, albeit often informally. In particular, we would like to express our thanks to Farouk Al-Kasim, Bengt Hope, Erik Jarlsby, Per Kulseth, Marilda Rosado, and Richard Wilfred Rwechungura. The good offices and support of Statoil, Petrad, and the Association of International

Petroleum Negotiators (AIPN) are also appreciated, in providing us access to experts, information, and potential colleagues.

As for the text itself, we thank David Đukanović for his research assistance early on in the project, along with Natalia De Lima Figueiredo and the anonymous reviewers at Oxford for their useful comments on earlier drafts. Michael Alvarez and Omano Edigheji have been frequent sources of support over the years, and diligent readers of earlier drafts. For that we are both very thankful. On a more personal note, Jonathon would like to thank Omano for originally sparking his interest in the subject matter.

Of course, should any errors remain—despite the generous assistance and support from the aforementioned—the responsibility and blame is entirely our own.

Table of Contents

List of Figures

List of Tables

List of Boxes

List of Abbreviations

AF	The Academics' Collective Organization (*Akademikernes Fellesorganisasjon*)
APA	Awards in Predefined Areas (TFO in Norwegian)
API	American Petroleum Institute
AS or A/S	Limited company listed on the Norwegian stock market (*Aksjeselskap*)
ASA	Norwegian public company (*Allmennaksjeselskap*)
ASCM	Agreement on Subsidies and Countervailing Measures
Bboe	billion barrels of oil equivalent
bbl	barrel
BIS	Bank for International Settlements
BIT	Bilateral Investment Treaty
BNOC	British National Oil Company
BP	Beyond Petroleum (formerly British Petroleum)
BSE	Balassa-Samuelson Effect
CA	concessionary agreement
CEO	Corporate Executive Officer
CERES	Coalition for Environmentally Responsible Economies
CFP	Compagnie Française de Pétrole
CJV	contractual joint venture
CSR	corporate social responsibility
DNV	Det Norske Veritas
DSB	Dispute Settlement Body
DSO	The Collaborating Organizations (*de Samarbeidende Organisasjoner*)
E&P	exploration and production
ECHR	European Court of Human Rights
ECSC	European Coal and Steel Community
EEA	European Economic Area
EEC	European Economic Community

EFTA	European Free Trade Association
EHS	Environment, Health, and Safety
EIA	environmental impact assessment
EITI	Extraction Industries Transparency Initiative
ENI	Italian National Hydrocarbons Authority (*Ente Nazionale Idrocarburi*)
ERR	External Reference group for Regulations
ESA	EFTA Surveillance Authority
ESG	Environmental, Social, and Corporate Governance
EU	European Union
FCPA	Foreign Corrupt Practices Act (US)
FDI	foreign direct investment
GATS	General Agreement on Trade in Services
GATT	General Agreement on Tariffs and Trade
GDP	gross domestic product
GEO	Global Economic Order
GFU	Gas Sales Negotiations Committee (*Gassforhandlingsutvalget*)
GNP	gross national product
GPA	Global Procurement Agreement
GPF	Government Petroleum Fund
GPFG	Government Pension Fund, Global
GRI	Global Reporting Initiative
GSP	Generalized System of Preferences
HSE	health, safety, and environment
IBRD	International Bank for Reconstruction and Development
ICJ	International Court of Justice
ICSID	International Center for Settlement of Investment Disputes
IFC	International Financial Corporation
ILO	International Labor Organization
IMF	International Monetary Fund
IMO	International Maritime Organization
IO	international organization
IOCs	international oil companies
IOR	increased oil recovery
IPIECA	Global Oil and Gas Industry Association for Environmental and Social Issues
ISDS	Investor State Dispute Settlement

ISI	Import Substitution Industrialization
ISO	International Standardization Organization
ITO	International Trade Organization
JVs	joint ventures
JVC	joint venture corporation
JVP	joint venture partnership
KP	Kyoto Protocol
LCPs	local content policies
LNG	liquid natural gas
LO	Norwegian Federation of Labor (*Landsorganisasjon*)
LULUCF	Land Use, Land-Use Change and Forestry
MDG	The Norwegian Green Party (*Miljøpartiet De Grønne*)
MJ	millijoule
MMboe	million barrels of oil equivalent
MNCs	multinational corporations
MOSOP	Movement for the Survival of the Ogoni People
MPE	Ministry of Petroleum and Energy
NAF	Norwegian Employers' Association (*Norsk Arbeidsgiverforening*)
NAFTA	North American Free Trade Agreement
NBIM	Norges Bank Investment Management
NCA	Norwegian Coastal Administration
NCS	Norwegian Continental Shelf
NDH	Norwegian Directorate of Health
NEA	Norwegian Environment Agency
NENT	Norwegian National Research Ethics Committee for Natural Science and Technology (*Den nasjonale forskningsetiske komité for naturvitenskap og teknologi*)
NFSA	Norwegian Food Safety Authority
NGL	natural gas liquids
NGOs	non-governmental organizations
NHO	The Confederation of Norwegian Enterprises (*Næringslivets hovedorganisasjon*)
NICs	newly industrialized countries
NIEO	New International Economic Order
NIOC	National Iranian Oil Company
NOAA	National Oceanic and Atmospheric Administration (US)

NOAF	Norwegian Operators Employers' Association (*Norske Operatørselskapers Arbeidsgiverforening*)
NOC	national oil company
NOCO	Norwegian Oil Consortium
NOFO	Norwegian Clean Seas Association for Operating Companies (*Norsk Oljevernforening For Operatørselskap*)
NOK	Norwegian krone
NOPEF	Norwegian Oil and Petrochemical Trade Union (*Norsk Olje- og Petrokjemisk Fagforbundet*)
NORSOK	NCS Competitive Position (*Norsk sokkels konkurranseposisjon*)
NOU	Official Norwegian Reports (*Norges Offentlige Utredninger*)
NPC	Norwegian Petroleum Council
NPD	Norwegian Petroleum Directorate
NRGI	Natural Resource Governance Institute
NRPA	Norwegian Radiation Protection Authority
NSHD	Norwegian Social and Health Directorate
NTNU	Norwegian University of Science and Technology (*Norges teknisk-naturvitenskapelige universitet*)
OBM	obsolescing bargaining mechanism
ODA	Official Development Assistance
OECD	Organisation for Economic Co-operation and Development
OECD GME	OECD Guidelines for Multinational Enterprises
OfD	Oil for Development
OFS	Federation of Oil Workers' Trade Unions (*Oljearbeidernes Fellessammenslutning*)
OGP	International Association of Oil and Gas Producers
OHS	Occupational Health and Safety
OK	Swedish Association of Oil Consumers
OLF	Oil Industry's National Association (*Oljeindustriens Landsforening*)
OPEC	Organization of the Petroleum Exporting Countries
OSC	oilfield service companies
OSH MS	Occupational Safety and Health Management Systems
OSHA	Occupational Safety and Health Administration (US)
PCA	Pollution Control Act
PDO	plan for development and operations
PDQ	production, drilling and quarters
PDVSA	Venezuela's NOC (*Petróleos de Venezuela S.A.*)

PPP	purchasing power parity
PPR	The Petroleum Price Board (*Petroleumsprisrådet*)
PSA	product sharing agreements
PSA(N)	Petroleum Safety Authority (of Norway)
PSC	production sharing contract
PWYP	Publish What You Pay
R&D	research and development
RNNP	Trends in risk level in the petroleum industry (*Risikonivå i norsk petroleumsvirksomhet*)
ROV	remotely operated underwater vehicle
RTA	Regional Trade Agreement
RVK	Compliance in Rules and Regulations for the Petroleum Industry (*Regelverkskompetanse*)
SAFE	Norwegian Union of Energy Workers (*Sammenslutningen av Fagorganiserte i Energisektoren*)
Sm^3	standard cubic meter
SC	service contract
SDFI	state's direct financial interest
SDR	Special Drawing Rights
SDT	special and differential treatment
SfS	Working Together for Safety (*Samarbeid for Sikkerhet*)
SFT	Norwegian Pollution Control Authority (*Statens Forurensningstilsyn*)
SMEs	small and medium-sized enterprises
SRWG	Social Responsibility Working Group
STAR	Stolen Asset Recovery Initiative
SV	Socialist Left Party (*Sosialistisk Venstreparti*)
TBU	Technical Calculation Committee for Wage Settlements (*Det tekniske beregningsutvalget for inntektsoppgjørene*)
TEC	Treaty Establishing the European Community
TFEU	Treaty on the Functioning of the European Union (aka Treaty of Rome)
TFO	Awards in Predefined Areas, see APA (*Tildeling i Forhåndsdefiniterte Områder*)
TRC	Texas Railroad Commission
TRIMs	Trade Related Investment Measures
TRIPs	Trade Related Aspects of Intellectual Property Rights
UDHR	Universal Declaration of Human Rights
UNCBD	UN Convention on Biological Diversity

UNCITRAL	UN Commission on International Trade Law
UNCLOS	UN Convention on the Law of the Sea
UNCTAD	UN Conference on Trade and Development
UNDP	UN Development Program
UNEP	UN Environment Program
UNFCCC	UN Framework Convention on Climate Change
UNGP	UN Guiding Principles on Business and Human Rights
WBCSD	World Business Council for Sustainable Development
WEA	Work Environment Act
WICE	World Industry Council for the Environment
WTO	World Trade Organization
WWI	World War I
WWII	World War II
YS	Confederation of Vocational Unions (*Yrkesorganisasjonenes Sentralforbund*)

1

Introduction

It's like winning the lottery. For decades, international oil companies (IOCs)[1]—desperately seeking to expand their access to proven reserves (and in response to an insatiable demand)—scoured the earth in search of new resources. In their wake, a number of countries discovered that they rested atop pools of unimaginable wealth.

Petroleum reserves are ambivalent about territorial sovereignty. From the perspective of a political scientist, their distribution seems random (at best), or downright antidemocratic (at worst). Since the year 2000, the petroleum industry has discovered 52 giant oil and 68 giant gas fields.[2] These conventional fields are estimated to contain proved plus probable (2P) reserves of 248.62 billion barrels of oil equivalent (Bboe),[3] and are scattered across the globe: 28 of these fields are located in the Middle East; 19 are offshore of Brazil; 13 are offshore of East Africa; 9 are in Central Asia; 8 are in Australia; 6 are offshore of West Africa; and 6 are found in the Gulf of Mexico (Bai and Xu 2014). As can be seen in Figure 1.1, very little (3 percent) of the new giant reserves are located in North America or Europe.

It will not surprise the reader to learn that the largest oil and gas reserves are found in the Middle East. But beyond that region there is still an astounding geographic spread in today's global oil and gas reserves—most of which are well off the beaten path. As the legendary oil-well firefighter, Boots Hansen, once said, "They don't drill them oil wells downtown anywhere. It is always out in the jungle or in the ocean, or in the desert, you know. That's just the way it is" (NRK 2009, III/27:34).

When we study the location of these valuable resources, we find that most of the world's proven (conventional) petroleum reserves lie buried beneath states that are not democratic. Indeed, of the 25 leading petroleum countries (measured in terms of proven gas and oil reserves) listed in Table 1.1, only four of these states are fully democratic (Australia, Canada, Norway, and the United States). Another three states have less developed forms of democratic regimes (Brazil, Indonesia, and Mexico). Thus, if we employ a relatively

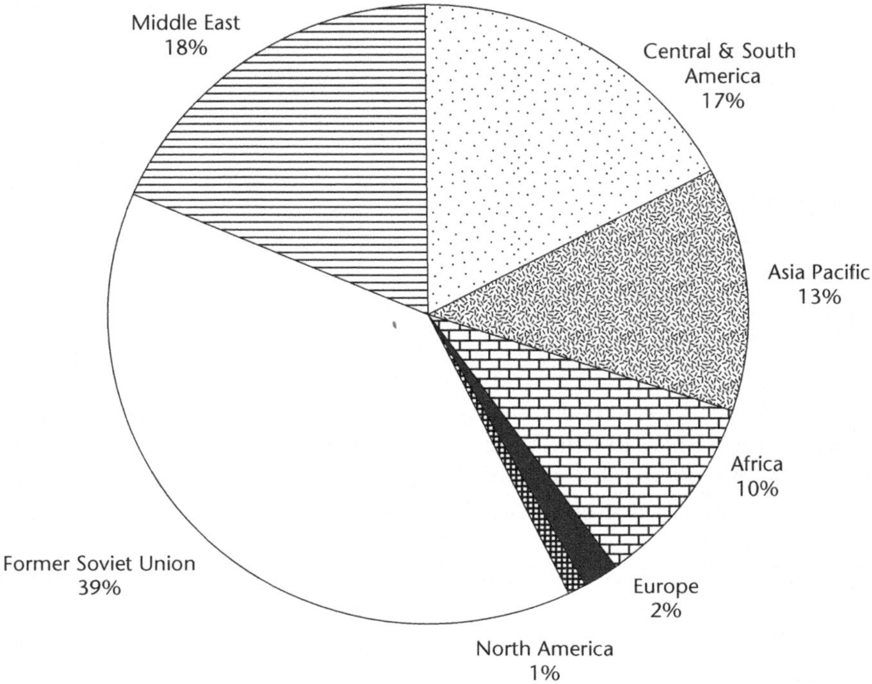

Figure 1.1 Distribution of new giant reserves[a]

[a] Adapted from Bai and Xu (2014, Table 1). Discovery of new giant reserves of oil and gas, between 2000 and 2010. Percentage of total. See text for definition of terms.

loose operationalization of democracy,[4] we can see that almost three-quarters (18/25) of the world's proven resources lie in non-democratic states with relatively underdeveloped economies.

In all likelihood, the world's future oil lies in the developing world. While much of the new production over the past three decades came from states in the Industrialized West (some 40 percent), this trend may not continue into the future (the joker lies in unconventionals). Back in 2007, the International Energy Agency projected that more than 90 percent of new hydrocarbon supplies will come mainly from the developing world in the next 20 years (Baker III Institute 2007, 1–2). Beneath the Persian Gulf, Latin America, Russia, and Africa lie large and still undeveloped oil and gas fields.

At first glance, this might not appear to be a problem. Indeed, the reader might find some hope in recognizing that many of the countries that enjoy rich underground reserves also have relatively poor populations. This is, at any rate, the most common reaction to citizens of states that have recently discovered wealth in natural resources; finally, their ship seems to have come in.

Table 1.1 Leading petroleum nations, by regime type[a]

Country	Gas Reserves	Oil Reserves	Polity Score	Description
Venezuela	5.57	298.35	4	Open Autocracy
Saudi Arabia	8.23	265.85	−10	Autocracy
Canada	*2.02*	*174.32*	*10*	*Fully Democratic*
Iran	33.78	157.00	−7	Autocracy
Iraq	3.59	150.00	3	Open Autocracy
Kuwait	1.78	101.50	−7	Autocracy
United Arab Emirates	6.09	97.80	−8	Autocracy
Russian Federation	31.25	93.03	4	Open Autocracy
Libya	1.55	48.47	0	Closed Autocracy
United States	*9.34*	*44.18*	*10*	*Fully Democratic*
Nigeria	5.08	37.14	4	Open Autocracy
Kazakhstan	1.53	30.00	−6	Autocracy
Qatar	24.68	25.06	−10	Autocracy
China	3.27	18.08	−7	Autocracy
Brazil	0.45	15.59	8	Democratic
Angola		12.67	−2	Closed Autocracy
Algeria	4.50	12.20	2	Open Autocracy
Mexico	0.35	11.08	8	Democratic
Norway	*2.05*	*8.68*	*10*	*Fully Democratic*
Ecuador		8.19	5	Open Autocracy
Australia	*3.68*	*3.96*	*10*	*Fully Democratic*
Egypt	1.85	3.90	−4	Closed Autocracy
Turkmenistan	17.48	0.60	−8	Autocracy
Indonesia	2.93	3.74	8	Democratic
Azerbaijan	0.88	7.00	−7	Autocracy

[a] Adapted from BP (2014) and Polity (2014). Proven reserves in 2013. Tables ranked by size of their proven oil reserves. Oil measured in thousand million barrels; gas in trillion cubic meters. The Polity Index stretches from −10 (Autocracy) to +10 (Full Democracy). The descriptions are derived from the index, where "Fully Democratic" states score 10; "Democratic" states score between 6 and 10; and "Open Autocracy" states between 1 and 5; "Closed Autocracy" between −5 and 0; and "Autocracy" states between −6 and −10. Fully democratic states are marked with italics.

The optimists among us can hope that this natural resource abundance will be used to help poor inhabitants of these countries, creating opportunities for economic growth and development. As developing economies suffer from an abundance of labor and a scarcity of capital (by definition), the money realized from resource wealth might be put to very good use—encouraging investments that can better utilize their abundant stocks of labor. Much of this money might end up in the state treasury, making it easier to provide public goods, such as education, health care, and infrastructure—each of which brings with it important economic benefits. As economic growth and development tend to encourage the growth of more democratic regimes, we might further hope that this natural abundance will facilitate a more democratic development over time.

Indeed, as residents of one of the four above-mentioned democratic countries (Norway), we are accustomed to being reminded that our national economic bounty is the result of fortuitous circumstance. Norway's enormous gas and oil reserves are frequently employed to explain our country's enviable economic and social record. How else could the country top the United Nations Development Program (UNDP) Human Development Indexes year after year, or manage to steer clear of the Great Recession?

These are the most common perceptions of how oil wealth affects a country's economic and political fortunes. Unfortunately, these common perceptions are frequently wrong. Most countries that discover significant petroleum resources come to suffer what we now call a "Resource Curse": their economies are handicapped, and their polities easily corrupted, by the discovery and extraction of vast natural resources. In light of a growing literature on the "Paradox of Plenty" (Karl 1997), we might say that Norway's enviable economic record has been secured despite, not because of, its petroleum wealth.

This book aims to introduce the sundry and significant challenges, both economic and political, that petroleum-producing countries face. Our intent is to help the reader learn about the pitfalls that policymakers face in the aftermath of a major resource discovery, and what they can do to protect their countries from the most adverse consequences. These lessons are derived from two very different sources: (1) the broader—if still underdeveloped—social science literature that examines the Paradox of Plenty in its disparate forms; and (2) the experience of a country that has successfully managed its natural resources over several decades. As a small country on the margins of Europe, Norway has stood up to powerful international interests in one of the world's most powerful industries. This small country has exerted sovereign control over its natural bounty, and exploited these resources in a way that has delivered significant wealth to all of its citizens.

By drawing on the Norwegian experience, we do not wish to imply that every country can (or should) follow Norway's example. Too much of the Norwegian success can be attributed to rather unique characteristics in the country's economic and political history, and/or the particular context in which Norway discovered and developed its resources. Rather, our ambition is to introduce a challenging academic literature that has studied many facets of the Resource Curse. In light of this literature, readers will be able to better understand why the Norwegian experience was successful (when it was successful) and why it sometimes failed (or needed to be changed). By juxtaposing general lessons with particular experiences in this way, readers will be well positioned to develop specific and targeted policies to fit the needs and circumstances of today's petroleum-producing states.

The Paradox of Plenty

The difficulty of harnessing natural resource wealth has many names. The "Paradox of Plenty" refers to a 1997 book with the same title, by Terry Lynn

Karl, which compared the disappointing economic records of several petro-states (Venezuela, Iran, Nigeria, Algeria, and Indonesia) in the wake of the 1970s' oil booms. The "Resource Curse" was first coined by Richard Auty (1993) to describe how countries rich in natural resources were unable to use that wealth to boost their economies and how, counter-intuitively, these countries had lower economic growth rates than countries with scarce natural resources. The notion of a "Rentier State" was first elaborated upon by Hussein Mahdavy (1970), and is commonly used to refer to states that depend on an external sources of income (e.g., resource rents); in so doing they can refrain from levying domestic taxes, and become less accountable to the societies over which they govern. Finally, "Dutch Disease," first used by *The Economist* magazine in 1977, refers to the decline of the Dutch manufacturing sector in the wake of a natural gas discovery off that country's coast in the late 1950s.

These different terms are sometimes used interchangeably, and it is no easy task to separate the effects of Dutch Disease from that of the Resource Curse and/or the Paradox of Plenty. One useful way of relating these descriptive terms is to separate the political from the economic challenges of resource wealth. Generally, then, the literature on Dutch Disease focuses on the economic challenges of resource wealth, especially as they relate to the international competitiveness of the economy in question. By contrast, the literature on Rentier States tends to focus on the political consequences of relying too heavily on an external source of government revenue. The literature on the Resource Curse is somewhat broader, as it includes references to both the economic effects (e.g., on economic growth) and the political effects (e.g., on corruption, violence) of that wealth. "The heart of the resource curse," writes Paul Collier (2008, 42), "is that resource rents makes democracy malfunction." A Paradox of Plenty can be used to describe the full spectrum of (economic and political) challenges.

While each of these terms is relatively young in origin, the economic and political challenges to which they refer are familiar to most historians and social scientists. For centuries, political, legal, and economic experts have contrasted the development trajectories of Europe's disparate monarchies, and noted how the source of a country's wealth matters for its subsequent chances at economic and political development. Although earlier studies didn't refer to these developments with reference to a Paradox of Plenty or the Resource Curse, we have long understood how the source of a country's wealth can influence the nature of its political and economic institutions (and hence its subsequent economic and political performance).[5]

Throughout most of the sixteenth century, bullion (especially silver) flowed from the New World into the Old. Spain was the main recipient of this wealth, and its bullion was used to fund one of Europe's most impressive military and administrative systems, centralized in the hands of an absolute monarchy.

On the surface, then, it appeared as though the Spanish authorities were able to use their new-found wealth to develop and extend their country's economic and political power. Because the Spanish crown could rely on external sources of revenue, it did not need to ask local elites for economic support. As a consequence, these local elites did not enjoy much leverage over the king: their parliament (the Cortes) was seen as relatively ineffectual in its dealing with the monarchy. Over time, the result was an institutional entropy that is often used to describe the decline of Spanish power in the seventeenth century.

By contrast, British monarchs didn't enjoy access to the same level of external financing. Britain's rulers were forced to negotiate for taxes with lords and commoners if they were to secure the resources need to fight wars and defend territory. These negotiations resulted in institutions (e.g., annual parliaments) that limited the monarch's authority and subsequently helped to guarantee property rights. The resulting accords also provided incentives for the public to ensure that the monies lent would be spent in accordance with the agreement. Consequently, the state apparatus became more transparent and efficient, and dependent upon (and responsive to) the local revenue providers (read residents).

These old lessons are reflected in a much newer and more vibrant literature on the Paradox of Plenty. In light of this contemporary literature, it is possible to focus on three main challenges for states that rely heavily on natural resource wealth. Although we will be examining the details of these particular challenges in subsequent chapters, it may be useful to begin the book with their brief introduction.

- **Decreased Competitiveness.** Countries that are rich in natural resources tend to experience decreased levels of international competitiveness. It is common to point to two sources of this decreased competitiveness:
  - *Real Appreciation.*[6] When a country begins to extract natural resources on a large scale, the resulting windfall will generate significant inflationary pressure within the domestic economy. This is the classic symptom associated with Dutch Disease, and it can have a dramatic effect on a country's ability to remain competitive in the global economy. This inflation drives an appreciation of the currency's exchange rate, making exports more expensive (and hence less competitive, internationally).
  - *Resource-Pull Effect.* This new money sloshing around in the domestic economy will increase the demand for skilled workers and capital investments, sucking them into the natural resource economy, pushing wages up across the economy, and crowding out investment in more traditional sectors of the economy. With time, the economy will become more dependent upon (and vulnerable to) the natural resource

sector; it will have fewer economic legs to stand upon. This makes the economy more susceptible to radical swings (like those that often characterize commodity prices), and can have a negative effect on the level of innovation and entrepreneurship in the domestic economy.

- **Poorer Economic Performance.** For many, this connection is the most difficult to fathom: how can increased wealth lead to poorer economic performance? One way to think about this is in terms of lottery millionaires; upon winning, lucky winners expect their life will become a piece of cake—they quit their job, buy whatever their hearts' desire, and pay less attention to the needs of the future. After all, more than one lottery millionaire has ended up bankrupt, broke, and alone. The same thing can happen to states that suddenly find themselves swimming in money. The effect of natural resource wealth on economic performance can flow through two channels.
  - *Lower Growth Rates*. The first of these channels has been the main focus in much of the Resource Curse literature: that a comparative advantage in natural resources is associated with lower rates of economic growth. For example, from 1965 to 1998, gross national product (GNP) per capita growth in member nations of the Organization of the Petroleum Exporting Countries (OPEC) decreased on average by 1.3 percent, while the per capita growth rate in the rest of the world was averaging at about 2.2 percent (Gylfason 2001, 848). A 2003 World Bank report found:

    > Data on real per capita gross domestic product (GDP) reveal that developing countries with few natural resources grew two to three times faster than resource-rich countries over the period 1960–2000. Of 45 countries that did not manage to sustain economic growth during this time, all but six were heavily dependent on extractive industries, and a majority of them also experienced violent conflict and civil strife in the 1990s.
    > (World Bank 2003, 12)

    This aspect of the Resource Curse has been made famous by Sachs and Warner's bivariate scatter plot, reproduced as Figure 1.2.[7] This figure shows a relatively strong inverse relationship between resource dependence and economic growth. More recent research (e.g., Ross 2012) suggests that this curse was more pronounced in the past—and that economic growth in today's petroleum states is neither faster nor slower than in the rest of the world.
  - *Bigger Government Expenditures*. Economic performance in resource-rich states can also be negatively affected by the growth of government expenditures. When large amounts of money flow into the government's coffers, it can be sprinkled over various projects of dubious economic value. In itself, public employment can be an attractive way for governments to gain greater public support. While this money may be useful for

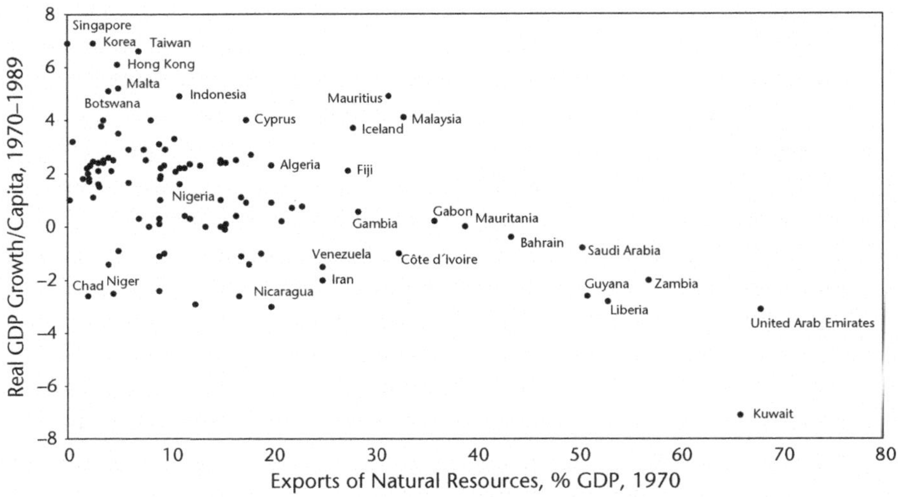

Figure 1.2 The original Resource Curse[a]

[a] Adapted from Sachs and Warner (2001, 829). Authors' reproduction of original scatterplot.

purchasing allegiance (for staying in power), the economic benefit from such spending is questionable. As a rule of thumb, resource booms tend to generate large public sectors and unsustainable budgetary policies. Worse, states can find themselves overextended as they expand public services when the price of oil is high, and governments are then left holding the bag when the price drops (the price of oil is notoriously volatile).

- **Rent-Seeking.** The third strand of work in this area considers the political effects of relying too heavily on resource extraction: the state's susceptibility to rent-seeking behavior, conflict, and even violence. Because officials do not need to rely on taxing their citizenry, they are less obligated to respond to the needs of that citizenry.[8] Instead, officials focus their attention on the IOCs, and the wealth and power that these can deliver. This leads to governments that are less responsive to domestic needs, more corrupt, and more susceptible to political violence, even civil conflicts. Indeed, the World Bank's (2011) *World Development Report: Conflict, Security and Development* considers the existence of natural resource wealth as one of the main stress factors leading to conflict and violence.

In short, there is a growing body of research demonstrating that countries that are rich in natural resources tend to:

- have national economies that grow slower than others;
- experience a real appreciation that undermines international competitiveness;

- become overreliant on the resource sector, as it sucks up domestic inputs of skills and capital;
- have politicians and bureaucratic elites that are less responsive to domestic/constituent needs, and less willing to provide public goods. As a result, these countries tend to have:
  - higher levels of corruption;
  - high levels of poverty, inequality, and unemployment; and
  - are more prone to internal conflict.

As a rule of thumb, then, resource-rich countries have great difficulty in leveraging their wealth into broader political, social, and economic development. We will examine these leveraging challenges in more detail in the chapters that follow. For now, we look briefly at how Norway's experiences seem to offer an exception to this rule.

Norwegian Exceptionalism

For a number of reasons, Norway has managed to avoid most of the problems listed in the previous section. Indeed, in Karl's (1997) book, Norway is highlighted as a unique case of a state that has largely managed to avoid the *Paradox of Plenty*: it has lots of oil/gas; it is becoming increasingly reliant on those resources; and yet its economic and political systems continue to function remarkably well.

Since discovering oil in the late 1960s, Norway's production of oil and gas climbed quickly and steadily throughout the 1980s and 1990s, and the country continues to maintain a fairly steady level of production (although increased gas production has made up for falling oil levels), as evident in Figure 1.3. Norway holds the largest oil and natural gas reserves in Europe, and supplies much of the oil and gas consumption on the continent. The US Energy Information Agency (EIA 2014) estimates that in 2013 Norway was the world's third largest exporter of natural gas (after Russia and Qatar), and the 12th largest net exporter of oil.

As a result, the Norwegian state and economy have become reliant on these petroleum reserves. For example, 21.5 percent of Norway's GDP could be attributed to the petroleum sector in 2013; as were almost half (48.9 percent) of the country's export revenues. Indeed, roughly a third (30.7 percent) of all investments in Norway were sunk in this sector, and 29.1 percent of the government's revenues in 2013 can be said to have come from the petroleum sector (NPD 2014, 12). Petroleum reliance does not often translate into an abundance of jobs, but in Norway it has been estimated that about nine percent of Norwegian employment is somehow related to the petroleum

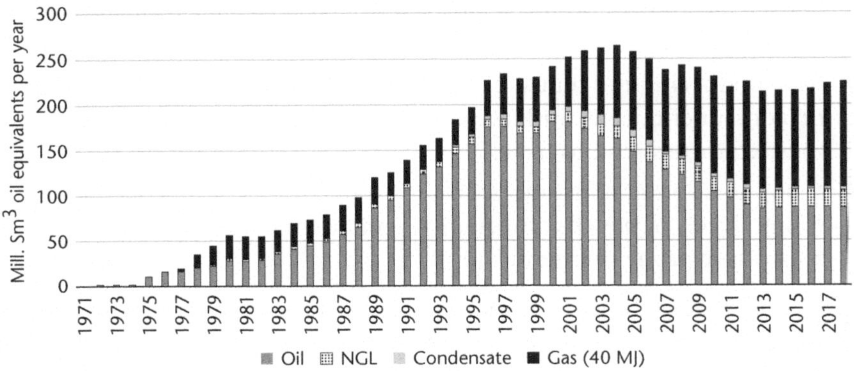

Figure 1.3 Production of oil and gas in Norway, 1971–2018[a]

[a] Adapted from NPD (2014, Figure 1.6). Sm^3 = standard cubic meter; MJ = millijoule; NGL = natural gas liquids.

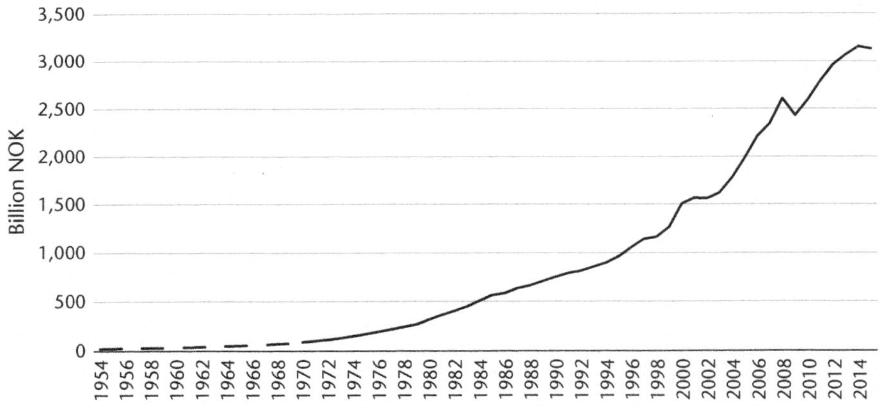

Figure 1.4 Norway's gross domestic product, 1954–2014[a]

[a] Adapted from NOS (1972, 14); SSB (2016, [9189-1]). GDP is in current prices. The pre-1970 data (dashed line) comes from NOS (1972).

sector (Cappelen et al. 2014, 31). In short, there can be no question that Norway has come to depend on its petroleum resources.[9]

Despite its reliance on oil, the Norwegian economy continues to deliver, and shows little sign of being affected by the most evident indicator of the Resource Curse: lower growth rates. Figure 1.4 shows the remarkable growth in Norwegian GDP. If we mark the transition to the oil economy in the year 1969, we can clearly see that the country's economy grew much slower before oil (dashed line) relative to the rate of growth after oil was discovered (unbroken line). We also know that the Norwegian economy has maintained

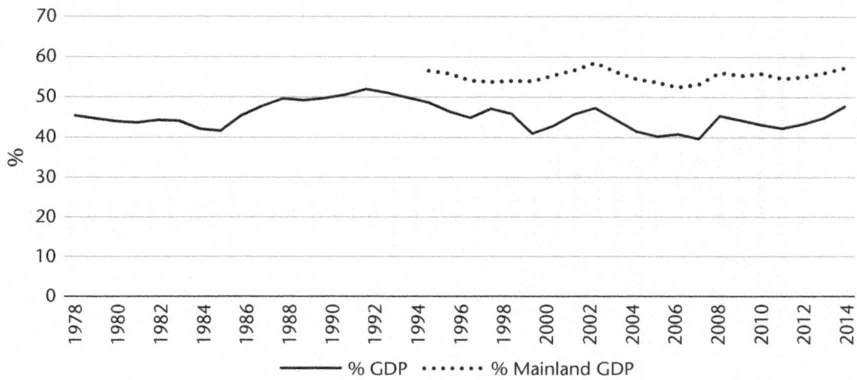

Figure 1.5 Total government expenditures in Norway, 1978–2014[a]

[a] Adapted from SSB (2016, [10909]; [09189]). % of GDP and mainland (non-oil) GDP. Total general government expenditures; GDP in market value current prices.

a higher pace of economic growth, relative to comparative countries (such as Sweden and Denmark) in the post-1969 period (Røed Larsen 2005).

This impressive rise in petroleum production and economic growth has not led to a government spending spree. If we look at government expenditures as a share of GDP, we see (in Figure 1.5) that Norway's general government expenditures remained remarkably stable over a long period of time, fluctuating between 40 and 50 percent of GDP. There is no sign here of government expenditures being overinflated by the new sources of revenue. The absence of negative effects is also evident in other areas as well. Since the end of the 1990s, Norway's inflation rate has remained very low (below 2.5 percent), as prescribed by the central bank's inflation target. In Transparency International's Corruption Perception Index for 2014, Norway is ranked 5th (out of 175 countries)—higher than any other petroleum-producing state (Transparency International 2014a).

By means of a responsive and responsible government, fortified with strong labor market institutions (as we shall learn more about in Chapters 3 and 9), this economic bounty has been distributed in a way that benefits a broad swath of the population. Throughout the entire postwar period, Norway has been a leader in securing full employment for its workers, and this commitment has not lessened after the discovery of oil. As evidenced in Figure 1.6, Norwegian workers enjoyed the lowest unemployment level in the entire Organisation for Economic Co-operation and Development (OECD) area, in 2014—well below both the EU and OECD averages.[10]

Of course, Norway's economic glory is not limited to full employment. During the recent economic crisis, Norway's GDP continued to grow (except

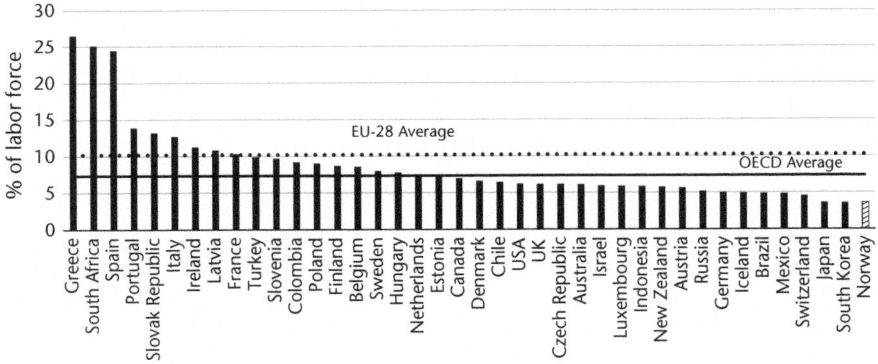

Figure 1.6 OECD unemployment rates[a]

[a] Adapted from OECD (2016a). Q3, 2014. Total percentage of labor force.

in 2009), under conditions of low inflation, an absence of debt, a surplus in the current account, and its government budget stayed in surplus (OECD 2012, 13). Just as importantly, these economic outcomes are distributed such that Norway's entire population benefits.

It is in large part because of Norway's full employment policies that the country suffers from relatively little abject poverty. Those Norwegians who are not able to secure full employment find protection from a developed welfare system that provides basic support (health, education, social security) and is designed to train them for quick re-entry into the labor market. Indeed, as shown in Figure 1.7, the level of wage inequality in Norway is very low, ranking seventh best in the OECD (as measured by the Gini coefficient). While this is itself impressive, the distribution of net income in Norway is even more egalitarian after taxes and transfers are taken into account; Norway then rises to third in the class (behind Slovenia and Denmark). This is the effect of Norway's famous welfare state.

It is Norway's generous provision of public services that secures the country a leading position on the UNDP's annual Human Development Index, which ranks countries by life expectancy, education, and income records; for the past 12 years, Norway has been ranked first. Similarly, Norway was ranked third (after Switzerland and Australia) in the Economist Intelligence Unit's 2013 "Where-to-be-Born" index (*The Economist* 2013). Norway's high scoring on social development indices such as these provides evidence of the opportunities available to the broader Norwegian population—opportunities generated by a well-managed economy and an extensive network of public services and support.

Instead of being cursed, the Norwegian economy has been blessed by the discovery of oil and gas off its coast. Much of the resulting wealth has ended up in a sovereign investment fund—the world's largest—stocked away for some

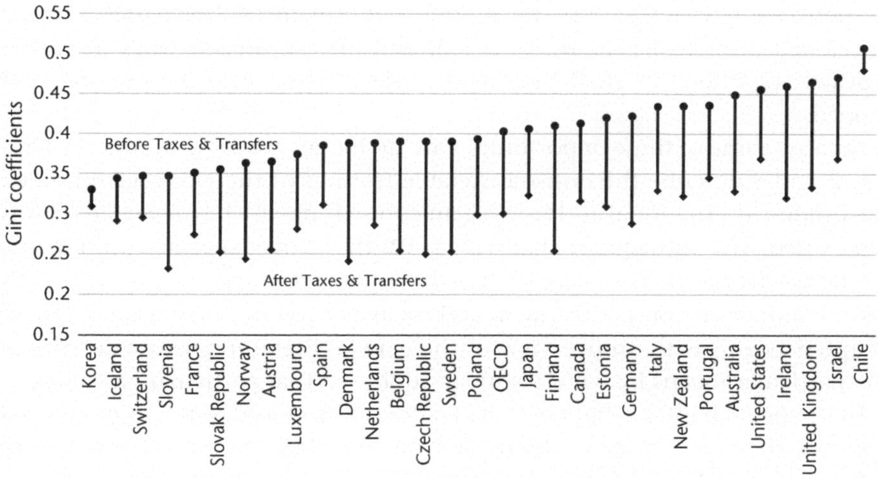

Figure 1.7 Inequality index[a]

[a] Adapted from OECD (2012,11).

future rainy day. Norwegian engineering firms and supply companies have not only grown in response to domestic demand, but established themselves in global markets so that they can sustain the wealth and experience necessary to thrive—even when Norway's petroleum reserves dry out.

We believe that many useful lessons can be drawn from the Norwegian experience, and that these lessons should be shared with policymakers from developing countries with newly discovered natural resource wealth, and those that have not effectively utilized their resource wealth for national development. But we hasten to add that these lessons need to be placed in a context of the larger, more general (and non-Norwegian) literature that informs the Paradox of Plenty. The question most relevant to policymakers is not "What did Norway do?," but "What can Norway's experience teach us about *what we should do,* given today's circumstances?"

To answer this more appropriate question, we need to break down the Norwegian model into its component parts, and show how these different parts respond to the basic challenges inherent to the Paradox of Plenty, as introduced in the section "The Paradox of Plenty". The remainder of this chapter provides an overview of these component parts, as each is described in its own chapter.

Chapter Overview

The success of the Norwegian model rests on three foundations. It rests partly on (1) the particular international context in which Norway first developed its petroleum resources; (2) the unique political institutions that Norway developed prior to its discovery of oil; and (3) the diverse ways in which Norway responded to challenges as they developed, once the resource was tapped.

Because some of these opportunities are no longer available to states, Chapters 2 and 3 describe the non-transferable features of the Norwegian model: the unique international and Norwegian contexts in which Norwegian oil was discovered. The subsequent chapters (4–10) then cover specific components of the Norwegian experience, such as how it has secured the government's take, maintained competitiveness, and regulated the industry. Each of these chapters begins with an overview of the general literature, before describing the particular means by which Norway addressed that particular challenge.

In the next chapter, Chapter 2, the reader is introduced to the international context, in which Norway discovered oil. In 1969, the world was a very different place. The late 1960s were a time of rising political turmoil, characterized by much inflammatory rhetoric and designs for a New International Economic Order (NIEO). This was a time when many developing states were breaking free from colonialism—when the balance of power between IOCs and sovereign states turned toward the latter, as exemplified by the rise of OPEC. In this volatile context, Norway became a very attractive investment site for international oil companies; but the same context also provided greater leeway for autonomous policies.

Chapter 3 turns to describe the broader political context in which Norway found oil, that is, a democratic *rechtsstaat*. In particular, this chapter begins by introducing a general framework to think about the nexus of politics and markets in the context of development. It then provides a short history of Norway's experience with foreign ownership and natural resource management (with respect to waterfalls and hydroelectricity) and introduces some of the most important institutional features in the Norwegian economy. It is these institutional features that allow Norway to adapt quickly to economic shocks. Norway was both unique and lucky in that it already enjoyed a stable and effective democracy—and was relatively wealthy—*before* it discovered oil offshore.

The rest of the book explores the different ways in which Norway has responded to the challenges of natural resource wealth. These lessons are much more transferable to today's resource-rich states. In Chapter 4, we discuss the need to respond to the political challenges of petroleum wealth, that is, those associated with the Rentier State. In particular, we consider what a state can do to manage the threat of corruption, and to leverage its bargaining position with respect to more powerful IOCs. This chapter begins very broadly, by focusing on the relationship between politics and economics and between states and firms. It then turns to describing Norway's unique institutional solutions for countering the threats that arise from resource dependence, that is, a three-part division of administrative responsibility that is the core of the Norwegian model.

In Chapter 5 we consider the size and nature of the government's take. This chapter introduces the challenge of how to decide between different ownership and fiscal structures, and how these decisions were made in Norway. Here, too, we start broadly with a theoretical discussion about the terms of trade and its relationship to ownership structures, before focusing in on the particulars of the Norwegian experience. Three main points are covered: (1) the different contracting systems that exist, and their consequences (joint ventures, production sharing contracts (PSCs), service contracts, and so on); (2) the nature of Norwegian allocation/licensing rounds; and (3) the organization and role of specific state interests in Norway (e.g., Statoil, SDFI, Petoro, Gassco).

Chapters 6 and 7 discuss the challenge of maintaining a macroeconomic balance and international competitiveness. In Chapter 6 we explain the need to minimize the inflationary consequences of large oil revenues, and to protect the exposed (tradables) sector. In particularly, we look at the role of Norway's highly organized labor movement (and the Scandinavian inflation model) and the country's willingness to control the pace of extraction to limit the chance of a real appreciation. Chapter 7 focuses on two other ways that Norway maintains its macroeconomic balance: its sovereign wealth fund (formally, the Government Pension Fund, Global, or GPFG) and its adherence to a firm budgetary rule (*handlingsregel*).

The Norwegian model is also admired because it encourages and facilitates the establishment of Norwegian firms and local competence. Chapter 8 considers the need to develop local content, and how securing this need has become more difficult in a global economy. Norway maximized local content in the offshore industry by employing diverse incentives that favored Norwegian suppliers/contractors. Two instruments attract special attention: (1) the need to control the distribution and the pace of development (so that the state stays in the driver's seat); and (2) the use of concession strategies that involve technology agreements (profit-sharing/carried interest clauses, and so on) and the political controls placed on development (e.g., §10 clause on Statoil).

In Chapter 9, we discuss the development of regulatory regimes that protect workers and the environment. Not only are these regulations necessary to protect locals from the profit motives of powerful international companies, but regulations can play a useful role in securing local content. This chapter focuses on the early division of authority in the Norwegian model and the need to establish an autonomous regulatory regime offshore. A central part of this regulatory regime was developing the principle of internal control (in 1976), the development of national standards (the NCS Competitive Position, or *Norsk søkkels konkurranseposisjon*, NORSOK), and the embrace of international standards and guidelines. In so doing, we emphasize the role of the Norwegian labor movement in bringing about many of these changes.

In Chapter 10 we consider the ethical challenges associated with developing a natural resource-based economy. This chapter focuses on international oil companies as global political actors with significant economic and political power and examines the concept of corporate social responsibility (CSR). In particular, we encourage readers to consider whether there are ethical limits to petroleum production and where these limits lie with respect to the choices that both states and firms face. In other words, we ask: to what extent do international oil companies have social and political responsibilities in the countries where they operate? Chapter 11 concludes.

Conclusion

While we are accustomed to writing academic books and articles on the subject, this book promises to be very different. Rather than a traditional textbook that introduces general themes and models, our aim is to provide much more practical and hands-on advice to policymakers, engineers, lawyers, and other actors in the petroleum sector.

To do this we build a bridge between the academic work on the Paradox of Plenty and the particular instruments and policy choices made by a state (Norway) that has successfully avoided this paradox. We hope this combination will make it easier for those that are not trained in the social sciences to better understand the management challenges that petroleum countries face. Given the varied background of our readers (engineers, lawyers, social scientists, policymakers), we have included a glossary at the end of the book.

While the following chapters draw extensively from the Norwegian case, these examples are used to illustrate more general points. Indeed, we have no intention of limiting ourselves to Norwegian examples. Examples from other countries are often used as a counterbalance to the Norwegian case, and for the purpose of elaboration. But in order to understand which elements of the Norwegian model can be transferred to other countries, and other times, we first need to consider the more unique and non-transferable aspects of the Norwegian petroleum management model. This is the subject of Chapters 2 and 3.

2

The International Context

Norway first learned that it had substantial oil deposits on the night before Christmas Eve in 1969. Over the decade or so that followed, the country developed a network of institutions, regulations, and policies for governing a nascent petroleum industry. Though the country had little relevant experience in the oil industry, and many still doubted the size and scope of its reserves, Norwegian policymakers did not lack self-confidence in their capacity to regulate this new industry in line with the Norwegian people's interests. Most importantly, Norway could exploit an opportune moment in history—it discovered oil at a time when the power of the state was at its apex, and markets, especially energy markets, were in turmoil.

This chapter describes that unique international context. Our objective is to show that many of the options which Norway chose then are no longer available to states today; the world has changed in ways that have made it more cumbersome for states to govern their economies.

Our younger readers may not realize how much the world has changed since the late 1960s. At the time, a generation—frustrated with the norms and values of its parents—created a "New Left" from disparate student and civil rights organizations to counter the dominant (and repressive) social values of the 1950s: social conformity, war, colonialism, racism, and conspicuous consumption. Patience for an unpopular war in Vietnam had begun to run out (the war lasted from 1955 to 1975), and young people were becoming increasingly radical. In both the United States and across Europe, protest groups took to the streets to challenge the war in Vietnam, deep-seated racism, and the injustices of capitalism.

The political shockwaves from these events reverberated all the way to Norway, shaking the foundations of several governments. Between the winter of 1971 and the September 1973 elections, four governments held power in Norway. Each change in government signaled a move further to the left. Part of this instability was generated by the same radical pressures emanating from Europe and the United States: rising dissatisfaction with a status quo that encouraged rampant materialism, while condoning racism and the war in Vietnam.

There was, however, another important driver to Norwegian radicalism at the time. Norway's political elites wanted the country to join the European Economic Community (EEC),[1] with a referendum set for September 1972. The campaign over this referendum split the country in two, but not along traditional party lines; it separated the political elites (who supported EEC membership) from their rank and file (who were mostly skeptical).[2] After a vigorous and bitter campaign, the Norwegian people narrowly decided against (53.5 percent) joining the EEC (as they would do so again, in 1994). When given an opportunity, Norwegians chose to prioritize democratic sovereignty over market access.

This radicalism was also evident in the international community, where a number of newly independent (postcolonial) states began to flex their political muscle, and demand a more just international economic order. In Africa, this was a time of radical political change, as 32 countries gained independence from European colonial powers. At the same time, there was a growing recognition that the international trading system—then organized as a General Agreement on Tariffs and Trade (GATT)—was not benefiting poor countries. In response, we find a growing demand for state intervention in the economy—both globally and within nation states.

This, in a nutshell, describes the ideological climate at the time when Norway discovered oil. The balance of power between states and markets was briefly leaning in the direction of states, and the domestic political climate in most countries was growing increasingly radical—with growing demand for more interventionist and just policies. In this political climate, Norwegian policy could reflect domestic values and influence: elected officials could do what they wished (more or less), within the context of a few (very general) rules of the game—for example, respect for property rights and contracts. International constraints on policy were mostly limited to market-related forces: corporate strategies, technological change, demand, and price.

Since the mid-1980s, however, Norwegian petroleum policy has been curbed by the development of an expansive network of supranational laws, regulations, codes, and guidelines designed to facilitate free trade and global markets. To attract market players, governments are expected to limit their sovereign control over those markets. In effect, democratically elected officials are no longer masters of their own economies; they must sacrifice their policy autonomy to satisfy international norms and rules, imposed by the likes of the EU, the World Bank, the International Monetary Fund (IMF), the WTO, and a thicket of international treaties—and, as we shall see, the petroleum industry itself.

Understanding the Changing Balance of Power

How should we understand this changing tide of state power? While large states, such as the United States, wield sufficient power to change the international balance of power between states and markets, most states are what we call "policy takers"; they must accept the international context, as is, and work within it. In other words, most states need to adjust their domestic ambitions and strategies in light of the changing opportunities offered by the global context. This chapter introduces those changing global contexts to emphasize the constraints placed on today's states when they attempt to develop their domestic oil sectors.

Karl Polanyi's notion of the "double movement" can help us understand this changing balance of power, over time, between politics and markets:

> It can be personified as the action of two organizing principles in society, each of them setting itself specific institutional aims, having the support of definite social forces and using its own distinctive methods. The one was *the principle of economic liberalism*, aiming at the establishment of a self-regulating market, relying on the support of the trading classes, and using largely laissez-faire and free trade as its methods, the other was *the principle of social protection* aiming at the conservation of man and nature as well as productive organization, relying on the varying support of those most immediately affected by the deleterious action of the market—primarily, but not exclusively, the working and the landed classes—and using protective legislation, restrictive associations, and other instruments of intervention as its method.
>
> (Polanyi 2001, 138–9, emphasis added)

When markets are unregulated, they create economic and social hardships that fuel political radicalism. In his *The Great Transformation*, Polanyi (2001) described the rising popularity of both Communism and Fascism as reactions to the interwar period's futile attempt to establish a self-regulating market. After World War II (WWII), the architects of the postwar international order were very aware of the dangers of an unfettered economy, and its links to political radicalism. In 1944, a number of national representatives came together (under the leadership of John Maynard Keynes and Harry Dexter White) in Bretton Woods, New Hampshire to design a system that could deliver both economic growth and democratic accountability.

The Bretton Woods compromise created an international order of "embedded liberalism" (Ruggie 1982). It reordered society's priorities; the free market was used as a means to acquire more just political ends (rather than being an end in itself). Embedded liberalism secured political stability by providing elected officials with the tools that they needed to avoid massive unemployment. This is the international order in which Norway discovered oil; it is a forgiving framework, in that it provided states with more room for political autonomy.

At the same time, there was a growing realization that free trade was exacerbating global political and economic inequalities, and that developing countries needed to respond strategically to these global challenges. Much of the intellectual framework for thinking about these structural challenges was provided by two UN economists: Hans Singer and Raúl Prebisch.[3] The Prebisch–Singer hypothesis held that the terms of global trade benefited manufacturing exporters more than it did the exporters of primary goods.[4] In other words, developing countries could not expect to leverage their comparative advantage by exporting primary goods. These countries had to (1) industrialize, in order to reap bigger gains from trade; or (2) improve their terms of trade. The first response led to a strategy of Import Substitution Industrialization (ISI); the second response encouraged producer cartels.

The Prebisch–Singer hypothesis has been subject to extensive and vociferous debate. Empirical support for the hypothesis varies over time; there was strong support for it before WWII, it then waned until the late 1990s, and support has grown stronger since. Most recently, however, with the rise in global food prices, the relationship has weakened again.[5] More relevant than the empirical relationship is the *perception* of a problem; when people think that there is a "terms of trade" problem, it affects their attitudes about free trade, and how it should be structured.

Hence, by the end of the 1950s a consensus was beginning to develop that a free market economy did not benefit all countries equally. Increasingly, it was recognized that economic development required the protection of (and support for) infant industries, so that countries could better exploit the terms of trade, or to develop alternative solutions to deteriorating terms of trade. In short, there was a recognition that free markets were neither fair nor politically sustainable; the state needed to intervene in the market in order to secure a more just distribution of resources and power.

Since the mid-1980s, the global balance of power has shifted away from states, to the advantage of multinational corporations (MNCs) and international organizations (IOs). One way to understand this shift is in terms of Polanyi's "double movement"—that is, part of a pendulum-like response to the growing power of the state vis-à-vis markets. Another explanation is found in the work of John Stopford and Susan Strange.

Stopford and Strange (1991) focus on three important characteristics of the new global economy: accelerating technological change; increased capital mobility; and cheaper communications. These characteristics have changed the nature of politics. Because of them, political negotiations and outcomes are no longer confined to the diplomatic channels that link states in the world. In a global economy, political relations are now channeled through three different conduits: between states (traditional diplomacy); between states and firms; and between firms. The nature of the second relationship, between states and firms, is the most relevant for our purposes, in that a globalized economy has strengthened the position of firms. This is because firms (MNCs) have access to the skills and technology, global capital/finance, and global markets that states need in order to develop economically.

> The growth of global competition can be seen as moving the world towards a position where events are conditioned mostly by an emerging managerial technocracy than by traditional notions of state power. In this new technocracy, firms feature prominently but are only one component of a wider network that links them to the educational and skills infrastructure and the financial system. Competition is increasingly among different production and institutional systems and contrasting social organisations. Further challenges to the role of the nation state are provided by the Europe of 1992 and North American debates about a free-trade area to run from Canada to Mexico. Supra-national bodies would emerge to offset the growing power of the world market. We are not attempting to forecast from among these and other possible 'meta' structures, we merely indicate that new triangular relations will continue to evolve. (Stopford and Strange 1991, 22)

Because firms have better access to these three important components of the global economy, they have strengthened their relative bargaining position. Just as significantly, this changing balance of power forces developing states in competition with one another. Whereas it was once possible for developing countries to work together in forwarding a common political agenda (e.g., to overturn declining terms of trade), developing states now find themselves locked in a tooth-and-nail competition with one another to attract MNC investments.

The order fashioned in Bretton Woods began to unravel at the very time Norway was establishing its petroleum management regime. By the mid-1980s, the balance of power had clearly shifted to the benefit of markets and firms, and it became thoroughly entrenched in a neoliberal ideology. Whether it was because of Polanyi's double movement or the changed incentives brought about by globalization (to follow Stopford and Strange), the contemporary international context came to prioritize the role of markets over states. In effect, we have returned to a period of "disembedded liberalism," where firms are less beholden to (national) political regulation, and states have less room to manage the domestic economy to secure political stability. Today's global marketplace is ordered by a number of supranational authorities that protect market actors from overzealous states.

The remainder of this chapter describes this changing global context over time. The next section introduces the rise of a New International Economic Order (NIEO) and a global economic environment that nurtured and protected the Norwegian petroleum management regime. We then turn to examine the decline of the NIEO and the nature of today's Global Economic Order (GEO). It is in this light that we can see why many aspects of the Norwegian management model (from the 1970s) are no longer feasible or permitted.

The Rise of a New International Economic Order

The Norwegian petroleum management model was developed in a context best described as a New International Economic Order (NIEO). To introduce the NIEO, we begin with a short description of the early oil industry, and how it was dominated by a handful of large firms cornering the market. These firms used their market position to accumulate great wealth and power. We then describe how the balance of power between these market forces, and the states that hosted them, slowly changed over the postwar period. The final section describes the rise of OPEC, and the rising influence of national oil companies (NOCs), as a response to the dominance of the largest international oil companies (IOCs).

Firms in Control

For most Americans, the history of oil is intricately linked to John D. Rockefeller, his Standard Oil Company, and a short period of American "trust-busting." Before the end of the nineteenth century (and over a remarkably short period of time), the Standard Oil Company and Trust came to dominate almost all of the oil production, processing, marketing, and transportation activity in the United States. Rockefeller stood accused of using unfair and destructive practices to create his industrial empire, and then leveraging this unfair advantage to limit competition, control the market, and set prices accordingly.

Not surprisingly, this unprecedented concentration of power attracted the attention of political authority. A number of US states brought antimonopoly suits against the Standard Oil Company and Trust. As a consequence, the Standard Trust was dissolved in 1892, and its component shares were transferred to 20 companies. Nevertheless, control still remained with the same network of owners, now reorganized as the "Standard Oil Interests." Then, in 1906, the US federal government brought action under the Sherman Antitrust Act of 1890. By 1911, the company was ordered to divest its major holdings (33 companies in all). The offspring of this broken trust continued to dominate the oil market for another half a century.

After WWI, and as the market for oil became increasingly international, new attempts were made to corner the (now global) market. The most famous of

Table 2.1 Achnacarry participants

Principal participants
- Anglo-Persian (later BP)
- Royal Dutch/Shell
- Standard Oil of New Jersey (later Exxon)

Smaller participants
- Standard Oil of Indiana (later Amoco)
- Standard Oil of New York (later Mobil)
- Texas Company (later Texaco)
- Gulf Oil (later Chevron)

these was the 1928 meeting at Achnacarry castle in Scotland, which resulted in the infamous "As-Is Agreement."[6] In this Agreement, participating companies divided among themselves the markets in Europe and Asia, and the price to be paid for the crude oil traded in global markets—based on the companies' relative share in 1928. (The US domestic market was excluded from the Agreement, so as to avoid violating US antitrust laws.) As Table 2.1 reveals, many of the participants were heirs to the original Standard Trust.

By the end of the 1930s, a few major IOCs, operating within the framework of four incorporated joint ventures, had secured exclusive petroleum arrangements in the Gulf countries in the Middle East. Most famously, a group of seven IOCs—or the "Seven Sisters"[7]—came to dominate the global petroleum industry from the 1940s to the 1970s. The companies concerned are listed in Table 2.2; their dominance across the Middle East is shown in Figure 2.1.

By the end of WWII, the strategic importance of controlling oil was evident to everyone. States began to encroach on these markets and wrest power from the IOCs. This slow transfer of power can be seen in the declining concession terms that IOCs were able to secure when negotiating with host states. In the early, interwar, agreements, IOCs were able to secure access to oil without any host state participation and with remarkably low levels of government take. Over time, the host state came to participate more, and secure a larger share of the revenues. By the mid-1970s, the IOCs had lost "control over a large part of their production capacity and oil reserves outside the US, forfeiting forever their capability to dominate the international market and to control the prices" (Taverne 2008, 58).

States on the Rise

From the end of WWII until the mid-1980s, the global economy was ordered in such a way that states could manage their domestic economies in response to domestic pressures and the need to secure economic stability (read full

Table 2.2 The Seven Sisters

Company Name	Original Nationality	Eventually Becomes
Anglo-Persian Oil Company	United Kingdom	BP
Gulf Oil	United States	Chevron
Royal Dutch Shell	Netherlands/United Kingdom	Shell
Standard Oil of California	United States	Chevron
Standard Oil of New Jersey	United States	ExxonMobil
Standard Oil Company of New York	United States	ExxonMobil
Texaco	United States	Chevron

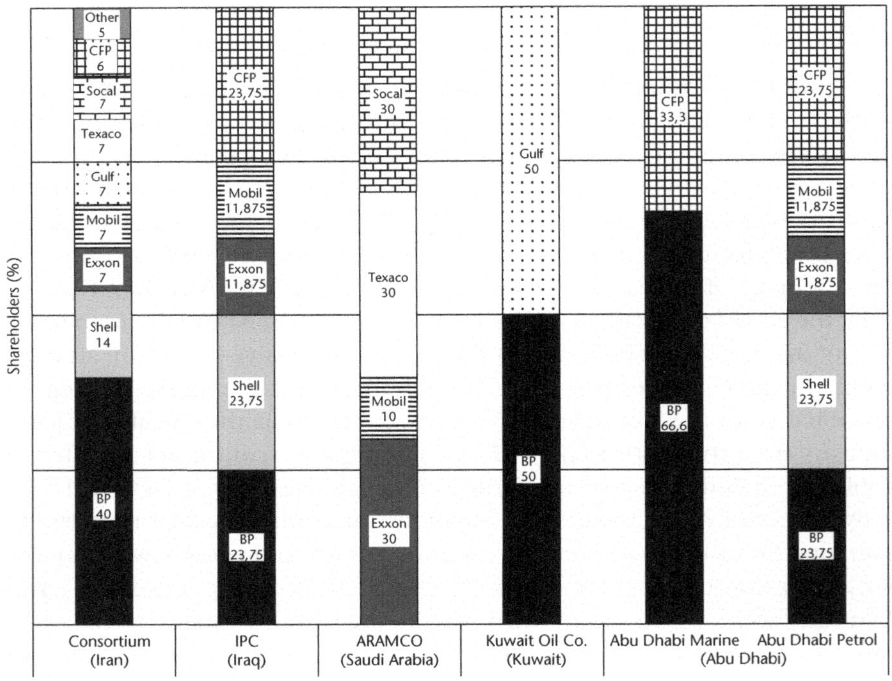

Figure 2.1 Joint oil production companies in the Middle East, 1972[a]

[a] Adapted from US Senate (1974, Part 5). CFP: Compagnie Française de Pétrole.

employment). The basic architecture of this system was agreed to at Bretton Woods in 1944.

In particular, economic growth was to be secured with free trade, by adopting a system of fixed exchange rates. In this system, all the major currencies were linked to the US dollar, and the dollar was linked to gold at a fixed amount. Democratic accountability was secured by providing states with autonomous monetary policies (this was carried out by limiting capital mobility across countries). In this context, central banks (now placed under the authority of elected officials) steered the money supply (and hence its price, the interest rate), with an eye at stabilizing the domestic level of employment and/or inflation.

To secure this managed system of exchange, three new international institutions were created: the IMF, the International Bank for Reconstruction and Development (IBRD), and the International Trade Organization (ITO). The first two of these institutions helped states secure finance for reconstruction (after the war), and to defend the fixed exchange rate arrangements (upon which the system depended). Together, these institutions provided an international context that gave states the freedom to pursue autonomous monetary policies, while encouraging free trade among themselves (from the price stability provided by fixed exchange rates).

The last institution, the ITO, was supposed to establish a free trade regime that would propel economic growth in the Bretton Woods system. The ITO Charter was successfully negotiated in March 1948 (in Havana), but proved too radical (read democratic) for the US Congress, which was unwilling to approve it. Consequently, responsibility for world trade fell on a temporary, fallback, agreement: the GATT. As can be seen in Table 2.3, the GATT met irregularly between 1947 and 1986, in eight different rounds of negotiation, to minimize tariffs among signatory states (who were growing rapidly in number).

Table 2.3 GATT bargaining rounds and outcomes

1947	**Geneva** (23 participants). GATT's founding countries established 20 tariff schedules that became an integral part of GATT. These schedules covered half of the world's total trade in goods (about 45,000 tariff reductions, relating to $10 billion in trade).
1949	**Annecy, France** (13 participants). Roughly 5,000 tariff concessions exchanged.
1950–51	**Torquay, England** (38 participants). About 8,700 tariff concessions exchanged, equivalent to a 25% cut in the 1948 tariff levels.
1956	**Geneva** (26 participants). About $2.5 billion's worth of tariff reductions.
1960–62	Geneva (26 participants). ***The Dillon Round.*** 4,400 tariff reductions, covering $4.9 billion's worth of trade. Agricultural and certain sensitive products excluded.
1964–67	Geneva (62 participants). ***The Kennedy Round.*** Introduces a formula approach to cut industrial tariffs by 35% across the board, staged over five years. Tariff concessions cover about $40 billion of trade. Separate agreements reached on grains, chemical products, and a code on antidumping.
1973–79	Geneva (99 participants). ***The Tokyo Round.*** Tariff reductions and bindings cover more than $300 billion of trade and lower the weighted average tariff on manufactured goods in the world's nine biggest industrial markets from 7.0% to 4.7%. Preferential treatment for developing countries is recognized and trade in many tropical products is liberalized. Revision of antidumping code and the establishment of GATT codes on subsidies, technical barriers to trade, import licensing, government procurement, customs valuation, dairy products, bovine meat, and civil aircraft.
1986–93	Geneva (117 participants). ***The Uruguay Round.*** Twenty-eight separate accords extend fair trade rules for the first time to agriculture, textiles, services, intellectual property, and foreign investment. Tariffs on industrial goods are cut by over a third; farm export subsidies and import barriers are substantially reduced. Creates the World Trade Organization (WTO), where trade disputes are settled by a single dispute procedure, with provisions for appeals and binding arbitration.

By the 1960s, it was becoming increasingly evident that GATT was a rich-man's club. Its tariff reductions were focused on manufactured goods, while agricultural products (the most relevant export market for most developing countries) remained largely unaffected. In 1964, developing countries responded to this imbalance by creating the United Nations Conference on Trade and Development (UNCTAD). The General Secretary for the UN's 1964 Geneva Convention on World Trade and Development was none other than Raúl Prebisch, and his plan for the new organization was clearly informed by a concern for declining terms of trade (Prebisch 1964). In particular, the UNCTAD was designed to meet four specific objectives (Johnson 1976, 3–4):

- To increase the amount of development aid;
- To introduce a unilateral preference system for developing-world manufactured exports (this eventually becomes the Generalized System of Preferences, or GSP);
- To reach an international commodity agreement that could stabilize commodity prices; and
- To discuss the possibility of creating a new international reserve currency (Special Drawing Rights, or SDRs), as an aid to developing countries.

While UNCTAD focused attention on the inadequacies of GATT, the broader global order was being placed under increased scrutiny. A decade later, in 1974, the UN General Assembly adopted a Declaration for the Establishment of a New International Economic Order, which was explicitly aimed at addressing the "problems of raw materials and development" (UN 1974a). This NIEO aimed to improve the developing world's terms of trade, increase development assistance, and reduce tariffs in the developed world—so that trade would be more beneficial for developing countries. Like UNCTAD, the NIEO was seen as a necessary counterweight to the GATT framework, a framework that was seen as unfair in the developing world. By the end of 1974, the UN adopted a related Program of Action and a Charter of Economic Rights and Duties of States (UN 1974b).

The NIEO generated a number of recommendations aimed at balancing the needs of poor countries with those of rich countries in the international system. In particular, the NIEO Charter established that each state had the freedom to choose its own political, social, and cultural system; it had the right to supervise multinational corporations; and it had the right to nationalize foreign property. For our purposes, there were four particular suggestions that are of interest:

- International trade should benefit developing and developed countries alike;
- Countries should be free to set up associations of primary commodity producers, to stabilize their price on the international market;

- Countries should be free to regulate and control MNC activities on their own territory; and
- Countries should be free to nationalize or expropriate foreign property on favorable terms.

For a number of reasons, the NIEO never gained significant traction. Indeed, most Western states voted against the Charter (or abstained from voting on it), and the GATT continued to expand its influence—becoming, eventually, the World Trade Organization (WTO). But the NIEO opened a small window of opportunity, through which many developing states entered.

Organization of the Petroleum Exporting Countries

In both the UNCTAD and NIEO charters, there was explicit recognition of the need to organize primary commodity producers in order to stabilize prices in the international market. This is in part because of Prebisch's role in founding UNCTAD, and in part because of the growing perception of a "terms of trade" problem in the international economy.

OPEC is the most evident and influential result of this push to create producer cartels. In creating OPEC, its members were embracing the spirit of an NIEO; it was an explicit attempt to turn the terms of trade advantage away from the developed world and to emphasize that foreign control of mineral resources was a vestige of colonialism.

The roots to OPEC can be traced back to Texas, with a long detour through Venezuela. The organization's father, Venezuela's Minister of Mines and Hydrocarbons, Juan Pablo Pérez Alfonzo, soon recognized the utility of getting producers to cooperate. He had learned this by studying the Texas Railroad Commission (while in exile in the United States). After first floating the idea with the Americans (who rejected it),[8] Pérez Alfonzo found a more receptive audience in the Middle East.

Pérez Alfonzo believed that the petroleum-exporting countries needed to unite in order to conserve their resources and to influence future price developments. This belief was informed by Raúl Prebisch's work and a recognition that oil-producing countries suffered from declining terms of trade (Rabe 1982, 159). To respond to these challenges, and to counter the power of the IOCs, Pérez Alfonzo encouraged oil-producing countries to organize in a producers' cartel. In September 1960, Venezuela joined with Saudi Arabia, Iran, Iraq, and Kuwait in a pledge to consider a system to ensure the stabilization of prices by—among other means—the regulation of production. The resulting agreement established a permanent intergovernmental organization with international status,[9] consisting of five original members. These founding, and subsequent, OPEC members are listed in Table 2.4.

Table 2.4 OPEC members

	Year of Entry
Founding	
• Venezuela	1960
• Saudi Arabia	1960
• Iran	1960
• Iraq	1960
• Kuwait	1960
Subsequent	
• Qatar	1961
• Libya	1962
• United Arab Emirates	1967
• Algeria	1969
• Nigeria	1971
• Ecuador	(1963–92, 2007)
• Angola	2007
Others	
• Gabon	Joined in 1975, withdrew in 1995
• Indonesia	Joined in 1962, withdrew in 2009, rejoined in 2016

OPEC was created in the spirit of the NIEO, to protect producer interests from two concomitant threats: (1) an international network of trade that benefited those that exported manufactures (read rich countries) over those that exported primary goods (read poor countries); and (2) to wrest power from powerful private interests, in the form of IOCs, and return it to the people who owned the oil (or, more accurately, their government officials). This link to Prebisch was clearly evident at the OPEC's 1975 Algiers meeting of the Sovereign and Heads of State of the Member States, where participants made a Solemn Declaration strongly supporting the NIEO (Taverne 2008, 96). An effort to extend greater political influence over IOCs had already been evident at the First Arab Petroleum Congress in Cairo (in 1959). In particular, an American lawyer, Franck Hendryx, presented an influential and controversial paper on the legal (and moral) right of sovereign states to change or to eliminate provisions in existing petroleum concessions—including the arbitration clause—should they prove contrary to the public interest (Hendryx 1959; Stevens 1959, 276–9; *The World Today* 1959, 247–8).

This sort of attention to public ownership and control over natural resources had been gaining legal traction since the 1958 Geneva Conventions on the Law of the Sea, which asserted that national governments had exclusive rights to exploit resources found on their respective countries' onshore and offshore territories.[10] These rights were confirmed by the 1982 Convention on the Law of the Sea, which also recognized that mineral resources (including petroleum) located under the deep seabed area are part of the common heritage of the human race (Taverne 2008, 120). Public ownership of land-based resources follows a similar trajectory, from a 1962 UN General Assembly Resolution to the 1992 Rio Declaration on the Environment and Development (see Al-Kasim 2015, 17–18).

Finally, the rise of state influence can be seen in the expanding influence of NOCs—a subject we will return to in subsequent chapters. Here, too, a door was opened by the NIEO, in that it recognized the legitimate and legal right of countries to nationalize or expropriate their natural resources, so long as fair compensation was provided. Since the early 1900s, only a handful of states had participated actively in the oil industry as part or full owners of companies. In the 1960s and 1970s, however, the tide turned, and NOCs began to explode in number and exert greater influence on the market (see Figure 2.2). By 2012, NOCs controlled between 73 and 95 percent of global oil reserves (Mahdavi 2014, 228). This new trend was propelled by the growing radicalism we have described.

Today's Global Economic Order

Since Norway developed its petroleum management regime, the world has become much more integrated, both politically and economically. This new Global Economic Order (GEO) prioritizes free markets by constraining the power and autonomy of states. With the rise of international capital mobility

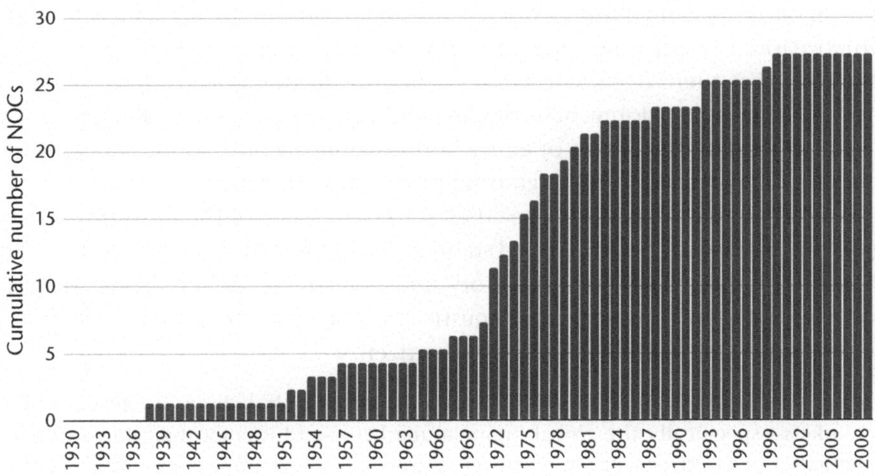

Figure 2.2 Growth in NOCs, 1930–2009[a]

[a] Adapted from Victor et al. (2012, 893).

and the demise of a fixed exchange rate regime, states have lost many of the tools that they once used to manage their domestic economies (Moses 2015). In this concluding section we describe the decline of the old, Bretton Woods, system and the rise of a new GEO, under the auspices of several international organizations and treaties.

The Collapse of Bretton Woods

There is no single factor that can explain why the Bretton Woods system collapsed. In the end, the United States pulled the plug in 1971, when President Nixon refused to defend the gold price of the US dollar. As the dollar was the key currency in the fixed exchange rate arrangement, Nixon's actions threw the international financial system into a tailspin. The Eurozone was one offspring of this market turmoil—a frantic attempt to recreate a fixed exchange rate system among EU member states. At the same time, international capital mobility was rising quickly with the advent of new electronic transfer and communication technologies, and the growth of offshore (unregulated) markets.

Together, the demise of the fixed exchange rate regime and the rise of international financial capital mobility made it increasingly difficult for states to manage their domestic economies. Monetary policies now had to be directed toward defending the exchange rate (rather than addressing the needs of the domestic economy); central banks were "depoliticized" in the sense that their leadership was isolated from democratic (political) pressures (that is, central banks became "independent" in the jargon of reform). The ease with which capital could now move around the globe made it more difficult to tax. The subsequent decline in tax revenues meant that governments had less money to spend on domestic programs and regulatory regimes. Finally, firms began to employ the threat of exit when negotiating with labor and governments; this increased their bargaining power substantially.

In short, with the arrival of the GEO, states lost some of the most important tools they had used to manage and stabilize their domestic economies. The aim of domestic monetary, tax, regulatory and investment policies were increasingly turned away from serving domestic constituents, and aimed at satisfying what Saskia Sassen calls the "economic citizenry":

> economic citizenship...does not belong to citizens. It belongs to firms and markets—specifically the global financial markets—and it is located not in individuals, not in citizens, but in mostly corporate global economic actors. The fact of being global gives these actors power over individual governments...
>
> (Sassen 1996, xiv)

The World Trade Organization

In 1995, the world got its first permanent organization for managing international trade: the WTO. Tellingly, the WTO grew out of the Uruguay Round of GATT negotiations (not UNCTAD), and the organization exhibits many of GATT's most evident shortcomings (from the perspective of developing countries, at least). In particular, the WTO extended free trade principles to a number of new areas, by including agreements on Trade Related Investment Measures (TRIMs), Trade Related Aspects of Intellectual Property Rights (TRIPs), as well as a General Agreement on Trade in Services (GATS). For several reasons, the WTO has become very popular, and the number of its member states continues to grow: there were 162 members in 2015 (WTO 2016a).

For the developing world, one of the most important constraints imposed by GATT (and subsequent WTO) agreements is that GATT prohibited discrimination based on national origin in its Article III (Sections 4 and 5 are most commonly used, but Sections 1 and 2 can also be raised to challenge local content measures). These provisions are designed to prohibit local content requirements that mandate local sourcing of goods and services (as they are seen to be discriminatory). Hence, states that have joined the world trading community are no longer allowed to pursue the sort of local sourcing requirements that most of today's developed states used to grow their domestic economies.

In contrast to the GATT, however, the WTO has a permanent dispute settlement mechanism, initially agreed to in the Uruguay Round (of GATT). This mechanism can be (and has been) used to challenge national regulations when they interfere with free trade. It works as follows: a member state can complain about an alleged violation of WTO rules before the WTO Dispute Settlement Body (DSB), which is composed of all WTO member governments. The DSB first tries to encourage governments to resolve the conflict on their own; if this fails, the DSB creates an Expert Panel (consisting of three experts in trade law) to investigate the complaint further. The Expert Panel reviews the evidence, meets with the parties, and prepares a final report that is submitted to the DSB. The DSB must accept the Expert Panel's final report unless *all* WTO members vote against its adoption (including the government that initially brought the complaint!). As this is a very high threshold to pass, the Expert Panel enjoys significant power to decide whether governments must change their regulatory policies.

This sort of international dispute mechanism achieves two things. First, it takes authority out of the hands of elected officials and places into the hands of technocrats who are not democratically accountable. In effect, this allows for a much more level playing field, in that the authority is (in a sense) independent of the interested parties. Second, the mechanism prioritizes trade. This is because the mandate and competency of these Expert Panels is limited to trade, even though trade may not be the dominant value underling a particular regulation or policy. For example, states might introduce regulations in order to protect the environment or workers' rights, even if these have a negative effect on trade—but the WTO's Expert Panel and the DSB are *only* allowed to asses polices in terms of their effect on trade flows.

The WTO is a multilateral trading agreement among 162 countries, covering a growing scope of trade issues. As can be seen in Table 2.5, several oil- and gas-producing countries are members or observers of the WTO—and this seriously affects their room to use protectionist measures to secure local economic outcomes and benefits. Significantly, however, the WTO partners have still not been able to agree on reducing agricultural tariffs. This is one of the reasons why the most recent, Doha, round of talks has run aground. In addition to the WTO, however, the world is littered with many smaller, Regional Trading Agreements (RTAs), as shown in Figure 2.3.

From 1948 to 1994, the GATT received just 124 notifications of RTAs (relating to trade in goods). Since 1995, however, over 400 other RTAs

Table 2.5 Oil- and gas-producing countries in the WTO

Developed Country Members	Developing Country Members	Least Developed Country Members	Developing Country Observer Status
Switzerland	United Arab Emirates	Angola	Yemen
Sweden	Malaysia	Chad	Azerbaijan
Singapore	China	Republic of Congo	Russian Federation
United States	Brunei		Kazakhstan
Germany	Tunisia		Syria
Japan	Oman		Libya
Finland	Kuwait		Sudan
Netherlands	Thailand		Iraq
Denmark	Indonesia		Iran
Canada	India		Algeria
Hong Kong SAR	Brazil		Equatorial Guinea
United Kingdom	Vietnam		São Tomé and Príncipe
Norway	Turkey		
Australia	Mexico		
Qatar	Romania		
Saudi Arabia	Colombia		
Italy	Peru		
	Egypt		
	Trinidad and Tobago		
	Philippines		
	Argentina		
	Ecuador		
	Ghana		
	Venezuela		
	Nigeria		
	Gabon		

[a]Adapted from Warner (2011, 35). © Greenleaf Publishing. First published in *Local Content in Procurement* by Michael Warner (2011). Reproduced by kind permission of Greenleaf Publishing.

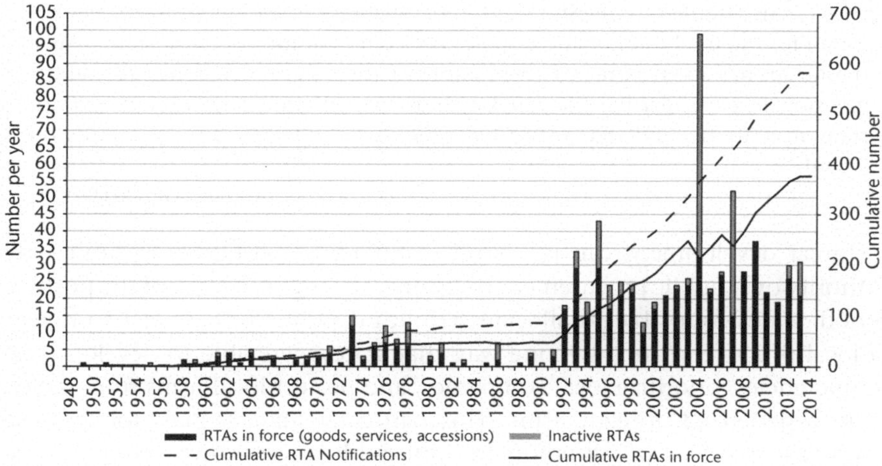

Figure 2.3 Growth in Regional Trade Agreements, 1948–2014[a]
[a]Adapted from WTO Secretariat (2014).

(covering goods and services) have been declared (WTO 2015a). Many of these RTAs include similar constraints on national sovereignty, in the form of binding rules and independent dispute mechanisms. For example, Chapter 11 of the North American Free Trade Agreement (NAFTA)—between Canada, the United States, and Mexico—includes an Investor State Dispute Settlement (ISDS) clause that allows investors/corporations to sue a government if its actions result in reduced profits.

ISDSs are found in a surprisingly broad array of international agreements. They are evident in various free trade agreements (e.g., they are currently being negotiated in the ongoing Transatlantic Trade and Investment Partnership (TTIP) and the Trans-Pacific Partnership (TPP) talks); they can be found in the International Center for Settlement of Investment Disputes (ICSID) at the World Bank; and they exist in a number of bilateral and multilateral investment agreements, such as the Energy Charter Treaty. Since 1997, 98 states have been subject to a total of 568 known treaty-based cases of ISDS (UNCTAD 2014, 1). By far the vast majority (3/4) of the states being sued are located in the developing world (ibid).

As we shall see in Chapter 10, this type of arrangements (and the use of stabilization clauses) is especially common for limiting a state's capacity to regulate investment activity in the energy sector:

> Given the needs for investment, the protections afforded to private investors through the global network of over 3,000 bilateral investment treaties (BITs) are particularly significant for those in the energy sector. Both the substantive

> protections afforded to investors and the availability of an independent, neutral forum for the resolution of any disputes between investors and host states play a significant role in encouraging these immense investment decisions. Oil and gas investors in particular have featured with disproportionate regularity in investor state arbitrations, and that sector has produced the largest arbitral awards on record.
>
> (Rivkin et al. 2015, 131)

A similar development can be found in international efforts to limit state authority over public procurement decisions. Throughout most of the postwar period, states were able to allocate significant public procurement projects to local producers. This was one way that states were able to develop local competencies, prioritize local companies, and protect local firms from larger, more experienced, foreign firms. This particular management tool was so important that all early attempts to limit public procurement policies were unsuccessful. Indeed, in the original (1947) GATT negotiations the idea of subjecting government procurement to GATT guidelines was brought up and rejected. Again, in the Tokyo Round of 1979, a similar proposal was rejected (but a "Code on Government Procurement" was achieved, to which a few countries agreed).

At the Uruguay Round (1994), several WTO members signed a plurality Agreement on Government Procurement (GPA).[11] The WTO sees procurement as a protectionist tool, which inhibits free trade, and it began to gather support for a multilateral agreement that would eliminate preferential treatment to national suppliers in procurement deals (Garcia-Alonso and Levine 2004). There are currently 15 parties, covering 43 WTO members, which have signed the GPA;[12] these countries have obliged themselves along two dimensions, when deciding on public procurement:

- *Substantial* obligations require that signatories avoid discriminating against foreign suppliers with regard to price. In short, the standard for evaluating a public procurement offer should be best value for money.
- *Procedural* obligations require member countries to build a procurement system that is transparent and encourages open competition.

Even states that have not signed the GPA may find that their ability to prioritize local suppliers is limited. This is because similar restrictions are included in the conditions set by several international financial institutions. The World Bank, the Inter-American Bank, and other regional developmental banks have played an important role in shaping generally accepted principles for procurement in connection with the projects they fund. Indeed, in 1986 these institutions encouraged the United Nations Commission on International Trade Law (UNCITRAL) to undertake work in the area of procurement—adopting the Model Law on Procurement of Goods, Construction and Services (Ssennoga 2006, 222).

This network of international trade agreements and organizations is designed to encourage free trade among signatory states. As tariff barriers to trade have been significantly reduced, the emphasis has now turned to reducing non-tariff barriers to trade. These "barriers to trade" are the very sort of regulations that countries have traditionally used to manage their domestic economies, while protecting workers, the environment, local communities, and so on. For this reason, we will touch upon them again in Chapters 8 and 9.

The European Union

As we saw in Figure 2.3, the number of RTAs has grown significantly in recent years. Each state, regardless of its geographic proximity, finds itself affected by this dense web of trade and investment agreements.

In the case of Norway, the most relevant RTA is obviously the European Union (EU). Even though Norway is not a member of the EU, it cannot avoid being influenced by decisions made on the other side of its border. Norway has twice held referendums to consider membership in the EEC/EU (in 1972 and 1994), and it has twice voted against that membership. The country's opposition to the European project rests in a population that wants to maintain sovereign control over important areas of policy.

Norway's relationship to Europe is complicated by the fact that it depends on the European market for its economic survival. For this reason, while negotiating the terms of its possible membership in 1994, Norway signed what was assumed to be a transitional agreement (in 1992), known as the European Economic Area (EEA) agreement (it came into effect on January 1, 1994). Tellingly, the EEA agreement does not cover the EU's common agriculture and fisheries policies, the customs union, the common foreign and security policy, justice and home affairs, or monetary union. Still, the areas covered remain extensive, and there exists a set of surveillance institutions (e.g., the European Free Trade Association (EFTA) Surveillance Agency (ESA) and the EFTA Court),[13] which ensure that the signatories respect their obligations.

Norway enjoys the possibility of vetoing any directive coming out of the EU, but—until very recently—it had been afraid to employ that veto for fear of the retributive consequences.[14] In large part, the EEA agreement secures Norwegian access to the EU's common market, but it does so at the cost of much political sovereignty. Because the EU forms a common market, whose explicit intent is to secure what Brussels likes to call the four economic freedoms (freedom of movement for goods, capital, people, and services), many of its laws are aimed at erasing any advantages that a member state might provide to local producers (and employers). Hence, the EEA has constrained Norway's capacity to manage its economy in a number of important ways.

These constraints are evident in the substantial reorganization of the Norwegian oil and gas market in 2001–02, but it is not always easy to see whether the impetus for change came from a growing liberal attitude among Norwegian policymakers or because of hard constraints imposed by the EEA. Statoil, the Norwegian NOC, was privatized in 2001 because of internal political pressures and concerns about its growing influence (see Chapter 4). But Statoil's privatization took place at the same time that the EU was demanding the sector be restructured (liberalized), and the Norwegian process was adjusted in response to this European pressure. In effect, the reorganization of Norwegian oil and gas activities was the result of three factors: the privatization of Statoil; pressure from the EU to make Norway adapt to EU competition law; and EU pressure to adapt to its gas directive (Austvik 2010, 119).

At a more general level, the EU's common market rules can be seen to affect Norway's management of the petroleum market in at least four policy areas:

- *Preferential Treatment.* On account of the EU's Concessions Directive (1994), Norway can no longer extend preferential treatment to domestic (Norwegian) suppliers.
- *Competition Policy.* Norwegian competition policy was gutted by the EEA agreement. For example, the country had to jettison its long-standing policy of choosing developmental concepts and materials specifically designed to fit the Norwegian supply industry (Arnesen 1995, 343–93).
- *Procurement Policy.* In the EU's main treaty, member states are prohibited from discriminating on the basis of nationality; they must grant the freedom of establishment, and they must establish the freedom to provide services.[15] In the mid-1980s, discriminatory public purchasing was seen as one of the last obstacles to completing a single EU market (Martin et al. 1997). Subsequent EU directives, such as the Procurement Directive and the Utilities Procurement Directive,[16] were aimed explicitly at establishing a non-discriminatory public procurement process: "Contracting entities shall treat economic operators equally and non-discriminatorily and shall act in a transparent way."[17]
- *Concessions Policy.* When Norway negotiated the EEA agreement in 1992, it made clear that its concessions policy was not to be affected by the agreement, as it did not extend to cover property rights issues. As we shall see in Chapter 3, Norway had developed a rather unique concession arrangement (*hjemfallsretten*) to maintain public control over waterfalls for hydroelectric generation. *Hjemfallsretten* provided the foundation for Norway's subsequent oil concessions regime.

In June 2007, the EFTA Surveillance Authority challenged Norway's position and asked that it jettison its long-standing concessions policy.[18] In particular, the ESA claimed that Norway's concessions policy discriminates between private and public power companies, and was therefore a breach of the EEA agreement. In the end, Norway managed to defend its concessions system, and the role of public ownership, but it was forced to make several adjustments.[19]

In effect, the Norwegian government was forced to choose between liberalizing its hydroelectric market (by allowing private and public actors to compete on an equal footing) and strengthening the public's ownership position. The government chose the latter; it eliminated the possibility of allocating new private concessions. These will only be given to public companies in the future (NOU 2012/2, 15.4.3).

Conclusion

Almost every state today is caught in a dense web of international trade and investment agreements that limits their sovereign scope. The constraints placed on Norway's policy autonomy can be found in several overlapping agreements. For example, Norway can no longer pursue procurement policies to the benefit of local (Norwegian) suppliers because of Norway's membership in the EEA and because the country has signed the WTO's bilateral GPA agreement. Similar webs of overlapping obligations can be found in the areas of trade and investment.

Developed and developing countries alike find themselves in a GEO where the room for policy autonomy is shrinking rapidly. These global rules of engagement are designed to protect investors and to encourage free trade. These are, of course, worthwhile objectives—and many economists believe that increased trade is the key to economic development. Our intent has been to emphasize how today's network of free trade and investment agreements have a less attractive flip side; they are designed in a way that limits the possibilities for democratic accountability and political sovereignty. This is a problem for any state that wishes to follow the Norwegian example of petroleum management.

Things were much different when Norway developed its oil industry. In the 1960s and 1970s, when IOCs began to harvest Norwegian oil, states enjoyed much more policy latitude and power in deciding how to develop and/or regulate their domestic industries. In Norway, the oil industry came to be managed and regulated in a way that reflected that country's long history of natural resource exploitation, and the powerful institutions and interests that had developed as part of that history. These unique national characteristics are the subject of Chapter 3.

3

The Norwegian Context

The aim of this chapter is to reveal Norway's economic and political backdrop, before oil was discovered. This is useful because Norway's petroleum experience was derived from this historical backdrop, and because it is quite common (albeit wrong) to assume that Norway is rich because of its oil, not despite of it. In fact, before the discovery of oil, Norway was very developed—both politically and economically. The discovery of oil prompted important changes in the way in which the Norwegian state managed its economy—but so too did changes in the surrounding international economy (as discussed in Chapter 2).

Although the political character of Norwegian economic management has evolved over the country's modern history—from a liberal democratic state (1884–1935) to a social democratic state after 1935—both types of state can be understood as forms of a *Developmental State*. In other words, the state played an active role in creating the institutions and policies that allowed Norway to catch up and modernize its economy (Moses 2005).

The chapter begins by introducing the general literature on Developmental States, as it provides an analytical vantage point for describing Norway's economic development. We then turn to describe the more important political factors for explaining that development: a tradition of state activism (with elites that embrace an ideology akin to developmental nationalism),[1] a strong labor movement, and a developed/capable bureaucracy. With this foundation in place, we explore the unique ways in which Norway managed its three factor markets: land (including natural resources), labor, and capital. This management regime was already established before Norway discovered oil. For many foreign (e.g., American) readers, this state-centered approach will appear rather alien. For this reason, it is important to describe it in some detail.

In employing an approach that prioritizes state-driven economic development, we have sidelined a more common approach that considers markets as independent of (or divorced from) political context. From a liberal (or neoliberal) approach, the state is granted only a cameo role in economic development, for fear of tilting (or severing!) the invisible hand. While academic economists and politicians may find utility in this sort of approach, students of policy and/or history tend to be wary of its highly stylized (and unrealistic) models.

For us, and following Chalmers Johnson (1982), there are at least three reasons why it makes sense to prioritize a state-centered approach that is market conforming:

1. A Developmental State approach provides a more realistic (and accurate) description of how states have successfully developed in global markets; its focus is on what states have done in practice, rather than what states should do in an ideal world. In the real world, markets are always saturated by politics, and the state often plays a central role in guiding the market's invisible hand. Worse, free market advocates often have something to gain by lessening the political constraints on market power: large private interests, such as IOCs, can leverage their power in more liberal contexts; and developed states often use free trade as a crowbar to access developing markets.
2. The market for natural resources is not like other markets. In particular, control over natural resources provides opportunities to capture economic rents—that is, a payment that exceeds the actual cost of bringing a primary factor into production. See Box 3.1. In most countries, the state controls the underlying natural resources (and hence controls the resource rents they generate), and their accompanying resource rents can be enormous. In this context, it is important to examine the state's role in controlling the resource and how it allocates those rents to different stakeholders.
3. A Developmental State approach is better suited for our intended audience: concerned policymakers and citizens in developing countries that hope to harness their natural wealth in a way that can benefit the entire country (rather than a handful of stakeholders). A Developmental State approach allows us to prioritize democratic objectives and develop strategies that can protect and enhance national sovereignty. It achieves this by paying attention to how markets are embedded in political and social contexts (Polanyi 2001 [1944]).

By employing this sort of approach, we do not mean to suggest that state intervention is always helpful or beneficial—quite the contrary. The

Box 3.1 RESOURCE RENTS

We think of rents as a prize, or an undeserved reward. Formally, rent is the surplus value, after all costs and normal returns have been taken into account. In other words, a resource rent is the difference between the price at which an output from a resource can be sold, and its respective extraction and production costs (including a normal return).

The notion of an economic rent can be traced back to Adam Smith (1776), who described rents as something going to landlords who "love to reap where they never sowed" (Smith 1776, I.6.8). David Ricardo (1971 [1817]) popularized the idea by making the simple observation that the amount of land was limited, and that it varied in quality. Poor-quality land was less productive than high-quality land. This difference in productivity levels generates rent, that is, rent arises out of the different productivity levels of different patches of land that are providing foodstuffs to the same market. The owner of the most fertile land receives a reward (a rent). The rent results from the quality of the farmer's land, not the result of her ingenuity, hard work, or other factors.

The same logic applies to oil fields. Consider an example from Daniel Yergin's award-winning book *The Prize*:

> [I]n the late 1940s, oil was selling for around $2.50 a barrel. Some grizzled stripper-well operator in Texas might only make a 10 cent profit on his oil. But in the Middle East it only cost 25 cents a barrel to produce oil. Deducting 50 cents for other costs, such as transportation, and allowing a "profit" of 10 cents on the $2.50 barrel, that would still leave a very large sum—$1.65 on every barrel of Middle Eastern oil. That sum would constitute rents. Multiply it by whatever the rising production numbers, and the money added up very rapidly. And who—the host country, the producing company, or the consuming country that taxed it—would get how much of those rents? There was no agreement on this elemental issue. (Yergin 2009, 414)

Why is this a problem? While both rent and profit are identical from an accounting point of view (they are "leftovers" after costs have been met), they play very different social roles. While profits play a role in promoting re-investment (and reflect that investment activity), rent has nothing to do with investment, innovation, or entrepreneurship. Rent is merely a windfall for the resource owner—a windfall that is derived not by effort, but by the accident of ownership. Hence, a fear of bankruptcy motivates the industrialists to re-invest her profits for more productive use, while the resource owner hoards his rents and spends it on non-productive trinkets. As a result, Ricardo (1971 [1817]) argued that rents constituted a brake on economic growth.

ingress of politics and a clumsy leviathan are frequent causes of market failure. Our point is *not* that the state can do no wrong. Rather, we intend to show that successful management of the petroleum economy is not possible in the absence of a strong, effective, and responsive state: a Democratic Developmental State.

The Developmental State

The Developmental State is a somewhat mushy and contested concept (see, e.g., Routley 2014); it is both disputed and evolving (Evans 2010a; 2010b).[2] Still, at the core of this approach lies an understanding that the state should play an active role in securing economic growth and enhancing human capabilities.

This understanding can be traced back to an older tradition of economic mercantilism—to the first US Secretary of the Treasury, Alexander Hamilton (1791), and to the German political economist Friedrich List (1885). This tradition contrasts what states actually say with what states actually do. Looking back over the history of economic development, we find that developing states have protected infant industries to keep them from getting trampled by stronger (more established) foreign competitors. It is only after developing economies were strong enough to compete that they came to embrace laissez faire policies.[3] The current popularity and utility of this approach are derived from the fortunes of Japan and the newly industrialized countries (NICs) in East Asia (e.g., Singapore, South Korea, Hong Kong, Taiwan)—whose growth is usually accredited to an active and interventionist state.[4]

Actually, the East Asian NICs do not offer a particularly useful point of departure. On the one hand, the Asian NICs all grew out of authoritarian contexts. On the other hand, they were all resource poor. What draws our attention to these states is the fact they were able to secure rapid economic development because their governments established capable state institutions and played an active role in promoting industrial polices (designed to assist strategic sectors), and they intervened directly to reduce the costly dislocations that accompany industrial transformation (from old to new industries).

At the most general level, a Developmental State approach is one that recognizes the importance of building state capacity. From this perspective, the state is seen as both *sufficiently embedded* in explicit synergistic relations with civil society and *relatively autonomous* from society's most powerful interests (Evans 1995). Peter Evans has zeroed in on the nature of state–society relations as the core of the Developmental State, and he describes this relationship in terms of "embedded autonomy." What he means by this is that the state's bureaucracy is not adversely influenced by interest groups, and yet remains close enough to them to be able to lean on them for advice as to how to ensure growth and (to a lesser extent) redistribution. To be sure, developmental states have three sets of capacities that set them aside from others, namely political capacity, organizational capacity, and technical capacity. It is these capacities that enabled developmental states to both formulate and implement policies to meet their development goals.

It is exactly these sorts of institutional characteristics that resource-rich states desperately need:

> Most mineral-rich countries in the developing world are the opposite of developmental states as they are marked by weak institutions; that is to say, the institutions of policy-making and implementation are generally weak and there is considerable lack of accountability and transparency. Furthermore, in most mineral-rich countries in the developing world, political leaders in particular and policy-makers in general rely on mineral wealth as the main source of state revenue. Consequently, they have limited incentives to create developmentalist institutions that will enable the state to intervene to diversity the structure of the economy. (Edigheji 2010a, 11–12).

This creates a more expansive image of the Developmental State: a state that plays an active role in guiding economic development, while using the resources of the country to meet the needs of the people. This is what Omano Edigheji (2010b) calls the Democratic Developmental State. From this perspective the state tries to balance economic growth and social development, using the state's resources and influence to attack poverty and expand economic activity (Edigheji 2007 and 2010b). In both the traditional view and the more expanded view, the state must develop the strategic, political, organizational, and technical capacities that it needs to play a role in subsequent development.

The Norwegian State

In the middle of the nineteenth century, Norway was a relatively poor country on the margins of Europe. By the time that Norway discovered oil (in the late 1960s), the country was already an established democracy enjoying a modest level of wealth. While a remarkable abundance of natural resources (not just oil, but also fish, timber, and hydroelectricity) can explain much of this development, it is the effective and just management of these resources that is key to Norway's success. This effective management can be attributed to three important political factors: (1) a long tradition of active state involvement in the economy; (2) the subsequent development of a strong labor movement; and (3) a highly developed network of political institutions.

Since the mid-1800s, the Norwegian state has played a central role in guiding and facilitating the country's economic development. This commitment to a strong role for the state can already be found in an 1886 speech to the Norwegian parliament (*Stortinget*) by A. M. Schweigaard, an influential parliamentarian, publicist, and professor of jurisprudence and economics. In that speech, Schweigaard notes the existence of a distinctly "Norwegian system," which can allow a small country like Norway to compete in an increasingly international environment.[5] In this system, government authorities were encouraged to use their superior credit rating to obtain capital from abroad, and supply technical supervision and general oversight over larger investment projects. Because the state enjoyed better terms of credit, and had more management experiences, it could help Norwegian industry gain a running start.

The simple idea behind the "Norwegian system" was that private initiatives should be supported by state loans, guarantees, or capital. Infrastructure (e.g., railway, telephone lines, telegraph, electricity) was to be established through private initiative, and run by private interests during the initial period. Over time, however, and because of the government loans and guarantees, the state would eventually take over the operation and ownership of these activities. For example, the Norwegian railway system was built by private firms, but its ownership eventually reverted back to the state.

The same sort of effort in asserting political control over economic activity is seen in 1899, when Norway passed its first law for regulating public procurement. This law gave priority to national companies in the public (municipalities and state) construction of electricity works. The law was updated in 1912 to strengthen procurement options as part of a larger industrial policy, providing a legal foundation for protecting "infant industry" (Wicken 2007, 29).

As industrialization began to take off, the Norwegian state helped to secure the necessary skills and competences that could attract further investment. For example, the Norwegian state actively encouraged the sort of research that could propel new industries (especially with regard to having access to cheap electricity). The Norwegian Technical College (NTH), which eventually became the Norwegian University of Science and Technology (NTNU), was established in 1910 with this in mind. At about the same time, the Norwegian state was introducing new institutions to promote research on natural resource-based development, for example The Raw Materials' Laboratory (*Råvarelaboratoriet*). As we shall see later in this chapter, in the section "Land and Natural Resources", the state also played an important role in securing public control over natural resources (waterfalls), to ensure that foreign interests couldn't abscond with the country's natural wealth.

Norway's Developmental State started from these humble beginnings. Even before the Labor Party came to dominate the political landscape, the country's more liberal and conservative governments were pragmatic in the sense of realizing that Norway had unique needs: a sparsely populated country, with little domestic capital and with little recourse to traditional (farming-based) sources of income. From across the political spectrum there was a recognition that the government needed to play an active role in encouraging investment, and building a costly infrastructure, if Norway was to industrialize and compete successfully.

This recognition of the value of state interventions was shared by elites of all political hues. Still, throughout most of the postwar period, Norway was dominated by the Labor Party (*Det norske Arbeiderparti*) in government, as can be seen in Table 3.1. Since the powerful red–green alliance formed by Johan Nygaardsvold in 1935, through a period of exile during WWII, and in the postwar reconstruction period under the leadership of Einar Gerhardsen, the Labor Party made an indelible mark on Norwegian politics. Indeed, the electoral power of the Labor Party was so strong in the period prior to Norway's discovery of oil that the historian Jens Arup Seip (1963) referred to the country as a "One party state" (the Labor Party) in the postwar period.

Table 3.1 Norwegian governments, 1935–present

Prime Minister	Took office	Left office	Party(s)
Johan Nygaardsvold	March 20, 1935	June 25, 1945	Labor
Einar Gerhardsen I	June 25, 1945	November 5, 1945	Labor's "Unitary" Government
Einar Gerhardsen II	November 5, 1945	November 9, 1951	Labor
Oscar Torp	November 9, 1951	January 22, 1955	Labor
Einar Gerhardsen III	January 22, 1955	August 28, 1963	Labor
John Lyng	August 28, 1963	September 25, 1963	Conservative, Center, Christian Democratic, Liberal
Einar Gerhardsen IV	September 25, 1963	October 12, 1965	Labor
Per Borten	October 12, 1965	March 17, 1971	Conservative, Liberal, Center, Christian Democratic
Trygve Bratteli I	March 17, 1971	October 17, 1972	Labor
Lars Korvald	October 17, 1972	October 12, 1973	Center, Christian Democratic, Liberal
Trygve Bratteli II	October 12, 1973	January 15, 1976	Labor
Odvar Nordli	January 15, 1976	February 4, 1981	Labor
Gro Harlem Brundland I	February 4, 1981	October 14, 1981	Labor
Kåre Willoch I	October 14, 1981	June 8, 1983	Conservative
Kåre Willoch II	June 8, 1983	May 9, 1986	Conservative, Christian Democratic, Center
Gro Harlem Brundland II	May 9, 1986	October 16, 1989	Labor
Jan P. Syse	October 16, 1989	November 3, 1990	Conservative, Christian Democratic, Center
Gro Harlem Brundland III	November 3, 1990	October 25, 1996	Labor
Thorbjørn Jagland	October 25, 1996	October 17, 1997	Labor
Kjell Magne Bondevik I	October 17, 1997	March 17, 2000	Christian Democratic, Center, Liberal
Jens Stoltenberg I	March 17, 2000	October 19, 2001	Labor
Kjell Magne Bondevik II	October 19, 2001	October 17, 2005	Conservative, Christian Democratic, Liberal
Jens Stoltenberg II	October 17, 2005	October 16, 2013	Labor, Socialist Left, Center
Erna Solberg	October 16, 2013	incumbent	Conservative, Progress Party

Although the entire political spectrum in Norway recognizes the need for an active state, the Labor Party has led the charge.

The dominant position of the Labor Party in Norway can be attributed to three important factors. First, Norway's mountainous and northern climes made it impossible to support large manorial estates. As a consequence, and unlike most countries in Europe, Norway managed to enter the era of industrialization without having experienced feudalism. In effect, Norway has always enjoyed a high level of equality, only it used to be an equality of poverty; the indigenous bourgeoisie or capitalist class has always been (relatively) weak and underdeveloped.

Second, the nature of Norwegian industrialization facilitated a strong and organized labor movement. Industrialization came late to Norway, and when it came it was concentrated in outlying areas, for example large hydroelectric installations (and the surrounding factories they supplied) and a number of fish-processing centers that dotted the coastline. By gathering together lots of men in this way, working far from home and family, it was easier to organize (and radicalize) the Norwegian labor force.

This leads us to the third factor: the strength of Norwegian labor—as both a political and an economic force. Indeed, both labor and capital organizations in Norway are highly institutionalized and accustomed to working with one another (in addition to working with the government). It was not always this way; this culture of cooperation grew out of a tumultuous interwar history, where Norway experienced numerous (and violent) strikes and lockouts. In an attempt to improve the political climate, the labor market partners agreed to a Main Agreement in 1935, establishing the ground rules for subsequent industrial cooperation (and shared growth). The end result was very high levels of unionization (which continue to this day, as evidenced in Figure 3.1) and a highly disciplined and organized workforce.

Labor union density rates have been falling in many neighboring countries over the past 15 years, but in Norway they have remained stable; just over half of the Norwegian workforce remains organized in unions. Although the overall level of organization has remained fairly constant, its makeup has changed significantly over the years, as seen in Figure 3.1. No longer is the Norwegian labor market organized in a hegemonic *Landsorganisasjon* (LO)—workers are now organized in a number of competing confederations, including the Academics' Collective Organization (*Akademikernes Fellesorganisasjon*, or AF), the Confederation of Vocational Unions (*Yrkesorganisasjonenes Sentralforbund*, or YS), Unio, and Akademikerne. This has clearly made it more difficult to deliver moderate wage demands. Still, the reach of these organizations is extensive, so that when the labor market partners sit down to negotiate collective wage agreements, the results influence the competitiveness of the entire economy; in 2008, about 70 percent of all Norwegian wage earners

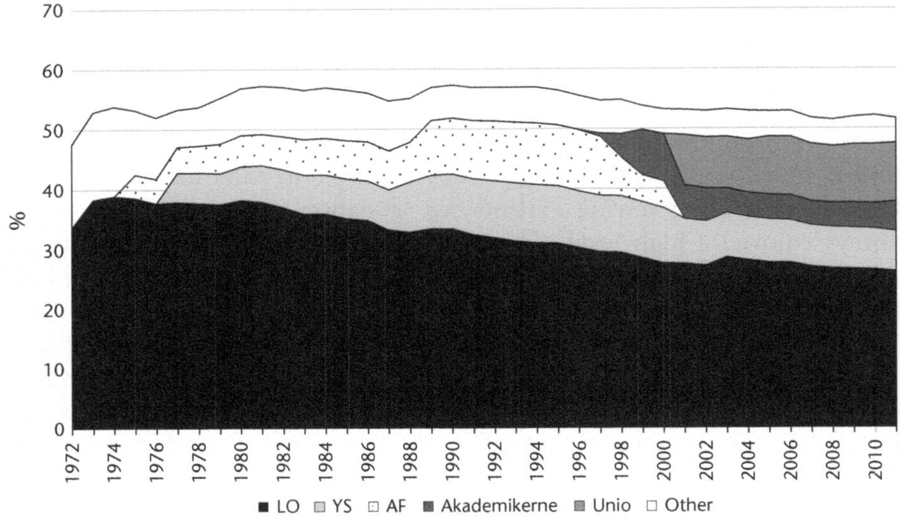

Figure 3.1 Norwegian union density rates, 1972–2011[a]

[a] Adapted from Stokke (2000) and Nergaard (2013). Percentage of workforce. LO = Landsorganisasjon; YS = Yrkesorganisasjonenes Sentralforbund; AF = Akademikernes Fellesorganisasjon; Unio = hovedorganisasjonen for universitets- og høyskoleutdannede; Other = FS (Frittstående).

were covered by these sorts of wage agreements, as all of the public sector workforce and just under 60 percent of the private sector were included (Bergsli 2013).

Because of the strength of Norway's labor movement, and because there are strong ties between the political and economic wings of that labor movement, the government has been able to secure cooperative incomes policies that have delivered full employment while maintaining international competitiveness. This form of "social corporatism" is a central part of the Norwegian approach to economic management (Katzenstein 1985), and will be described in more detail in the "Labor" section. For now, it is enough to note that Norway's economic outcomes are a product of a joint project of the state and the labor market partners *working together* to secure sustainable (that is, internationally competitive) outcomes.

State capacity is the third political factor that explains effective management in Norway, and the Labor Party in government has always been enthusiastic about building state capacity. For most of Norway's early history (1397–1814), the country found itself under Danish tutelage.[6] As a result, Norway was late in developing many of the institutions that we usually associate with nation building. The economic, political, and cultural capital of Norway was, for all intents and purposes, Copenhagen (in Denmark). For example, Oslo did not enjoy an independent university until 1813 and

Norway's central bank was first established in 1816. While Norway did not enjoy independence, it did benefit from a long and relatively peaceful tradition of respecting private property rights and the rule of law (under Danish rule).

This tradition can be seen in two important components of the Norwegian democratic tradition. The first is Norway's constitution. At the time of its writing, in 1814, Norway's constitution was considered to be one of the most democratic in Europe.[7] It remains a remarkably liberal document, clearly influenced by some of Europe's most progressive thinkers at the time; for example, it includes a suspensive veto (based on Rousseau's conception of popular sovereignty), a provision for an independent judiciary (traced back to Montesquieu), and guarantees of certain individual rights (from Locke). The remarkable character of this constitution is perhaps most evident in the fact that it is one of the oldest democratic constitutions still in use, second only to that of the United States.

The second important component of the Norwegian democratic tradition is the country's willingness to build up a strong, independent, and competent bureaucracy—one that was autonomous from important interests, to help monitor and manage the state's role.[8] This is, of course, a central component in Evans' notion of embedded autonomy. This characteristic was also noted by Terry Karl (1997, 217) as an important reason for why Norway avoided the Paradox of Plenty:

> In this exceptionally favorable environment, the prevailing policy style emphasized caution in the face of change, respect for standard operating procedures, segmentation according to issue areas, consensus building and egalitarianism.

Likewise, one of Norway's most celebrated political scientists, Johan P. Olsen (1983, 122), has described Norway as "a typical civil servant state." He notes that power in the Norwegian parliament is distributed in "a matrix like organization with the party caucus and the committees as building blocks" (Olsen 1983, 64).

This exceptionally favorable environment can be seen in the parliament's (*Storting*'s) permanent committees. In Norway, there is a strong tradition that every bill should be scrutinized by parliamentary committees, where Norwegian politicians are able to generate expertise in a context that relies heavily on expert input from the state's bureaucracy. This has the effect of distancing politicians from their (narrower) party interests (Rommetvedt 1998).

Norway's heavy reliance on a professional bureaucracy and parliamentary committees plays three important roles in the Norwegian Developmental State: (1) it provides expertise to policymakers that is not overly influenced by particular and vested interests; (2) it divides up political authority in a way that facilitates autonomy (and makes capture difficult); and (3) it provides

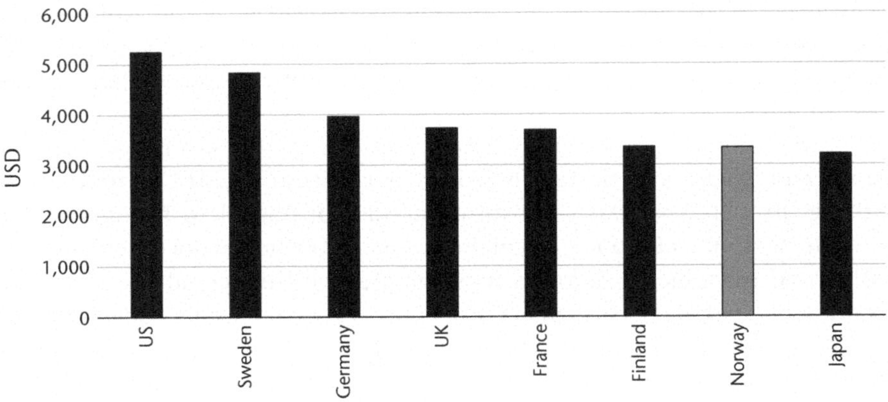

Figure 3.2 GDP per head of population, selected countries, 1970[a]

[a] Adapted from OECD (2016b). GDP per head of population in 1970. USD, current prices, current PPP.

access to alternative interests and an opportunity for these interests to contribute to the formulation of that policy. Together, these features make it easier to create consensus and to generate polices that benefit the interests of the community at large (not just narrow and particular interests of, say, the oil sector).

These three political factors (an interventionist state, a strong labor movement, and a highly competent and independent bureaucracy) combined to provide Norway with the capacity it needed to engineer a solid and broad-based economic development. This capacity was already evident before Norway found oil, as can be seen in Figure 3.2, where Norway's relative wealth in 1970 was similar to its neighbors, and greater than Japan's. In short, Norway enjoyed a distinct advantage relative to many other states that discover oil; it was already relatively well developed—both economically and politically.

Factor Management

Let us now look at the unique ways in which Norway has managed its three primary factors of economic production: land (including natural resources), labor, and capital markets. One key to successful economic management is a recognition of the central role played by these markets; factor markets are different from any other, in that each deals in a commodity that does not exist naturally. They are fictitious commodities, to use the expression coined by Karl Polanyi (2001 [1944]). After all, land is more than just real estate—it is nature; labor is more than wage earners—it is human life; and capital is a social invention needed to facilitate market exchange.

It is because these commodities do not exist naturally, that the community or state always needs to ensure their provision. In the words of Fred Bloc (2001, xxvi–xxvii):

> Real market societies need the state to play an active role in managing markets, and that role requires political decision making; it cannot be reduced to some kind of technical or administrative function. When state policies move in the direction of disembedding through placing greater reliance on market self-regulation, ordinary people are forced to bear higher costs. Workers and their families are made more vulnerable to unemployment, farmers are exposed to greater competition from imports, and both groups are required to get by with reduced entitlements assistance. It often takes greater state efforts to assure that these groups will bear these increased costs without engaging in disruptive political actions.

As we have already learned, rents are commonly associated with natural resources (and control over primary factors, more generally) and the challenge of their effective management. In Norway, an attempt at securing these primary resource rents was already in effect before World War I, when Norway was eking out its independence from Sweden. This effort rested on legal controls over ownership (concessions) and active support for producer organizations in each of the different factor markets.

Land and Natural Resources

Immediately after its independence from Sweden in 1905, Norway's Minister of Justice, Johan Castberg, began planning for the state's role in securing public control over Norway's natural resources. Castberg's thinking was influenced by the Progressive Movement in the United States, especially the work of Henry George (Ryggvik 2010, 14). In his 1879 work *Progress and Poverty*, George (1920) took Ricardo's theory (1971 [1817]) of economic rent and gave it a radical twist; he argued that any surplus or rent accumulated as the result of ownership of a particularly rich natural resource should belong to the public as a whole.

Castberg employed George's argument and made it a central component in Norway's new concession laws. In these laws (1906–17), the government intended to ensure public ownership over natural resources. While the laws were aimed mainly at protecting ownership over the hydroelectric powers inherent to waterfalls, the effects of the laws were wide ranging (and covered forestry, mines, farms, and eventually oil/gas fields).

The concession laws granted permission, for a limited time, to companies willing to exploit the resource. This provisional nature is known as *hjemfallsrett*, and is based on two legal principles: (1) the German principle of *heimfall* (which means that the land, or block, "falls" back to the state after a certain period of time); and (2) the common law tradition of *escheat* (where the property of a person who dies without heirs is transferred to the crown or state, so that property is not left in "limbo" without recognized ownership). In particular, the Norwegian concession laws ensure that all exploitation rights return to the state, without compensation, after a given period of time (for example, 60 years).

To understand the need for these laws, a little history might prove useful. Since 1888, foreigners needed permission to own property in Norway. Exceptions to this rule (concessions) were allowed, but they had to be sought out. This was done with increased frequency (and apparent ease) so that by 1906, three-quarters of Norwegian hydroelectric installations were owned by foreigners. This did not sit well with a newly independent country, trying to maneuver out from beneath the economic shadows of its two more powerful neighbors. By 1906, this foreign dominance became a political lightning rod, when half of the world's largest power station at the time (Rjukan) was sold to foreigners for a pittance (600 kroner!). The Norwegian government's response was swift, and became known as the "Panic Law."

The 1906 Panic Law introduced a temporary ban on all foreign purchase of Norwegian waterfalls (but foreign ownership was still possible, via concession). Two months after the law was introduced, its scope was extended to include forest and mining properties. It was also determined that a foreign company needed to establish an office in Norway if it hoped to be granted a concession. By 1907, all hydroelectric power fell under public control, and concessions were only granted for a limited period of time (between 60 and 80 years). After that, the ownership was to be returned to the state, free of charge.

The government's concession laws played a central role in securing the just and efficient management of the rents that were accrued from publicly owned resources. At the same time, the government encouraged other forms of (private) organization and cooperative control, to ensure that producers were able to secure a degree of stability over the markets that provided their livelihoods.

In the other (primary) markets, the main challenge came in the form of increased international pressure; the prices of primary goods became more volatile and exposed to international competition. This put fishermen, farmers, and forest workers—both in Norway and abroad—in a bind. The Norwegian response to these global pressures was to establish cooperative institutions, and allow them to steer the economy. These institutions had begun to take form as early as the 1880s, but grew to become much more influential in response to the economic threats of the interwar period.

During that time, primary producers faced a sharp fall in prices, owing to overproduction. The producer cooperatives first tried to regulate prices and competitive practices by controlling entry to the market, but they found it difficult to stop free-riders. Eventually, the government and parliament had to step in. For example, in 1928, a state-owned monopoly for the import of grain was reorganized to buy all domestically produced crops. The monopoly was then able to pay a better price to domestic producers (financed by selling imported grain at excess prices to consumers).

Rather than attempting to break up these collectives, the government encouraged them to regulate competition on their own—offering supervision and help in their efforts to corner the market. As a result, the cooperative movement made enormous strides in the interwar period; in 1922, the Norwegian Farmers' Association (*Norges Bondelag*) was born, followed by the Norwegian Fisherman's Union (*Norges Fiskarlag*) in 1926. In 1936, the Norwegian Herring Sales Union (*Norges Sildesalgslag*) gained a monopoly over the export of herring. In 1938, an unprocessed fish law (*Råfiskloven*) was passed that gave fishermen's collectives a monopoly over the sale of fish.

In effect, Norwegian "marketing centrals" were sponsored and promoted by the government under provisions of legislation passed in the 1920s and 1930s. These producer cooperatives covered very large shares of the country's dairy, fruit, and meat/poultry production, but also its fish and forestry products. By the end of the 1950s, producer cooperatives in fishing enjoyed exclusive rights to the sale of fish in Norway; virtually all dairy production and about 70 percent of meat and egg production were sold through cooperatives; a substantial percentage of private forest owners (holding over 60 percent of total forest acres) marketed their timber through the Forest Owners' Association; the marketing of grain was carried out by a government monopoly that set domestic grain prices; the sale of all potato starch was made through the Potato Starch Factories' Sales Association, controlled by the potato farmers; and so on (Leiserson 1959, 10–11). This domination by producer cooperatives continued through the twentieth century (and their remnants are still evident today). Each organization was obligated to accept everything delivered to it by the producers, so that prices couldn't be fixed at arbitrarily high levels. When prices rose too quickly, the state used a number of price control mechanisms to protect consumers.

Over time, these arrangements—both the concession agreements and the use of producer cartels—became both legitimate and important parts of Norway's approach to economic management. All new activity was naturally channeled through them. Thus, when new resources were discovered (e.g., oil, fish farming), they simply adopted an institutional and political context that was already established.

Labor

The market for labor is probably the most sensitive and protected of all markets. The Norwegian state developed a three-prong strategy to protect workers from the demands of the market (so-called decommodification): (1) an active labor market organization/incomes policy; (2) a network of state regulations; and (3) a broad-based and generous welfare state.

These three instruments work together. For example, the development of an extensive welfare state provides workers with a safety net, but it also strengthens their relative bargaining power at the negotiation table—as workers are no longer forced (out of desperation, at least) to accept any wage that is offered. The same thing can be said of the occupational regulation of the labor force: laws that prohibit child labor, limit the length of the work day/week (and require overtime pay), protect against occupational (safety) and environmental hazards—all of these regulations protect workers from unscrupulous employers. In so doing, they indirectly strengthen the relative bargaining power of workers vis-à-vis their employers.

Although our focus will be trained on the first of these instruments—an active incomes policy—an increased demand for all three instruments was forged in the fire of the interwar period. As was already mentioned, this was a period of massive unemployment and rampant industrial strife (numerous and large strikes and lockouts). Before WWI, some of the main institutional players came into being: the Labor Party itself was formed in 1887; the Norwegian Federation of Labor (LO) was formed in 1899, unifying labor unions in all sectors; and the Norwegian Employers' Association (*Norsk Arbeidsgiverforening*, or NAF) was founded the following year (1900), and renamed the Confederation of Norwegian Enterprises (*Næringslivets hovedorganisasjon*, or NHO) in 1985/6. It was at the turn of the twentieth century that we see the labor market players working out a system of industrial relations, which was first nailed down in the interwar period. In particular, the 1935 Main Agreement between the LO and the NAF became the de facto constitution of the country's labor relations, limiting strikes to periods of negotiation, coordinating the timing of negotiations, and specifying a framework for centralized bargaining.

To dive into the historical details of Norway's industrial relations is beyond the scope of this book.[9] Instead, we will focus our attention on how the Norwegian system is used to secure international competitiveness for a natural resource-based economy in a global economy. The main component in this strategy is corporatist intermediation, that is, tripartite bargaining between (1) the government and peak, encompassing, and hierarchical organizations representing (2) labor and (3) capital. Because Norwegian labor market partners are so well established, and they have built up a significant amount of trust among themselves (Høgsnes 1999), they are able to sit down together (with or without the government) to plan out a strategy for maintaining full employment and international competitiveness.

At the core of Norwegian incomes policy is something called the Aukrust (1977) model (or the Scandinavian inflation model, more generally), which espouses that the nation's wage growth can be influenced by means of a national incomes/wage policy that prioritizes the needs of the exposed (or tradables) sector. This model ties the national price level to the surrounding wage level with three important linkages: (1) wages in the exposed sector are linked to the profitability in that sector; (2) wages in the exposed sector are linked to wages in the sheltered sector; and (3) wages and prices in the sheltered sector are linked. Each of these linkages is secured by Norway's income policies: the labor market partners ensure that wages, productivity, and profitability levels move together (the first and third linkages), while the second linkage (between wage developments in the exposed and sheltered sectors) is secured by the way in which these negotiations are organized (the so-called "leading-branch model" [*frontfagsmodellen*]).

This negotiation framework dates back to 1967, when a Technical Calculation Committee for Wage Settlements (*Det tekniske beregningsutvalget for inntektsoppgjørene*, or TBU) was first established to set the stage for subsequent negotiations between the labor market partners. In order to facilitate collaboration, the TBU tries to establish, as much as possible, a shared understanding of the state of the Norwegian economy, and the challenges it faces. Primary among these are the expected rise in wages, prices, and productivity levels at home, and in Norway's main trading partners.

Hence, the TBU includes representatives from the government and the labor market partners to generate a common report that becomes the departure point for subsequent negotiations. More specifically, the TBU provides baseline wage information to help the exposed sector recognize the limits to its competitiveness. The resulting outcome is secured by allowing the partners in the exposed sector to settle first (hence, "leading branch" [*frontfag*]), and that the outcome of these agreements set the parameters for subsequent agreements (in the sheltered sector). In short, there is much pressure on the sheltered sectors to deliver wage demands that will not undermine the competitiveness of the exposed sector industries.

A final characteristic of the Norwegian corporatist arrangement is a commitment to wage solidarity. This plays out in the overall level of income equality in Norway, as we saw in Figure 1.7. This commitment has always been important to the Left, but it also has a significant effect on Norwegian productivity levels and investment patterns (and this makes it attractive to employers as well). Briefly, the existence of a narrow wage band forces firms in the least productive sectors to go bankrupt (as they have to pay higher wages than the market would support), while generating larger profits for more productive firms (who are paying wages below what would be the market rate). Bankrupted firms then have an incentive to invest in more productive sectors, while their laid-off workers are provided government support to help them retrain and move into these more productive jobs. Because of a broad and supportive welfare net, workers are able to move to more productive jobs without worrying too much about the consequences of falling between two jobs.

In short, the labor market partners in Norway had already developed a system of negotiated incomes policies, to ensure that wage growth in the sheltered sectors of the economy would not undermine the competitiveness of jobs in the exposed sectors. This system worked because of the highly organized and disciplined nature of the Norwegian labor market partners. As we shall see in Chapter 6, this institutional framework became a very important tool for protecting the Norwegian economy from Dutch Disease, once oil revenues began to pour into the economy.

Capital

Controlling capital markets has always been a central component of Norway's Developmental State, but it was a component that was largely facilitated by the international economy (in particular, the Bretton Woods regime, as described in Chapter 2). The main reason for this is the strength of the Norwegian labor movement and its desire to control social investment, for the benefit of the broader population. Norway has always relied on foreign capital, and the fear and influence of foreign capital have driven Norwegian policymakers to impose strict ownership regulations on Norwegian natural resources (the concession laws, as described in the section "Land and Natural Resources").

Despite this skeptical attitude, Norwegian governments have always recognized the necessity of attracting foreign capital in order to fund the domestic process of industrialization. This was especially true at the turn of the twentieth century—when building up its hydroelectric capacity—and in the 1970s—when developing its oil industry. Although this foreign capital could expect to receive a fair return on its investment, Norway's political authorities intended to channel this source of investment in ways that could serve the interests of Norwegian society.

As in many other countries, Norway's experience from the interwar period galvanized public and political attention on the need to maintain strong political control over the central bank, and the need to develop an international regime which fixed exchange rates and limited international capital mobility. In particular, after WWII, the Bretton Woods system imposed limits on capital mobility, and encouraged political control over the business cycle. This meant that the central bank, Norges Bank, was largely seen as an extension of government policy—and its control of the domestic monetary supply was used to complement broader economic policies (e.g., fiscal policy, and targeted investment policies).

As a general rule, the state intervened in the management of credit markets, with private banks cooperating alongside. The Norwegian policy rested on permanently low interest rates, brought about by regulating the quantity of money in circulation (Moses 2000). This provided the government with regular budget surpluses (that is, that the money was not being used to fund countercyclical Keynesian spending, as is often assumed), which could be used to fund its investment priorities (focusing on the export sector, to generate jobs and foreign exchange); and/or to build out the infrastructure (homes, education, roads, and so on).

Laws and institutions were used to regulate domestic access to capital (state banks, rationing systems that prioritized selected investment purposes, and so on). Indeed, for most of the postwar period, Norges Bank was a branch of the (social democratic) government, with little autonomy.

> What has come to be called the "iron triangle" of Norwegian economic policies consisted of the tight networks between the University of Oslo Economics department, Statistics Norway and the Ministry of Finance, with the latter as the primus inter pares. Norges Bank had little autonomy vis-à-vis the latter.
>
> (Mjøset and Cappelen 2011, 173)

Although social democrats had already decided against the nationalization of banks in the 1950s, the volume and price of Norwegian capital was highly regulated. The state had a credit budget (part of the national budget), and employed strict quantitative regulations (e.g., reserve requirements). The bond market was similarly managed. The state prioritized the needs of the competitive exports industry, along with the need to secure a housing stock and infrastructure for the benefit of the people. Special banks and credit arrangements were set up to help these prioritized sectors.

In effect, the existence of international controls on capital mobility allowed the Norwegian authorities to prioritize political objectives (not just economic objectives, such as maximizing returns) when allocating capital to the domestic economy. These objectives were mostly financed by budget surpluses and they facilitated a lower interest rate policy.[10] The result is a system that induced banks and insurance companies to run something like a price cartel (Mjøset and Cappelen 2011, 173), or what Knutsen (2007, 567) calls "strategic capitalism."

A low interest rate policy was preferred because it could deliver social and political objectives. Low interest rates meant that capital would get a smaller (and workers a larger) percentage of the national income (Aukrust and Bjerve 1945, 252). Interest rates were not to be used as a tool for countercyclical

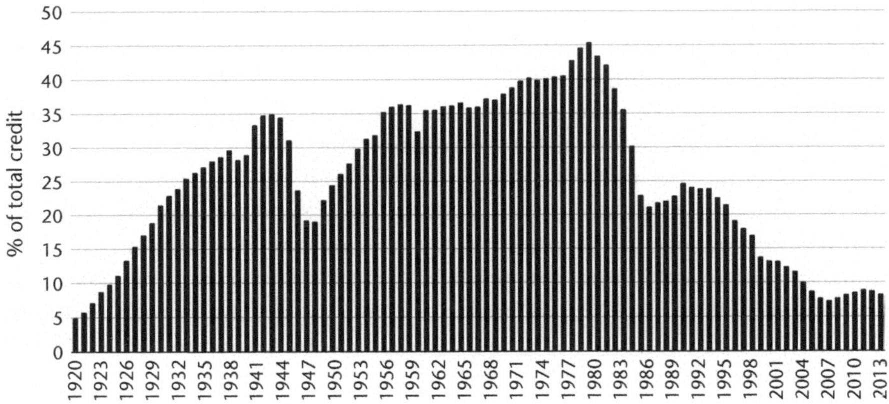

Figure 3.3 The share of Norwegian state banks, 1920–2014[a]
[a] Adapted from Norges Bank (2012).

economic management; rather, they were used to encourage long-term stable investments in strategically important areas (such as home construction). A large state bank network was created for the same reason: to have political control over significant distributional and allocation instruments. As can be seen in Figure 3.3: state banks were providing a significant share of the country's lending activity throughout the 1960s and 1970s, when the Norwegian economy begins to transition to an oil economy (see also Fagerberg et al. 1990, 67).

As in the country's other factor markets, the Norwegian capital market was heavily regulated and managed in a way that delivered politically desirable (and democratically responsive) outcomes. When that capital market was liberalized in the mid-1980s, the country fell into a deep and systemic banking crisis (1988–1992). Once again, the government was forced to intervene and resolve the crisis—its response is recognized as one of the most successful in history (Anderson 2009). In particular, the central bank offered very cheap (below market value) loans in exchange for stricter regulations on the banks. This allowed Norwegian banks to continue functioning, despite the crisis, with very little real effect on the country's productive economic activity (Moe et al. 2004).

Before concluding, we would like to add a word about fiscal conservatism. It is commonly (if incorrectly) assumed that socialist or social democratic governments are spendthrifts. Left-wing governments are accused of pursuing unsustainable economic policies out of a desire to tax and spend, and/or exploiting Keynesian-style, countercyclical macroeconomic policies to flatten out the business cycle. While the Norwegian state has played an important role in providing a broad welfare edifice, active retraining schemes, and policies that facilitate rapid adjustments to the changes emanating in the global economy, it has done so from a position of fiscal orthodoxy.[11] This is because it would be very difficult for a small open economy, like the Norwegian one, to maintain its external balance or international confidence in its economy (and with it, the exchange rate) if it were to run large, or long-lasting, budget deficits.

Conclusion

This chapter has focused on some of the non-transferable, domestic, components to the Norwegian approach to economic management. In particular, we have emphasized three unique aspect of Norway's experience:

- Norway was already wealthy and industrialized before oil was discovered;
- Norway already enjoyed a well-established democracy, and an efficient and professional bureaucracy prior to discovering oil. Indeed, the state enjoyed a long tradition of active state involvement in economic management; and
- The Norwegian economy was reliant on natural resources before oil (e.g., fish, forests, hydroelectricity), and had developed institutions to deal with the pressure of competing as an export-based, natural resource economy.

Hence, Norway did not need to build up its political and economic institutions from scratch, while struggling with how to deal with the inflow of money and increased economic activity that came with oil. The Norwegian economy was already fairly developed, and Norwegian industry had developed competencies in a number of sectors that would become useful to the oil industry later (e.g., concrete technology, dam building, maritime operations, and advanced methods of numerical analysis).

In short, Norway had a well-functioning democracy, and a vibrant economy, managed by institutions that were both efficient and legitimate. It is into this context that the Norwegian oil boom was introduced.

Of course, countries that have recently discovered oil do not usually enjoy this sort of history. Today's oil-rich countries are more likely to suffer from several shortcomings: they have only recently come out from under the shadow of colonialism; their states and economies remain relatively underdeveloped, feeble, and (often) corrupted; and they cannot draw on a strongly organized civil society (e.g., a well-organized labor movement, with significant influence on policy), or a broad-based economy. But it is important to recognize that there is hope for these countries, and this hope is found in the lessons of those states that have successfully managed the resource. In the chapters that follow we examine these more transferable lessons in detail.

4

The Norwegian Petroleum Administration

How can states avoid the Paradox of Plenty? The answer to this question is complex, in that it depends on a country's response to a host of choices, both political and economic in nature. In this chapter our focus is trained on the broader political choices available to states when they establish a system for administering the petroleum sector, and the challenges inherent to those choices.

It is widely accepted that good resource governance requires a strict separation of functions—and this separation is usually operationalized in terms of autonomous and independent institutions (for each function). For example, Thurber et al. (2011, 1) hold that "in research and technical assistance projects throughout the world, advisors from international institutions and donor governments—including Norway itself—treat a strict separation of functions as something of a *sine qua non* of effective oil sector governance." In short, foxes shouldn't be put in charge of henhouses. But what is the fox and what is the henhouse in the context of petroleum management?

Every government that is rich in natural resources faces three important administrative tasks: policymaking, regulation, and developing the operational competencies needed to extract those resources. The latter is usually achieved by encouraging diverse commercial activities. Each task is important in its own right—states depend on good policies, a professional cadre of administrators/regulators, and a vibrant commercial sector—but each task is also critically dependent upon the others. Governments need to balance these sundry tasks, and the costs of imbalance can be high. Balancing a country's regulatory, policymaking, and operational tasks is difficult for any state, but it is especially difficult in countries where the state's administrative capacities are not yet fully developed.

The heart of this chapter describes the institutional evolution of Norway's petroleum management model. Institutional reforms, of the type described here, remain a real and important option for states hoping to manage their petroleum resources in a more democratic and efficient manner to achieve a developmental goal. The Norwegian model is frequently held up as an example, because it separates and balances the country's policy, regulatory, and operational functions (it keeps the fox away from the henhouse). Our description of the Norwegian experience reveals a state that is continually adjusting to new conditions—a state whose autonomy and power are constantly being challenged, and subject to changing balances of interests and influences.

The Tripartite Model

If you are a policymaker in a state abundant in natural resources, where can you turn to learn about appropriate administrative design?

Most of the work in this field is frustratingly vague and/or conducted at a level of aggregation that provides little utility to policymakers; governments are encouraged to employ "best governance" practices, respect the rule of law, protect private property rights, and so on. This literature can be divided into two schools: one by political economists, the other by international consultants. Since Sachs and Warner's (1995) early work, political economists have used sophisticated econometric techniques to track down the institutional features that explain the Resource Curse.[1] The second school relies more on case studies and common sense, and tends to be supported by international consulting groups associated with the likes of the International Monetary Fund (IMF), Chatham House, the Natural Resource Governance Institute (NRGI), and so on.[2] Both schools offer reasonable, well-meaning advice, but the sort of institutions being advocated (e.g., rule of law, "good governance," protection of property rights, parliamentary vs. presidential systems) are both vague and very difficult to change, as they reflect underlying balances of power, and vested interests.

Nearer to the ground lie a handful of studies that focus on institutional design in petroleum states. Here the focus is on three critical and interdependent tasks: policymaking, regulation, and operational expertise. These studies recognize the need for states to develop an institutional context that allows policymakers, regulators, and commercial actors to balance their need for autonomy and interdependence. Because it focuses on three tasks, we call this the Tripartite Model of petroleum administration.[3] In achieving a sort of institutional balance, governments can monitor the commercial sector, while reducing potential conflicts of interests (either among commercial actors; between commercial actors and the state; and/or between the competing needs for protecting workers and the environment while encouraging production). Examples from this literature include an earlier (2009) version of the Natural Resource Charter, which referred to the utility of separating commercial from regulatory responsibilities,[4] and a growing number of case studies,[5] many of which take the

Norwegian example as a point of departure (e.g. Al-Kasim 2006a; Boscheck 2007; Thurber and Istad 2010; Thurber et al. 2011).

The easiest way to think about this Tripartite Model is from the perspective of a government in a country that has recently discovered natural resources. The first thing such a government must do is consider the best way to extract its resource rent, to the benefit of the country's population.[6] This is a question of *policy*, where policies tend to be developed in legislative bodies and are executed by governments in power (usually in the form of a dedicated Department or Ministry of Oil/Energy). Hence, the institutions of policymaking can be spread across government, but they depend on information and competencies that rest with commercial actors out in the field, that is, policymakers need to be near the commercial actors that hold operational expertise. At the same time, however, these policymakers need to be shielded from the powerful influences that operational/commercial actors can have (as their interests do not always coincide with those of the larger population, for whom policy should be designed). In short, policymakers require an institutional setting that can produce just, transparent, efficient, and sustainable policies that are beneficial for the entire country, but be insulated from the very powerful vested interests associated with the oil industry (recall Peter Evans' (1995) concept of embedded autonomy, from Chapter 3).

One of the first policy choices facing a government in this context is to decide how it will encourage the *operational expertise* needed to develop its resources. This operational expertise is usually linked to a commercial strategy (that is, what kind of companies will develop the operational expertise needed to develop the resource). Linked to these operational decisions are a number of related policy concerns, such as encouraging local content and employment, expanding the government's take, and so on. A country can meet its operational needs in any number of different ways, using commercial entities that range from foreign to domestic, and/or private to public.[7] The nature of the relationship between the state and these operational actors is complicated—for example, there can be one national oil company (NOC), or many; there can be monopolistic control over infrastructure, or there can be a plethora of (private) domestic firms competing with international oil companies (IOCs). For this reason, we describe the commercial challenge in more detail in the following section. For the time being, we will assume that the state's operational interests are concentrated in an NOC.[8]

Finally, the government has a responsibility to protect its people and their environment from these operational/commercial activities, and it does this by monitoring and *regulating* the industry. This is an extremely important task, as the government needs a competent body that can not only regulate, but also optimize, the country's resources and resource activities. The institutional placement of this regulatory body is controversial, as the required competencies often lie in the NOC, while the regulatory mandate lies with the government or relevant ministry.[9] While the safety and environmental consequences of ineffective regulations are clear, we often forget that an ineffective regulatory framework can also discourage commercial actors, and limit the opportunities (and revenues) available to states.

> Regulatory frameworks determine the revenues and responsibilities for actors involved in the industry, including the private sector and the government. By defining certain requirements regarding exploration, field development activities, technologies applied, the number of operators, the role of NOCs, and monitoring opportunities, this framework sets the scene for the industry's activities. Regulation determines the opportunities for the sector to function efficiently, and whether this works for the benefit of society at large. (Al-Kasim et al. 2008, 8)

To facilitate this careful balancing act, the Tripartite Model assigns regulatory responsibility to an institution that enjoys significant autonomy from both policymakers and the NOC (e.g., an independent Agency, Commission or Directorate).

Thus far, we have considered the importance of each function in a static fashion, and linked them to particular institutions; for example, policymaking is the responsibility of a Ministry of Energy; regulation is based in a Regulatory Commission; and operational competence is located in an NOC. But the Tripartite Model draws our attention to the way in which these three functions (and their requisite institutions) *depend upon one another* (and yet require protection from one another).[10] To understand this delicate balance, consider the constellation of tasks in the Tripartite Model, as depicted in Figure 4.1.

The policy function is placed at the top of the diagram to emphasize the fact that policy must be privileged over both operational and regulatory matters. When push comes to shove, policymakers determine the scope for the

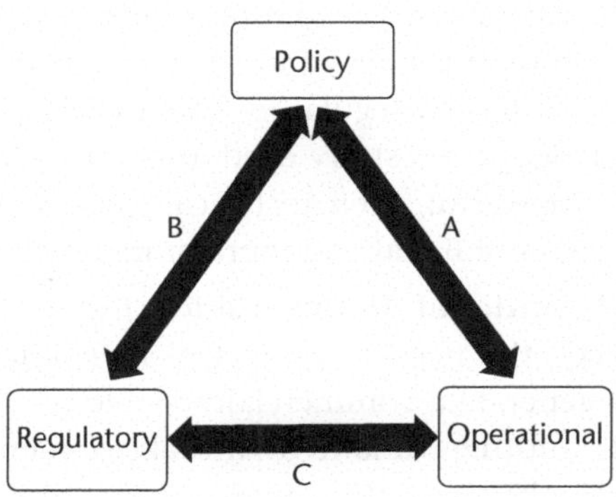

Figure 4.1 The tripartite model of petroleum administration

government's operational (or commercial) and regulatory reach. But we now need to shift our focus from the particular function (policy; regulation; operations) to the nature of the relationship between functions (that is, on the lines connecting these three functions—A: Policy–Operational; B: Policy–Regulatory; and C: Regulatory–Operational). Each relationship is symbiotic, even if the policy dimension will necessarily dominate, and each depicts a tug-of-war from the history of social science.

- **A: Policy–Operational.** This relationship reflects a classic tension in political economy: the difficulty of balancing the need for justice (in the form of just policy) with that of efficiency (in the form of commercial viability). After all, a government's petroleum policy aims to secure outcomes that benefit the entire community (for current and future generations). This objective might clash with economic imperatives and/or commercial needs of those that hold the operational expertise. Although it is important that policymakers are ultimately responsible for determining the scope of the state's operational and commercial activities, when determining this scope, policymakers must be careful to avoid undermining the commercial viability of the sector. Poor policies can deter commercial actors (and sufficient competition), undermining the state's potential revenues and opportunities. On the other hand, an NOC (or even private firms, for that matter) can easily become so powerful that it functions as a "state within a state," acting independent of government control, and acting in a way that can undermine a government's standing.
- **B: Policy–Regulatory.** This relationship reflects a classic tension between means and ends. Here policymakers should design the overall objectives/goals for the government's petroleum activities, while the regulators should ensure that this policy is carried out efficiently. Policymakers should be responsive to moral, ethical, and democratic appeals—they should be free to set the country's overall objectives. Regulators, by contrast, are administrators, who employ more technical and rational means to bring about the policymakers' ends. If policymakers receive poor information from their regulators, the resulting polices can be ineffective and impractical, even starry-eyed (utopian). On the other hand, if regulators are too powerful, their technical myopia can lead to policies that undermine larger political and moral imperatives.
- **C: Regulatory–Operational.** In this relationship we find another classic tug-of-war: between the need to protect workers and the environment, and the need to generate a commercially viable sector with operational expertise (this is another variant of the above-mentioned justice/efficiency trade-off). Although this tug-of-war is most evident when it comes to protecting workers and the environment, it is important

> to emphasize that the state's regulatory capacity is also involved in technical operations that ensure efficient operations with respect to optimizing government returns. If the state's operational arm (the NOC) is too powerful, it will secure a laxer regulatory environment, in order to increase profits. This is the fox and the henhouse argument, noted in the introduction to this chapter. On the other hand, the government's regulatory authority relies on information provided by commercial actors (from out in the field), and it is often in competition with these actors over the same limited talent pool of workers. In short, the relationship is bound to be contentious. Worse, by imposing an overly burdensome regulatory regime, government regulators can restrict commercial opportunities and operational expertise in the country.

For a state to manage its petroleum resources effectively, it needs to balance these complex relationships. This approach to petroleum administration, filled with checks and balances, is widely heralded as a means to secure fairness, integrity, and transparency in government decisions (and in their relationships with commercial actors). The nature of the first relationship, between policymakers and operational actors in the petroleum sector, is perhaps the most contentious of these (and the most interesting), so we will describe it in more detail in the following section.

Commercial Negotiations

Thus far, we have assumed that the government wants to maintain control over operational matters, so it creates a commercial entity in the form of an NOC. This is not an unreasonable assumption, given the ubiquity of NOCs. As Ralf Boscheck (2007, 370) has suggested, one would have to be historically agnostic to take a market-driven approach to petroleum management at a time when 90 percent of the world's hydrocarbon resources are controlled by state-owned enterprises.

Still, the growing presence of NOCs has not made IOCs irrelevant or impotent. Petroleum states still rely on IOCs, especially in more complex fields/contexts, and need to develop strategies for dealing with them. When a country first discovers oil, it is usually because of an IOC, as few domestic companies (or countries) have the capital, stomach for risk, or expertise required to develop the resource. We can assume that every government hopes to develop and privilege domestic firms, but it will always take time before those (domestic) firms are competent enough to assume the task. In the meantime, governments will need IOCs. The resulting relationship between IOCs and their host governments is a negotiated one, and it revolves around two axes: (1) the division of financial gains; and (2) the control over investment, exploration, and production activities (Andersen 1993, 2). In entering into such negotiations, the state finds itself bargaining with some very powerful adversaries.

We are already familiar with Stopford and Strange's (1991) argument that MNCs in a global economy enjoy increased influence (vis-à-vis states), because they have access to skills/technology, capital, and global markets (see Chapter 2). This influence is especially strong in IOCs, as the oil industry is extremely capital intensive and is able to wield remarkable technologies to entice the oil/gas out of the ground and into markets on the other side of the globe. While countries may own the resource, they still need to negotiate with IOCs in order to get it to the market.[11]

Although an IOC starts from a position of enormous influence, its power slowly erodes as the company begins to invest in the host country's economy. Ironically, the more an IOC invests, the weaker its bargaining power, as the country knows that the company is unable to walk away from its "sunk" investments, and the host country becomes less dependent on the IOC for expertise. This recognition of the changing balance of power, over time, is known as the "Obsolescing Bargaining Mechanism" (OBM), a term first introduced by the eminent economist, Raymond Vernon (1971). While the OBM is not limited to the oil industry (it affects many types of foreign direct investment, or FDI), it is especially important in this industry, given the scale of risks and investments involved.

In an OBM framework, we distinguish between pre-investment and post-investment phases. During the pre-investment phase, states and IOCs negotiate over ownership shares, tax rates, the size of the proposed investment, and so on. Here the partners haggle over the terms of extraction, and the outcome will depend on the two parties' relative demand for the other's resources; the nature of the stakes in play; the degree of interest overlap; and the bargaining skills on each side of the negotiation table (e.g., Fagre and Wells 1982; Grosse 2006; Kobrin 1987). During this round of negotiations, the levels of risk and uncertainty are high, while the potential rewards can be enormous. An IOC can exploit its previous experience in dealing with the commercial uncertainty involved; the host country does not yet have the necessary expertise, technology, and capital to develop the resource. This constellation of risk and reward allows the IOC to secure very favorable terms in the pre-investment phase.

After the IOC begins to invest, the balance of power shifts as the host country can hold the IOC's sunk investments hostage. The size of the required investment (which was once a crucial, ex ante bargaining chip for the IOC) becomes a liability once the company's costs are sunk.

Imagine a country where oil was just discovered. Suddenly, there is certain money to be made and the large rewards associated with the uncertain risk (in the pre-investment negotiations) now seem exorbitant (even the initial risk now appears as exaggerated). As the IOC has already invested a great deal of money in the exploration and production (E&P) facilities (which it cannot quickly pack up and abscond with!), the government can renegotiate the original terms of the agreement, improving its bargaining position (and resulting share).

> Uncertainty is reduced, and the old doubts are forgotten. The host government gazes out at a profitable operation, carrying off resources the country was sure it had all along, in which a large part of the revenue is flowing away to foreigners. The price paid to the foreign investor seems, in retrospect, too high, and the government in power (or its opponents) point out that the country is being cheated. The foreign company, on the other hand, has sunk its capital and won its bet, and has an attractive mine [or platform] that it will now continue to operate up to some point even though its share of the revenues may be reduced. Empirically, few large natural resource concessions in underdeveloped countries remain long unchanged. The terms of the original agreement are tightened in favor of the host country. (Moran 1974, 9)

Aware of the OBM, both states and firms design their bargaining strategies accordingly. IOCs are able to secure more attractive terms in the earlier stages of negotiations, while states benefit more in the post-investment phase. As a consequence, the longer the state can draw out the bargaining process, the stronger its relative position. This sort of renegotiation may strike the reader as unfair or unjust, but it is remarkably common—as illustrated by the recent case of Chad, described in Box 4.1. Finally, it is important to note that the balance of power between the host state and the IOCs can shift for other reasons as well (beyond the OBM)—for example, after a large discovery is made, or in a context of falling prices, regime instability, and so on.

The discussion so far implies an adversarial relationship between oil companies and their host countries. After all, the two parties tend to prioritize different objectives, interests, and time frames. In its negotiations, the IOC focuses on the short term; it wants to maximize early and high rates of production, early profits to shareholders, and (overall) project profitability. The state, by contrast, tends to focus on the larger, long-term picture—with an emphasis on securing a high rate of recovery and long-term benefits for the people at large, such as the need to maximize revenue, competence, employment, and general welfare. Because of these contesting priorities, we should expect to find some tension between the host state and the IOC when they negotiate over the terms of their cooperation.

Having said this, however, it is easy to exaggerate this tension, and to forget how both sides must necessarily rely upon one another. Indeed, the state and

Box 4.1 CHAD AND THE OBM

Chad is a poor, landlocked nation in Western Africa. In 2000, a consortium of IOCs, led by Exxon, began building an oil-production and pipeline project (over Chad, across Cameron, and to the Atlantic coast) that promised enormous wealth. Given its tumultuous history, Chad was not a particularly attractive investment environment, and the country had to accept some very hard initial conditions from the project developers and their sponsors. Most notably, the World Bank became involved, and was responsible for an independent oversight committee, to guarantee that a substantial portion of the resource rents ended up supporting local social and economic development schemes (and not in the pockets of corrupt officials or troublesome rebels). This arrangement was unique, in that it imposed a blatant constraint on Chad's sovereign powers (see Box 10.3).

After the IOC consortium had sunk its investments in these two countries, and the oil began to flow from Chad, and across Cameron, the government of Chad invoked the OBM. The struggle began in April 2006, at first with the World Bank, over the issue of oil-revenue monitoring. Chad's government then threatened to stop oil production if the consortium did not promptly deliver the $100 million it allegedly owed. By August (2006), two of the three consortium members—ChevronTexaco of the United States and Malaysia's Petronas—were forced to leave the country for failing to pay their taxes (both firms vehemently denied these charges). Chad's President, Idriss Déby, declared that his government would itself take control of the remaining oil reserves, and he demanded a majority stake in the project. In the end, Chad softened its demands, and the conflict ended when Chevron and Petronas agreed to pay some $280 million (Jakobsen 2012, 78–9).

This example exhibits all the signs of a classic OBM: a poor, desperate country is encouraged to undertake a large natural-resource development project; the IOCs secure a very lucrative initial deal; they then invest enormous amounts of money in a number of immobile assets (a 1,070-kilometer pipeline; oil production installations; and all the necessary infrastructure); and after production commences, and the commercial risks have evaporated, the host country begins to press for new terms—it asserts that the terms of the deal primarily benefit foreigners and curtail the sovereignty of an independent country. Leveraging the OBM, the host government can force IOCs (and here, the World Bank), to agree to new terms, demanding a significant share of ownership and a larger portion of the profits.

IOCs share many common interests, including the need to minimize costs, and to collect and manage data. This is one of the most important lessons of the Tripartite Model, the need to recognize the symbiotic and sensitive nature of these relationships.

While governments tend to rely on IOCs during the early stages of development, they hope to transfer that responsibility over to an NOC (or to domestic firms, and/or a combination of both) as quickly as possible. NOCs represent a popular option as they provide governments with a more direct route to promoting and controlling economic development, redistributing income, and advancing national pride. As we shall see in subsequent chapters, NOCs will share the government's long-term perspective. They can also be used to dispense jobs to favored groups and provide useful products and services, which governments can readily control for political purposes.[12] Finally, NOCs are popular because governments hope to use their direct ownership to control what is likely to be the most lucrative source of revenue in the country.[13] After all, many policymakers are afraid that foreign ownership could result in lost rents and less control over the pace of resource extraction, and even threaten their rule. While it is possible to control this source of revenue without direct ownership, most countries lack the administrative competence to ensure reliable control over private enterprises (especially IOCs). By contrast, it is often believed that NOCs are comparatively easier to tame (Grayson 1981; van der Linde 2000; and Victor et al. 2012, 9).

While most governments are eager to establish NOCs, they seem unaware of their potential threat. NOCs have a tendency to begin small, but they grow quickly. When they do so, they can wield significant economic and political power, and represent a serious challenge to government authority.

> Most NOCs were initially given, or grew into pursuing, a much broader set of tasks. These typically include contracting or undertaking infrastructure development, awarding and monitoring licenses, collecting hydrocarbon-related taxes, subsidizing energy and non-energy consumption, creating and maintaining employment and skill-building in related and less related areas, as well as representing the country in international fora. As its remit broadens, the distinction between operation and policy making begins to blur and the NOC gradually becomes a "state within a state." As a result, conflicts of interest seem "unavoidable," bear on operational and regulatory efficiency and raise concerns about the true location of political power. Now, who should monitor the NOC and, ultimately, based on what? Will multiple regulators and stakeholder groups deliver effective checks or be divided and conquered by the state-owned operator? (Boscheck 2007, 374)

This difficulty in maintaining political power over commercial interests with operational expertise—whether in the form of IOCs or NOCs—is the biggest challenge in petroleum administration. The power, money, and influence associated with these commercial interests can easily influence (and pollute) the policy and regulatory environments. The Tripartite Model is designed to help keep these commercial interests in check—to ensure that the operational experts within these commercial interests collaborate with policymakers and regulators. This is also the main lesson of the Norwegian Model, to which we now turn.

The Norwegian Petroleum Management Model

There are two faces to Norway's petroleum administration: one that looks outwards in its negotiations with IOCs, and one which looks inwards and considers how best to manage the resource for the benefit of the country (and its people). The remainder of this chapter describes these two faces.

International Constraints

Given the enormous capital investments involved, and the even greater risks, the Norwegian authorities were reluctant to play an active role during the early, exploratory phase. Indeed, Norwegians had little knowledge about the petroleum industry, and the country's authorities had little faith that the IOCs would discover oil;[14] at the time the government explicitly discouraged Norwegian industrial and shipping interests from entering the oil industry, citing uncertainty and high risk to investments (*St. meld. nr. 11 (1968–9)*, 7).

Given this uncertainty and ignorance, Norwegian officials had little to lose in courting the IOCs, and Norway's first licensing system provided companies with more flexibility (that is, bigger blocks during the exploration phase), lower fees and royalties, and more lenient relinquishment requirements, compared to the (competing) British system (Nelsen 1992). Not surprisingly, representatives of the IOCs were, at first, pleasantly surprised:

> Those of us coming from abroad thought of Norway as a small country, with little relevant legislation or practical experience. Worse, the country was governed by a socialist government and had a reputation for being bureaucratic, heavily–taxed and offering few possibilities for commercial expansion. But what did we find? A country that was able to assemble in record time, the most progressive and modern petroleum legislation; a country that met us with an uncommon openness and honesty—one that was willing to explain to us why things needed to be done in a particular way, or that was willing to take another look at the tax problem; and a country that managed to distribute its North Sea concessions in a uniquely short period of time and in remarkably thorough way.[15]

In fact, IOCs were beginning to lose faith in the Norwegian Eldorado; by the start of 1969, 16 wells had been drilled on the Norwegian Continental Shelf (NCS), but only one (Cod) showed any trace of hydrocarbons—and even this find proved to be commercially questionable. At this rate, each barrel of oil, on average, cost about 20 million kroner (Skule and Grytli 1997, 26)! It was only in the spring of 1970, after the impressive Ekofisk discovery, that the situation (and Norway's bargaining position) changed significantly. In four bold steps, the Norwegian authorities improved their position, relative to the IOCs. Each step leveraged the OBM.

First, and most evidently, Norway quickly rescinded the terms of its first licensing agreements. The ink had hardly dried on the official announcement to the second allocation round of licenses, when the government changed its mind on two of its most important components. In light of the (1968) Cod Field discovery, the authorities demanded higher taxes and a larger role for the state in future licensing arrangements.[16] The first came in the form of a new net profit sharing arrangement that resembled an income tax. In short, IOCs were required to pay a fixed percentage to the government (e.g., Esso was granted four blocks at 17.5 percent net profit participation rate).[17] The latter came in several different forms, including the use of a carried interest clause.[18] This is an arrangement where the government's interest is "carried" by the partners during the exploration phase; however, when petroleum is discovered, the government is given an option to participate. In other words, the government would not have to pay for the exploration costs, but could decide to contribute its fair share, only after a commercially viable discovery was made.

Just as importantly, by the mid-1970s, Norway was developing its own NOC: Statoil. The existence of a domestic competitor allowed the government to sharpen its demands to the IOCs, as it could now rely on a ready alternative, should the IOCs make good on their threat of withdrawing from the Norwegian sector (Ryggvik 2010, 52).[19] This was the second step along Norway's path to more autonomy; the government developed national competencies that could compete with, and eventually challenge, the IOCs.

Third, the government pushed to widen its influence and competence in order to secure control over the most important bottlenecks in the industry, such as pipelines and refineries. Following Rockefeller's earlier example in the United States, Statoil's first Chief Executive Officer, Arve Johnsen, pressured the government to secure control over how Norway's oil and gas were to be delivered to market (Ryggvik 2010, 26). Originally, the Norwegian Ministry of Industry wanted the IOCs to take responsibility for these delivery systems. State regulators were convinced that Norwegian industry (and Statoil, in particular) was not yet competent enough to deal with the challenges of developing an underwater delivery network. Government committees had been established in 1968, 1972, and 1974 to study the potential (with regard to the Cod, Ekofisk, and Frigg fields, respectively)[20] and each had concluded (on strong advice from the IOCs) that Norwegian competence was insufficient. Still, Arve Johnsen and Statoil persisted; they continued to push the state to control any subsequent trunk lines. Then, in the late 1970s, Statoil secured responsibility for developing the main (gas to land) trunk line to service the Statfjord Field. This was an especially challenging task, as the technology did not yet exist to lay pipe across the (remarkably deep) Norwegian Trench. By means of an impressive set of political maneuvers, Johnsen and Statoil wrested control of the project from Mobil; in 1981, the parliament approved the start of *Statpipe*; and by 1983 it was operational (with Statoil in control).

Finally, the Norwegian government pursued a deliberate go-slow policy of extracting and developing its petroleum resources. The Norwegian parliament made it clear from the very beginning that the pace of Norway's field development was to be slow and steady—in order to avoid undue hardships and bottlenecks while Norway developed its own competencies. By stretching out the timeline, the government was—in effect—strengthening its bargaining position (through OBM).[21] The government could be confident that it would approach future negotiations from a position of even greater competence, knowledge, and strength. Once Norwegian companies had become internationally competitive, the government lifted these constraints, and the pace of extraction reached breakneck speed (where it has remained).

In taking these four early steps, the Norwegian government was able to improve its relative bargaining position vis-à-vis the IOCs, and create more room for an autonomous petroleum policy, as described in the section that follows.

Domestic Components

Once Norway had secured better terms with the IOCs, the government turned its attention to developing a regime for managing the industry in a democratic fashion. Although it would take some time for this regime to develop, Norway's political elites, from across the political spectrum, clearly signaled their intention to maintain democratic control over developments. Consistent with the Tripartite Model, the Norwegian government allocated policy, regulatory, and operational responsibilities to three distinct (and relatively autonomous) government institutions.

This division of power is the result of a special committee that was appointed by the government to organize the new sector. In its brief to the committee, the government made it clear that it may consider exercising the options of state participation, which had been obtained through negotiations with some licensees (Industridepartementet 1971). Having studied how other countries had divided their administrative responsibilities, the committee proposed a three-way division of tasks within the Norwegian petroleum sector.

Similar to the Tripartite Model (Figure 4.1), responsibility for Norwegian policy rests primarily with the Ministry of Petroleum and Energy (MPE). Regulatory responsibility is based in an independent Norwegian Petroleum Directorate (NPD), and operational and commercial interests are vested in Statoil.[22] As Al-Kasim (2006b) points out, by separating the more political and regulatory functions from the commercial, the model promotes fairness, transparency, and integrity—to the benefit of both NOCs and IOCs. Al-Kasim's depiction is reproduced as Figure 4.2.[23]

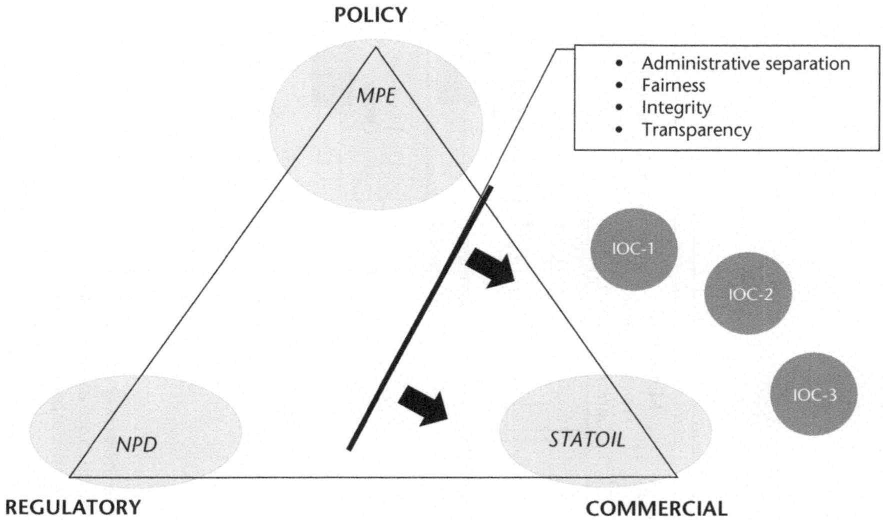

Figure 4.2 The Norwegian model of petroleum management[a]

[a] Adapted from Al-Kasim (2006b). MPE is the Ministry of Petroleum and Energy; NPD is the Norwegian Petroleum Directorate; and IOCs are the international oil companies.

Obviously, this is a simplification (as all models are). There is more to Norway's oil policy than what is carried out in the MPE. As we shall see, the parliament demanded a central role in the development of Norwegian petroleum policy, and continues to play an important role in deciding where (and when) an oil field can be opened for exploration. The same thing can be said of the regulatory and commercial sides of the triangle. Although the lead regulatory body is the Norwegian Petroleum Directorate, the complexity of the oil industry requires that the NPD must coordinate these regulatory activities across several other ministries, as shown in Figure 4.3. After all, the oil industry can affect many facets of a government's regulatory responsibilities: worker safety, environmental protection, and so on. For this reason, the Ministries of Environment, Labor, Transport, and Finance are all involved in regulating the industry.

Finally, to secure operational competence, the government needs to cultivate a healthy commercial environment, where firms are encouraged to develop the fields and necessary contracts, and be responsible for operational and marketing activities. To gain insight into and experience with these activities, the Norwegian government established Statoil. While Statoil began as a full-fledged NOC, it has subsequently morphed into a hybrid creature: a privately owned IOC, but where the Norwegian state is the majority shareholder. Recently it has been joined by two other companies (Petoro and Gassco) which resemble more traditional NOCs in their relationships to the government.

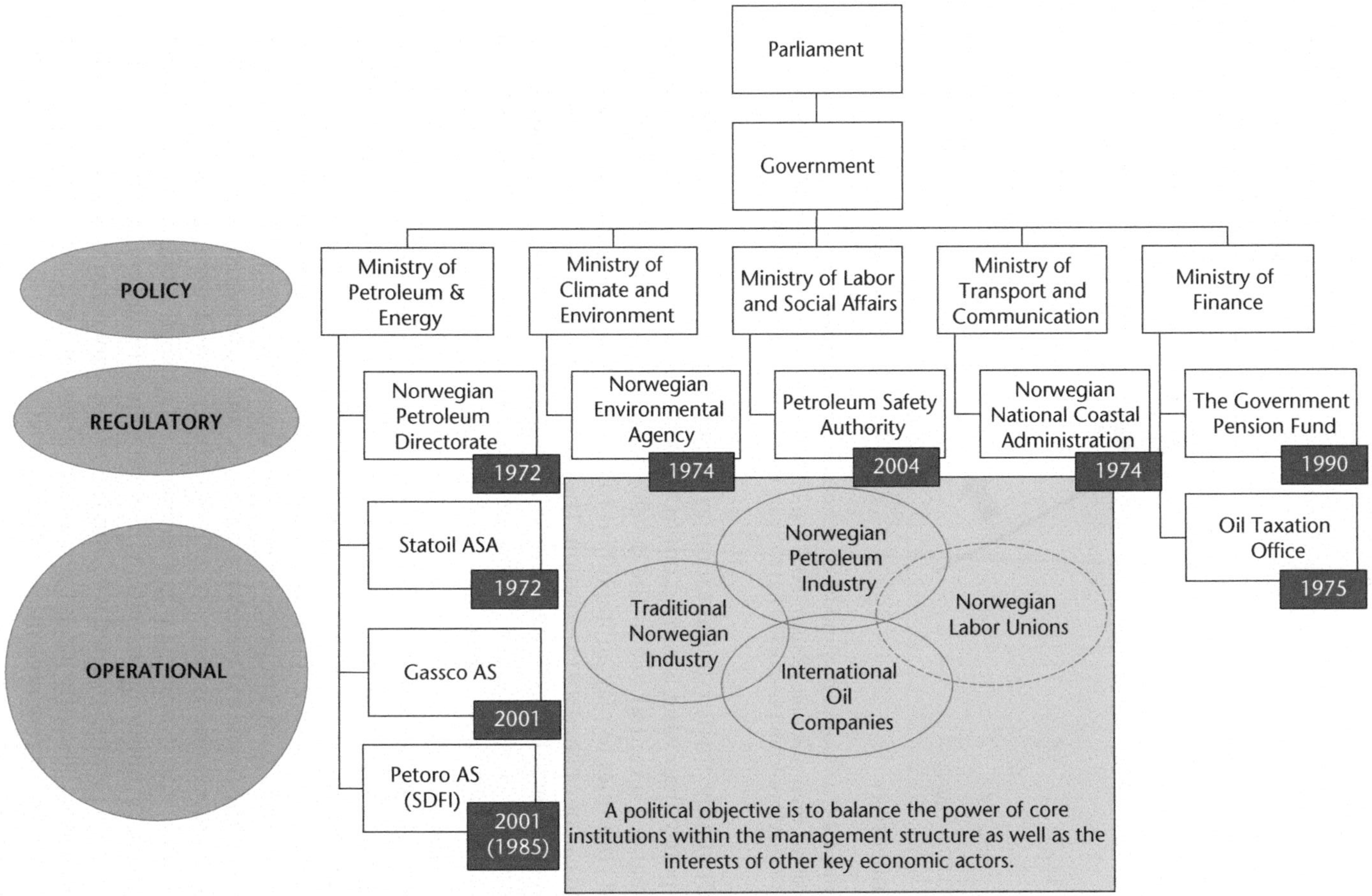

Figure 4.3 Institutional components to the Norwegian model[a]

[a] Adapted from NPD (2014, 31).

But perhaps the biggest shortcoming of this model (and most approaches to Norwegian petroleum management) is that it ignores the important role that a strong labor movement played in securing the norms, institutions, and outcomes that we associate with the Norwegian petroleum administration. We will try to rectify this shortcoming in the chapters that follow, but it is important to emphasize the significance of the labor movement and the corporatist nature of political decision making in Norway.[24]

The remainder of the chapter takes a closer look at each of these three main components of the Norwegian administration regime: policy, regulatory, and commercial interests.

POLICY

There was never any doubt about the need for a strong democratic component to Norwegian petroleum policy; the main outlines of that policy were hammered out in three founding documents, all from the 1970s.

The first of these policy instructions came in the form of an industrial committee that reported to parliament in June 1971 (*St. meld. nr. 76 (1970–71)*), in which the government was advised (among other things) to secure government control over all petroleum activities; encourage the establishment and support of a domestic oil industry; and create a state oil company that could manage Norway's economic interests. It is important to remember that these initial decisions about the nature of Norwegian petroleum policy were being made at the very time in which the Norwegian population was debating EEC membership (see Chapter 2). In other words, the merits of democratic control and national sovereignty were central components of the period's political vocabulary.

The list of demands in that government white paper came to be known as the Ten Oil Commandments (see Box 4.2) and many of these "commandments" became central components of the Norwegian petroleum administration, as we shall elaborate upon in subsequent chapters. In many ways, the Ten Commandments provide an explicit account of fundamentally Norwegian values, as they apply to the petroleum sector.

Two years later, in another parliamentary report, the Royal Norwegian Ministry of Finance (*St. meld. nr. 25 (1973–74)*) again emphasized the need to maintain national control over the resource. The chapter entitled "Democracy and Control," begins:

> Democratically elected institutions must have full control of all important aspects of petroleum policy: exploration, pace of extraction, safety measures and localization. It is important to have public direction and control of the exploitation of resources. (*St. meld. nr. 25 (1973–74)*, 9)

Box 4.2 THE TEN OIL COMMANDMENTS[a]

The Ten Oil Commandments are a declaration of principles underpinning Norwegian oil policy, submitted by the Standing Committee on Industry in a government white paper of June 14, 1971. These principles clarify how to ensure that the oil activities would "benefit the entire nation":

1. National supervision, control of all activity on the Norwegian Continental Shelf (NCS) must be ensured.
2. The petroleum discoveries must be exploited in a manner designed to ensure maximum independence for Norway in terms of reliance on others for supply of crude oil.
3. New business activity must be developed, based on petroleum.
4. The development of an oil industry must take place with necessary consideration for existing commercial activity, as well as protection of nature and the environment.
5. Flaring of exploitable gas on the NCS is only allowed in limited test periods.
6. Petroleum from the NCS must, as a main rule, be landed in Norway, with the exception of special cases in which socio-political considerations warrant a different solution.
7. The state involves itself at all reasonable levels, contributes to coordinating Norwegian interests within the Norwegian petroleum industry, and to developing an integrated Norwegian oil community with both national and international objectives.
8. A state-owned oil company should be established to safeguard the state's commercial interests, and to pursue expedient cooperation with domestic and foreign oil stakeholders.
9. An activity plan must be adopted for the area north of the 62nd parallel which satisfies the unique socio-political factors associated with that part of the country.
10. Norwegian petroleum discoveries could present new tasks to Norway's foreign policy.

[a] Adapted from *St. meld. nr. 28 (2010–2011)*, 8

The main point of this report was to emphasize the need to develop strong democratic foundations for Norwegian petroleum activities, and that the wealth from oil should be used to develop "a qualitatively better society" in Norway. In addition, however, this report signaled two other objectives. First, it noted that the pace of extraction needed to remain moderate:

> Wishing to prioritize a long-term perspective in the exploitation of resources, and after a comprehensive evaluation of its social aspects, the Government has concluded that Norway should take a moderate pace of the extraction of petroleum resources. (*St. meld. nr. 25 (1973–74)*, 6)

Second, the report signaled a need to plan for the threat of Dutch Disease, by hinting at the possibility of something like a future petroleum fund. The parliament was urged to be cautious as to how it spends the incoming oil revenues:

> [A] significant share of the public revenues from the petroleum activities must be invested abroad. It is essential that this be done in such a way that it will be useful, even after the petroleum activities have ended and will contribute to reducing the problems that can accompany a phasing out of such activity.
>
> (*St. meld. nr. 25 (1973–74)*, 18)

This advice seems to have been ignored at the time; it was not taken up in the subsequent debate, and it did not seem to affect actual policy in the subsequent decade. We will return to this subject in Chapter 7.

By the end of the decade, in 1978–79, the third foundational document was put into place (*St. meld. nr. 53 (1979–80)*)—in which the Labor Party took a more activist approach to managing the country's oil resources. *St. meld. nr. 53 (1979–80)* "marked a milestone in the development of Norwegian oil policy" (Al-Kasim 2006a, 65), in that it consolidated Norway's approach, by emphasizing four fundamental principles:

- maximize Norway's economic gain from the resource;
- contribute to the country's social and economic development;
- develop and maintain strong environmental and safety standards; and
- develop the resource in a slow, careful, and deliberate manner.

More specifically, this report proposed that Statoil, on behalf of the state, should receive large shares in future licenses and assume a more active role in decision-making (and operations). These four principles became the core political objectives of the new management regime that developed in the mid-1970s.

At the start of the Norwegian oil age, policy development was left in the hands of a very small group of experts, with remarkably little legal or bureaucratic support. Most of Norwegian petroleum policy can be traced back to a few people in the Oil Office at the Ministry of Industry (Gundersen 2007). In these early years, policy was the responsibility of something called the Norwegian Petroleum Council (NPC)—founded in 1965, with Jens Evensen as its chair. The NPC functioned as an advisory body regarding the exploration and extraction of subsea petroleum deposits on the NCS, and it was the licensing authority in the first licensing round. Because the circle of policymakers was small, and the regulators were largely unfamiliar with the industry, Norwegians could (and needed to) adapt to—and learn quickly from—changing conditions on the ground. As the workload grew, a separate Oil Office was established in 1966 at the Department of Mines in the Ministry of Industry. This Office quickly expanded into the Division of Oil and Mining, before a new, independent Ministry of Petroleum and Energy (MPE) was established in 1978.

The internal organization of the MPE changes with each government, but we can gain a general idea of its mandate by looking at its most recent organizational chart, as depicted in Figure 4.4.[25] While the MPE also has a number of responsibilities that lie beyond the oil and gas sectors, our focus is trained on the fact that the MPE is responsible for all petroleum operations on the NCS; it ensures that these operations are pursued in accordance with the guidelines set by parliament.

As early as 1972, the parliament made it clear that it wanted a say in how Norwegian petroleum policy would develop. It required the government to submit all of its field development plans for parliamentary approval, in order to influence the state's investments in joint ventures; to assess the economic and social impacts of the development plans; and to make sure that the resources were being managed efficiently. This level of active political management declined over time, along with the government's ability to control Statoil. As Norwegian commercial interests became more competitive (and outward looking), the Norwegian authorities unleashed the pace of petroleum development and allowed Norwegian and foreign companies to compete on a more equal footing. Still, overall policy responsibility remains with the MPE, the parliament continues to decide on the opening of new areas for petroleum activities, and the government still awards the licenses.

It is important to note that Norway did not rush to pass comprehensive petroleum legislation. Its first petroleum law (1963) was mostly aimed at declaring sovereignty over the country's offshore oil fields. Until the mid-1980s, oil legislation in Norway focused on establishing the most essential rules for regulating activities on the NCS. The legal basis for activities came in the form of royal decrees, which were designed to address issues/requirements that needed regulation at specific stages of development. Only with the Comprehensive Petroleum Law of 1985 was a more all-encompassing legal framework for the industry introduced—*after* the authorities had gained substantial experience with regulatory practices.

REGULATION

Regulating how a country develops its petroleum resources requires commercial acumen, broad-ranging competencies, and political finesse. Managing these resources requires adequate information (and data), as well as the knowledge and experience necessary to analyze that information in a way that can best serve the long-term interests of the country. This information and expertise are necessary to manage the resource effectively, but having this information can also strengthen the government's hand vis-à-vis the IOCs when the two meet to negotiate.

The Norwegian model relies on an administrative middle man, a single institution that can house the (petroleum) expertise and competence that the country needs when deciding what is possible, or how best to proceed. For this institution to function, the regulatory authorities need to have a good working

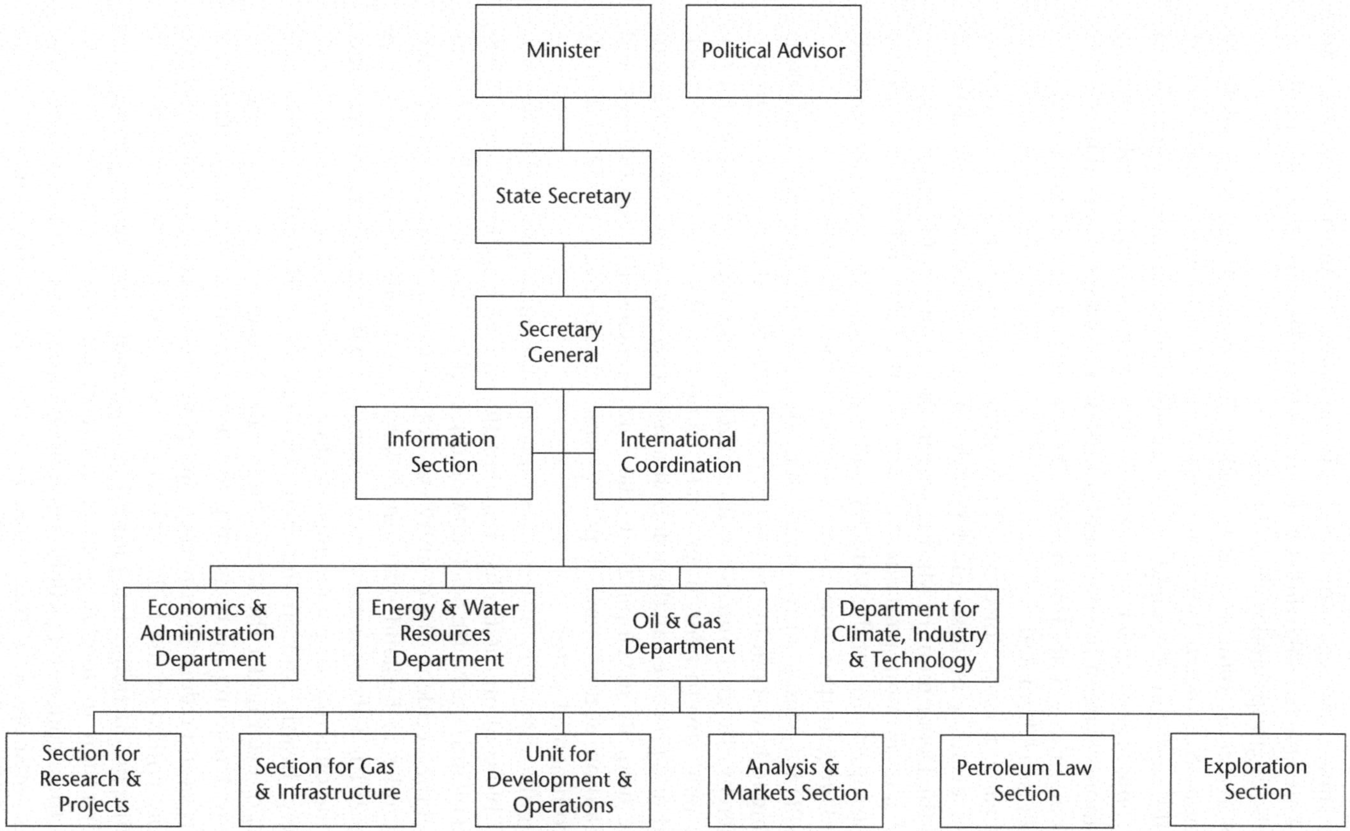

Figure 4.4 The Norwegian MPE[a]

[a] Adapted from MPE (2013).

relationship with both policymakers and the commercial players in the field. At the same time, these regulators need to be far enough removed from each, so that they are able to make difficult decisions that can have dramatic ramifications for both groups (and society at large).

The Norwegian solution was to collect this competence in a relatively autonomous Norwegian Petroleum Directorate (NPD), which was to report to the MPE (see Figures 4.3 and 4.4). Over time, however, it was necessary to slice off some of the regulatory responsibilities from the NPD, as policymakers became more aware of the potential conflicts of interests involved in regulating an economic activity that has significant safety and environmental consequences (see Chapter 9).

In the early days of Norway's oil adventure, when the regulatory requirements were rather modest, most petroleum activity was monitored by an office in the Department of Mining within the Ministry of Industry. On July 14, 1972, a unanimous parliament established the NPD and placed it directly under the supervision of the Ministry—although it had its own (independent) Board of Directors (this board was abolished in the early 1990s).

The NPD was designed to ensure that the country's laws and regulations were being followed out in the field. But the NPD was also seen as an institution whereby the government could collect technical competencies for the Ministry (and government, in general). To do this, the Directorate mapped, collected, and processed all the information and geological material coming in from the field. By developing internal competencies, the authorities could develop a more autonomous oil policy—they did not need to rely so heavily on the commercial sector.

Two significant events rattled this simple administrative model. The first was the 1977 blowout on the Ekofisk Bravo platform, which spilled about 200,000 barrels of oil into the North Sea (NOAA n.d.). Luckily, there was no loss of life as a result of this accident. The country was not so lucky in 1980, however, when the Alexander L. Kielland platform collapsed and sank, claiming 123 lives (Melberg 2010). Following these two accidents,[26] Norway reflected more critically on the difficulty of combining resource management (on the one hand) and safety, health, and environment protection (on the other) under one Ministry, or even under the same Directorate. As one might expect, the NPD was opposed to losing any of its authority; it argued that a separation of the two functions would reduce its efficiency. But policymakers and the public at large were more critical of the NPD's capacity to wear two hats at the same time (promoting production, while protecting workers and the environment).

The first step in resolving this conflict of interest was taken in 1979, when matters concerning safety, working environment, and emergency preparedness offshore were transferred from the NPD to the MPE and to the Ministry of Local Government and Labor (now Labor and Social Affairs).

From that time forward, the NPD had to report to more than one ministry. The second step came in 2004, when the Safety Division was extracted from the NPD and became the Petroleum Safety Authority of Norway (PSA(N)) (see Figure 4.3).[27] This was done to ensure that the supervision of safety and the working environment would be separate from the management of the country's petroleum resources. The resulting safety and environmental regulatory regimes are considered in much greater detail in Chapter 9.

Today, the NPD has a staff of over 200 people, and is responsible for maximizing the greatest possible value for society from Norway's petroleum resources. Its contemporary role is complex and manifold. It is responsible for the data collected from the NCS (NPD 2015, see also http://www.diskos.no/); it uses this data and its expertise to advise the MPE in a way that will realize the resource's full potential (that is, focus on long-term solutions, upside opportunities, economies of scale, and so on); and it works with the other authorities to ensure that the government's wishes are carried out in the field (NPD 2011).

COMMERCIAL INTERESTS

Since issuing the Ten Commandments, Norway's political elites have been clear about their interest in forming an NOC. After all, this intention was made explicit in Commandment 8, that:

> A state-owned oil company should be established to safeguard the state's commercial interests, and to pursue expedient cooperation with domestic and foreign oil stakeholders.

This should not be surprising; the early 1970s were a time of increased radicalization, and several countries were moving to control this strategic resource under the auspices of a national company. In fact, Norway was one of the few European countries that did not yet have direct state involvement in the oil industry: the United Kingdom had BP; Italy had ENI; and France had Elf/Total. In 1974, the UK had formed another state-owned oil company, the British National Oil Company (BNOC); and in Sweden, the Swedish Association of Oil Consumers (OK) ran gas stations and refineries as a cooperative. At the same time, it was recognized that a Norwegian NOC could support the state in realizing its nine other commandments. In this light, one could argue that the establishment of Statoil was the most important single instrument used by the state to ensure that oil activities would "benefit the whole nation" (Austvik 2012, 324).

It was much easier to develop a political consensus about forming an NOC than it was to decide what it should look like. As the birth of Statoil occurred at a time when the country was being torn apart over membership in the European Community, varying political constellations had competing ideas about what the NOC could (and should) look like. Worse, the scope for an active petroleum policy was not at all clear, at least until after the referendum was decided.

Even today, the operational component is the most controversial in the triangle used to depict the Norwegian petroleum management model (Figure 4.2). For example, Norway's Oil for Development (OfD) program focuses only on the first two components (policymaking and regulation), as it navigates around the political reefs associated with the commercial interests:

> The key example of this danger was a set of articles in the *Financial Times* in May 2006 accusing Norway of supporting Bolivia's nationalization of its oil industry as part of the OfD initiative. This concerned Norwegian officials considerably as it both questioned the independence of OfD, but also could potentially damage the reputation of Norwegian private companies. A policy of ensuring a "firewall" between Norway as a development partner and Norwegian commercial interests was thus established. (Norad 2013, 19, n. 1)

Many political duels have been fought over the nature and size of the Norwegian NOC, and Statoil itself has changed dramatically over time (as a result). Indeed, Statoil's earliest champions would probably not recognize the company's current form (not to mention its international ambitions). In Chapter 10 we return to the subject of Statoil's role in the world.

The first attempt at creating a Norwegian NOC came from the center-right Borten government. In late 1970, this government commissioned Hambros Bank in London to secretly secure 51 percent of the shares in Norsk Hydro, on behalf of the Norwegian state. The idea was for the government to buy a controlling share of an existing private company, and use it as an NOC. This initial attempt was stopped dead in its tracks.[28] When a new, Labor Party government came to power in March 1971, it quickly changed directions: Labor wanted to build a new NOC from the ground up (Johnsen 1988, 9). Fifteen months later (June 14, 1972), Parliament unanimously agreed to the creation of a state-owned oil company, and by September 18 (a week before the EEC referendum), Statoil was holding its founding general assembly.

Established as a state-owned limited company,[29] "the Norwegian State Oil Company" (*Den norske statsoljeselskap a.s.*), or Statoil, was subject to the same legislation and organizational structure as a private limited company, with a General Assembly and a Board of Directors. It differs from private companies in that the state (here the MPE) owned all the shares and could thus determine the Board's makeup. The Board was responsible for choosing (and, if necessary, firing) the company's Director.

The supporting white paper that launched Statoil contained mechanisms that were intended to assure parliament of adequate political controls. Statoil was designed to be the keystone to Norwegian petroleum policy (*St. meld. nr. 25 (1973–74)*, 9). In particular, Statoil was required to provide an annual report to parliament on "significant issues relating to principles and policy." This "§10 plan," became the basis for wide-ranging oil debates in the Storting in the 1980s. These sorts of democratic control on the NOC are typical in the Norwegian petroleum management model (in the early years of its existence):

> In this way, both the composition of the board and the parliamentary debates represented important democratic control mechanisms. The wide-ranging oil policy debates which these arrangements made possible were by no means purely symbolic or a distraction. The insight which they enabled into the company's strategic choices contributed to raising the level of knowledge about oil questions among politicians and other interested parties. Thus in many cases the debates became an expression of genuine popular democratic involvement in questions of oil policy. (Ryggvik 2010, 100)

Over the decades that follow it is possible to discern an important cleavage dividing the Left from the Right over the nature of Statoil. The Labor Party came to embrace a large and strong Statoil—one that was big enough to compete directly with the IOCs. Toward that end, Statoil was to become a fully operative state-owned company, with a mandate that included all phases of operation, both upstream and downstream.[30] The Conservative Party, by contrast, was continually worried about the lack of control over Statoil within the Norwegian system. From this (Conservative) perspective, Statoil should be constrained to act as a holding company that managed state ownership shares, without developing operational capacity. In short, the Norwegian Right wanted to limit Statoil's size and influence, so it would not become a state within the state (Osmundsen 1981; Høyre 1981).[31]

Once the company was formed, the Norwegian state did everything it could to further its interests. Political control was strictly maintained by the responsible minister, who appointed the Board and set the groundwork for Statoil's policy. Parliamentary control was secured by introducing the §10 constraint (an annual report to parliament) by the Conservative Party (in opposition), with support from an unlikely ally in the form of the radical Socialist Left (SV) Party.

Statoil clearly benefited from receiving advantages that the state, as landlord, could grant through its licensing policy. In particular, Statoil enjoyed high initial shares in the various licenses (between 1973 and 1985 Statoil received 50 percent or more of licenses),[32] and it could use its voting power as the clear majority shareholder; the right to be carried during exploration; the option to increase state participation by high percentage increments in accordance with the maximum daily production (sliding scale shares); and the promise of IOC help in developing its expertise and capacity. We will elaborate on the use of many of these instruments in subsequent chapters. For now, our point is simply to show how Statoil was designed to function as an instrument for furthering the Norwegian state's interests.

This policy of favoritism was incredibly successful, such that Statoil's economic position in the 1980s was becoming very large, relative to Norway's overall GDP. In 1983, Statoil conducted a study predicting that it would be responsible for 70 percent of Norway's total oil and gas production in 2000, while Norsk Hydro and Saga together would capture less than 10 percent (Skule and Grytli 1997, 34). In effect, Statoil had become too big for its host country.

Cutting Statoil down to size involved two major reforms.[33] The first of these was to spin off some of the company's gas wealth and expertise. In 1983, the MPE appointed a Gas Committee to coordinate the country's gas activities. This move was followed in 1985 by another: the establishment of the Gas Sales Negotiations Committee (*Gassforhandlingsutvalget*, or GFU)—made up of Norway's three largest oil companies: Statoil, Norsk Hydro, and Saga. GFU was established to reduce Statoil's powers and to develop expertise in gas negotiations in more than one Norwegian company (Al-Kasim 2006a, 71); it was intended to offset downstream market power and it excluded foreign companies (because of their alleged downstream stakes).

The second major reform (*St. meld. nr. 73 (1983–84)*) was more noticeable, in that it literally split the state's involvement in oil activities in two: one part went to Statoil, the remainder went to the State's Direct Financial Interest (SDFI). SDFI was created on January 1, 1985 as a separate juridical entity (yet, originally, still managed by Statoil) to collect and manage the country's oil rents. The intent was to keep Statoil from being able to control (by itself) a decision-making majority in any given license group, and hence control the development and operation of the field. As part of the deal, Statoil was allowed to keep its share of fields which were already in production or close to completion. At the same time, Hydro and Saga were allocated more fields, to develop their competence and market share: Hydro was given Oseberg, while Saga received Snorre (Skule and Grytli 1997, 34).

From 1984 until 2001 (when Statoil was privatized), Statoil was not allowed to use its SDFI shares for the purpose of voting within license groups, unless it received written instructions to do so from the MPE. SDFI was organized as a state holding company, without operational activities, and it was managed by a small administration (in practice, at first, an office in the MPE). Still, it controlled a great deal of money; in 2000, SDFI was managing oil reserves that were about three times larger than Statoil's (Ryggvik 2010, 99).

At the end of the 1980s, Statoil began to think of itself more as an oil company, and less like a *national* oil company. Having enjoyed the protections of an infant industry, the company now wanted to be freed from its political obligations and compete in the world beyond Norway's borders.

Toward that end, Statoil's leadership worked actively to free itself from the §10 debates in parliament, arguing that its reporting requirement was revealing information to competitors about strategic decisions within the firm. By the 1990s, the §10 report was reduced to a short account which rarely provided a basis for any real debate.

In 2001, Statoil's links to the state were severed further, when the company was fully privatized and made a public limited company, listed on both the Oslo and New York Stock Exchanges. At the time of the privatization, the government retained 81.7 percent of the shares, but that share has reduced over time. Today, the Norwegian state remains the largest shareholder in Statoil, with a direct ownership interest of 67 percent—and this interest is managed by the MPE. But Statoil is not an NOC in the sense that it promotes government policy—the government uses it more as a milk cow than an instrument of policy. Indeed, the Norwegian government is reluctant to use its majority shareholder position to affect policy decisions, for fear of threatening its market value (which determines the size of its dividends). To underscore the importance of appearing apolitical, privatization was also used as an excuse to repeal the §10 constraint.

That same year, in 2001, the EU introduced its Gas Directive—and this affected the way in which Norway could organize its economic activities in the petroleum sector (via the European Economic Area (EEA); see Chapter 2). EU internal market provisions have always affected Statoil's activities and its role in the Norwegian energy sector. The GFU, Statoil's 50 percent share, the "carried interests," and the "sliding scale" have always been in clear violation of the EU's antidiscrimination standards. For over ten years, the Norwegian government fought the EU on these issues, and protected its political autonomy. But in 2001, the Norwegian government decided that Norwegian industry was strong enough to compete, and it embraced the EU Gas Directive, privatized Statoil, and abolished the GFU. As a result, Norway's petroleum policy relies less on Statoil, which is now a private company, and more on two other companies: Petoro and Gassco.

When Statoil was privatized, the SDFI became a state holding company, Petoro, which was tasked by the MPE with three main responsibilities. Most importantly, Petoro manages the state's holdings in its sundry joint ventures. Although Petoro (like SDFI), doesn't have any operational responsibility, it represents the largest owner on the NCS (in addition to Statoil). As such, it is tasked with safeguarding the state's interests in these joint ventures by protecting assets and creating value. Petoro's second responsibility is to monitor Statoil's sale of the petroleum produced from the SDFI. Finally, Petoro is tasked with the financial management of the SDFI, including responsibility for its accounting activity. Here the company strives for the highest possible value creation from the SDFI (that is, economic objectives are prioritized over political, social, or ethical objectives). In 2011, this work resulted in a net cash flow to the government of 100 billion NOK, or about a third of the government's revenues from the petroleum sector (Petoro 2015).

The EU Gas Directive also signaled the end for the GFU (Claes 2002, chapter 5). Prior to June 1, 2001, all Norwegian gas was sold through the GFU. While many different companies (including the SDFI, now Petoro) own Norwegian gas, GFU was run by the three aforementioned Norwegian companies. In other words, instead of letting the different (international) producers compete, GFU would set a price for all Norwegian gas available for purchase. The EU claimed that this arrangement violated fair trading practices, and it threatened Statoil (and Norsk Hydro) with huge fines and legal action. As a result, a new state-owned company, Gassco, was created to operate all of Norway's natural gas pipelines (since January 1, 2002).

Like Petoro, Gassco is 100 percent owned by the Norwegian government. It does not own any gas reservoirs or infrastructure, but instead administers access to an extensive gas delivery regime. As such, Gassco operates all the leading pipelines for landing gas and the most important land-based facilities.

Conclusion

This chapter has introduced the Norwegian approach to petroleum administration, and showed how it is an example of a Tripartite Model, where the state's policy, regulatory, and operational interests are coordinated, yet anchored in relatively autonomous institutions. This model was most evident in the early years, when Norway was first establishing its relationship with the IOCs and establishing its own (domestic) industry.

Over time, Norway has become less active in steering the petroleum sector. There are several reasons for this. First, and foremost, the Norwegian petroleum industry has become internationally competitive, and it sees its future in the world beyond Norway's (dwindling) resources. We return to this theme in Chapter 10. The Norwegian model was useful for protecting this infant industry when it was first getting established, but it is less necessary today, when Norwegian companies have become powerful international market actors. As a backdrop to these changes, it is important to remember that the world oil market, and the surrounding political contexts, has changed a great deal since the 1970s—when this market was first being established.

We wish to close this chapter with a caveat and a plea. First, the caveat: although the Norwegian model is frequently touted as a set of important reforms that states can adopt to improve performance and enhance transparency (e.g., Collins 2003; Al-Kasim 2006b; Nore 2009), there is no scientific basis for these claims. One reason for this is the difficulty in trying to operationalize institutional independence; another reason is the difficulty of extending lessons from Norway to developing countries.[34] Thus far, the literature has focused mostly on a handful of cases to illustrate the role that independent institutions can play—and the lessons from these case studies are anything but clear. To our mind, the best example is Thurber et al. (2011), who focus on the role that the Norwegian model can play in improving performance. They conclude their study by noting "Reformers in oil need context-specific guidance as to when the Norwegian model is or is not a good idea" (Thurber et al. 2011, 11).

More research is clearly needed, and this brings us to our plea. We need to develop better indicators of institutional autonomy that can travel across varying contexts (cross-national and temporal). For example, it is possible to conceive of a political indicator, which considers how the decision-making body in each institution (policymaking, regulatory, operational) is selected/appointed, or an economic indicator (that considers where an institution gets it funding). But there are many other institutional elements that vary across contexts—for example, some countries have multiple NOCs (Norway has these now, as does China); some countries open the sector up to international and domestic competition (again, Norway does this; as does Brazil), or privatizing the NOC (again, Norway and Argentina in the 1990s). There is much work to be done, and your authors are currently working to develop these sort of cross-national indicators and tests.

More scholarship has been directed on how states manage to secure their revenues from the commercial sector, and how the size of the government's take varies across regime types. It is to this important issue that we now turn.

5

Commercial Frameworks

This chapter considers the sundry ways that a government secures its share of the wealth from petroleum. This government share is determined largely by two sorts of frameworks: (1) the nature of the underlying contractual/licensing regime; and (2) the nature of the government's revenue or fiscal system. It is a common, but inaccurate, belief that the former determines the latter—that is, that the choice of the contracting regime will determine the size of the government take. In practice, there are so many variants and instruments available—within each type of contracting arrangement and revenue system—that it is difficult to generalize about this relationship. The devil, as always, lies in the detail. We begin by providing an overview of the alternatives available to different states (on both the contractual and fiscal fronts), and then describe the changing nature of Norway's licensing and fiscal regime.

Government Take

We start by recognizing that in most countries *the people* (more accurately, their government), not private entities, own mineral rights—both onshore and offshore, regardless of who actually owns the property on the surface. This view came to dominate after the 1962 UN Resolution on Permanent Sovereignty over Natural Resources; and the principle was subsequently enshrined in the constitutions of many newly independent, mineral-rich countries. At the same time, a series of UN Resolutions encouraged that change—the most salient was Resolution 2158 from 1966, which called upon oil-producing countries to gain full control of their natural resources (UN General Assembly 1966). Although there are exceptions (and they are important)—such as the United States and some parts of Canada, where private ownership is possible—in most countries, mineral rights are owned by the state. Because the state controls these underlying mineral rights, it must decide on the most effective, judicious, and cost-effective means to get the petroleum out of the ground and into markets, while maximizing the government's share of the resulting revenue.

On discovering it may have gas/oil, a government usually begins by dividing up the potential development area into blocks of territory (offshore or onshore) in a systematic fashion. These blocks can vary significantly in size, but tend to be quite large (in the North Sea, they range between 100,000 and 125,000 acres). It then has to decide how to develop these blocks.

The first, and perhaps most important, decision concerns whether the contract or license should be granted by way of an auction to the highest qualified bidder, or whether states should use a more discretionary approach. Auctions are likely to deliver better economic results, and they tend to be easier to administer (this is a distinct advantage in states with underdeveloped competencies). But the results of these auctions are fairly predictable: international oil companies (IOCs) will end up with a preponderance of the licenses/contracts. For most states, this is politically untenable; authorities hope to encourage a diversity (large, medium, small) of companies, and especially domestic companies, to develop the resource. Discretionary approaches allow the government to achieve this, as they can employ other criteria (than cost) when allocating contracts, and use these criteria to prioritize local interests (of course, it is this sort of discretion that is often abused by corrupt governments).

The government would prefer that several companies enter into a competitive bidding process to secure a license to explore and develop the block—as competition increases the chance of a higher return for the government. But states are not always able to entice several (competing) bidders.[1] To begin with, countries are competing with one another to attract the attention of a limited number of IOCs (and their resources). Newcomers to this competition may not get the attention they desire/expect, as their terms may be too demanding. After all, the level of IOC interest (or the willingness to invest or enter a competitive bidding process) depends critically on the initial explorations—usually a few seismic surveys, or a geological interpretation of a nearby wildcat. Hence, the IOCs may not want to engage in what appears as a risky project, and the list of uncertainties can be long and intimidating, as they relate to geology, prices, costs, production, investments, technology, strategy, and politics. Their initial bids will reflect that underlying risk and uncertainty.

Contract Types

The government's next step is to decide on the nature of the underlying contractual system—that is, how licenses for exploring and developing the blocks will be distributed. The scope of these licensing arrangements can vary significantly, but they all determine how rents and costs will be distributed between the host government (and/or its national oil company, or NOC) and any interested IOCs. It would be surprising if governments and IOCs did not favor different forms of contracts. Governments tend to prefer familiar contract forms (e.g., similar to those they have already used in the mining or hydroelectric sectors), while IOCs often place a premium on owning proven oil reserves, that is, "booking barrels"—as this allows them to add assets to their balance sheets, and hence improve their market value.[2] In addition, the government needs to decide on the scope of the contract. Some contract agreements may require the use of local content/producers; others might instruct IOCs to support and carry the NOC. In countries that lack extensive administrative/regulatory traditions, governments may even use the licensing agreement to extend regulatory authority over environment and health standards to the contracting IOC. These particular requirements are discussed in subsequent chapters.

Most licensing arrangements today include a "carrying" clause, which provides the NOC with experience and the revenue needed to develop the nation's resources, while minimizing its exposure to risk. Formally, a carried interest refers to a context in which a working interest partner (usually an NOC) pays a disproportionately lower share of its costs and expenses (than its working interest share) during the exploration or development phase of a contract. Typically, NOCs are carried through the exploration phase, at which point the government takes up a working interest (or "backs in").[3]

Already at this level, before we describe the nature of the sundry contracts, we stumble across a serious challenge in petroleum management. It is common industrial practice to keep the terms of these contracts confidential. Both government and IOCs allude to the importance of protecting proprietary information. While a "model contract" is often distributed and discussed in public, the actual details of the contract remain out of the public's critical gaze. (We discuss a specific example of this, in Tanzania, in Chapter 10.) Because of this, there is enormous potential for corruption in these underlying contractual arrangements.[4] As companies have an incentive to make illegal payments to government officials in order to influence outcomes on lucrative deals, it is difficult to know whether a contractor was chosen because of its competence, its competitive bid, or because it had a beneficial relationship with an influential government official. Worse, the complexity of the contract system and the sundry fiscal tools that the state uses to secure its resource rent, makes it very difficult to police.

States usually draw from among four main contract types: concessions (royalty/tax) agreements (CAs), joint ventures (JVs), production sharing contracts (PSCs), and service contracts (SCs).[5]

- *Concessions (Royalty/Tax) Agreements.*[6] Under a CA, the government grants ownership of the oil, for a given period of time, in return for the payment of royalties and taxes.[7] At the wellhead, the IOC lifts its share of the oil—the total minus the royalty share (e.g., if the royalty level is set at 10 percent, then the IOC lifts 90 percent, and the state gets the remaining 10 percent). The IOC then pays an additional tax on the sale of its oil. These arrangements tend to be relatively straightforward (compared to the alternatives), and require less judicial/financial expertise on the part of the host country. At the time of bidding, however, there is usually inadequate information about the block's potential. This makes it difficult for the government to secure generous (and competitive) bids, as IOCs will be cautious when bidding on risky (as yet unproven) blocks.
- *Joint Ventures (JVs).* JVs vary significantly in form,[8] but each represents a form of partnership between two or more parties (usually between an IOC and an NOC, but they can also include several IOCs, or a combination of IOCs and NOCs). These partnerships are attractive in that they allow partners to share in the high levels of risk and capital that characterize the industry. JVs split ownership of the oil (e.g., 50/50) at the wellhead, but the IOC usually shoulders all the exploration risk (via carrying clauses). The IOC is often designated as the block/field operator, but governments can help with the management duties (e.g., via a joint management committee, or by approval of the work program and budget and so on). This makes the JV a useful instrument for transferring competence across partners (from the IOC to the NOC). JVs are a form of partnership more than a contract type; hence, JVs are found in other types of contract arrangements (e.g., in CAs and PSCs).
- *Production Sharing Contracts (PSCs).*[9] In a PSC, the state retains ownership of the resource, and the IOC is invited to develop the sector. These are probably the most predominate forms of contract in the world of petroleum. In these contracts, the IOC bears responsibility for exploration and production (E&P), and hence carries all the investment risk. When successful (that is, commercially viable), the IOC is reimbursed for the cost of production in the form of oil (cost oil). The remaining oil that is produced (profit oil) is then distributed between the government and the IOC (on the basis of negotiations).[10]
- *Service Contracts (SCs).* In SCs, as in PSCs, the government retains ownership of the resources in the ground. But the IOC's relationship to the petroleum is even more tenuous in SCs, as the IOC is simply hired as a subcontractor and paid a fee for its service (while in a PSC, the IOC gets a share of the oil produced). Originally, service contractors were limited to different stages of the production process (exploration, drilling, development), but hybrids

and integrated service contracts are more common now. SCs can vary in scope and the degree of risk involved: technical assistance contracts provide technical assistance to governments, and tend to be limited in scope; pure service contracts can be broader in scope, but entail little or no risk for the contractor, while risk service contracts can include some risk (and more substantial reward options).

Each contract form offers varying levels of control, risk, and reward. There is also much overlap among, and nuance within, contract types. While each of these contract types still exist, and each tends to predominate in different regions of the world, their makeup has slowly changed to reflect increased state power and influence (relative to the IOCs),[11] as we already hinted in Chapter 2. This change in the balance of power, and the development over time, can be seen in the type of contracts, as they vary from left to right in Figure 5.1, that is, from CAs to PSCs and (increasingly) to SCs.

As the international oil industry began in the United States, where concessions systems dominate, CAs were most common in the early years. These agreements were often set in a context of political dependency (colonies, dependencies, protectorates, empires, and so on), and hence reflected an underlying (unequal) balance of power. In CAs, the IOCs maintained exclusive exploration rights and controlled the timetable and process of development. This made it difficult for governments/NOCs to transfer relevant skills/experience to local firms and workers, and the level of government payments/royalties were relatively low.

In the early 1950s, host governments slowly came to replace or supplement these CAs with JVs and SCs. In response to demands for greater state

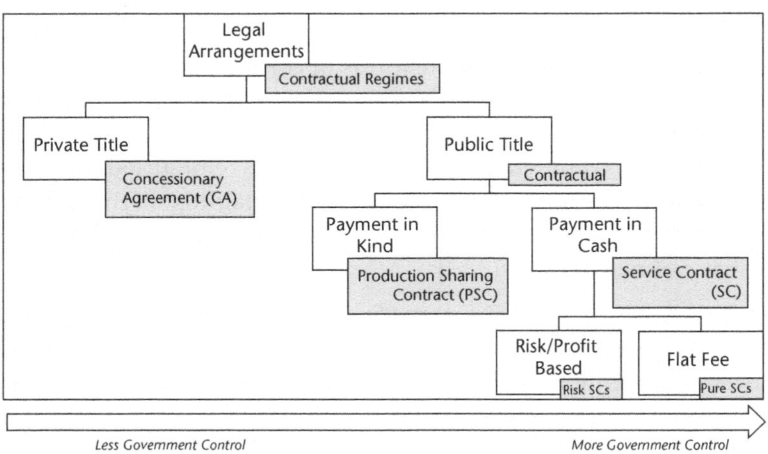

Figure 5.1 Contract classification

participation, many of the original CAs were amended to include JVs and so-called "Participation Agreements," which host governments used to secure better access to technology and knowledge, to influence production decisions, and to increase their government take.[12] Participation agreements increased government revenue and control, but they fell short of satisfying the host state's desire for sovereign control of the resource. Ownership and management still remained in the hands of the IOCs, because they usually held the majority interest in the resulting partnership. JVs were more promising in that they allowed for greater transfer of technical expertise, and hence better subsequent control (over operations, production, management, marketing). Better yet, from the perspective of governments, were the advent of PSCs (1966) and the extended use of SCs. Even greater control could be transferred to the host government with these contract types.

Today's JVs, PSCs, and SCs can take many forms, and are characterized by varying degrees of detail/coverage. The rise of NOCs has freed states from their heavy dependence on IOCs (and the nature of the relationship between host government and its IOCs is changing as a result). IOCs have been forced to adapt continually in order to remain relevant to NOCs. This has meant the rise of a number of oilfield service companies (OSCs), which take on a widening variety of contracting roles.[13] On the other hand, IOCs often try to avoid political risk by requiring very extensive and detailed contracts, where all eventualities are covered, including stabilization clauses that lock in existing regulations and terms.

Fiscal Systems

The contract arrangement establishes the parameters for commercial development of the sector. The contract will determine the level of state involvement, when the state/NOC will back in, and many of the fiscal particulars (e.g., how the state's revenues will be generated; the level of royalties; the cost recovery/ deduction framework; and how the state and the IOCs will divide any resulting profits). It is as this point that we wade into some of the essential details of a state's fiscal system.

Before we do, however, we need to remind ourselves of the unique nature of the petroleum market, in that it generates enormous rents. The resource rent is, in effect, a gift to the people of a country; in each of the contract types, the government tries to maximize the size of this gift.

When the oil/gas is sold on the market, the resulting revenue can be divided into three main components, as shown in Figure 5.2. Some portion of the revenue must cover the total *costs* (exploration, development, production, marketing, and so on). These costs will vary with the nature of the rent/gift; they will be lower in states where the oil is easy to access, and of higher

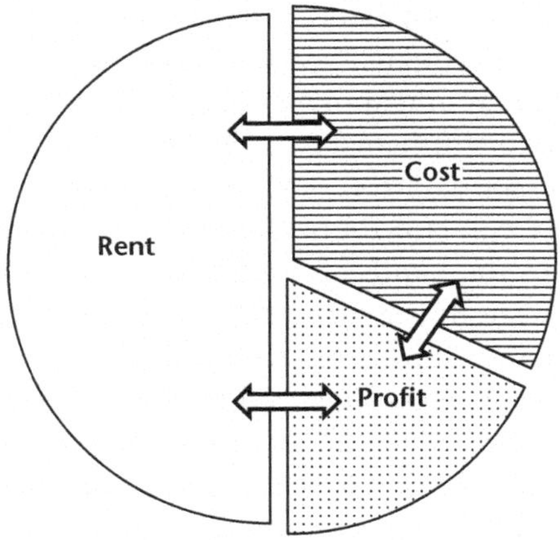

Figure 5.2 Major components of petroleum revenue

quality. The cost element can also be subject to negotiation (over what can be considered a legitimate cost, and subject to strategic decisions over how to develop the resource). The remaining revenue is split between a return (*profit*) for the IOC's investment/activities, and the resource *rent* (which belongs to the country). The size of this profit/rent trade-off is also subject to negotiations; the government seeks to offer sufficient profit to attract/reward IOCs, while pocketing as much of the remainder as rent. This rent can be secured in any number of different ways (e.g., royalties, taxes, fees, and so on). In short, the boundary lines separating rent from profit, profit from cost, and cost from rent are pliable and subject to negotiation.

The main task of the licensing system is to ensure that the IOC gets a fair return on its investment, but does not abscond with the underlying rent (which belongs to the country). How this is achieved varies across contract types. In some contracts, royalties will play a larger part; in others, states will secure their share in the form of taxes, rents, or participation. The overall cut is colloquially referred to as the "government take," although there is no agreed-upon means for measuring or comparing this across countries.[14]

FISCAL ELEMENTS

As there can be much variance within and across each contract type, it is perhaps easiest to look at the different types of fiscal tools that governments employ to obtain their rents. There are four main types of fiscal tools that the state can use to secure a bigger slice of the revenues, and the use of these tools varies across contract types.

- *Participation/Equity*. One important way for a state to secure a larger share of revenue is to ensure that its NOC plays a role in bringing that resource to market. By introducing PSCs, JVs, carried interest clauses, and/or diverse profit-sharing measures, the government (usually the NOC) is able to participate directly in the economic activity and earn its corresponding share of the return. With access to the profits, or the profit oil generated by the NOC, the government can use these revenues as it sees fit. Obviously, the way in which the government taps into these participation revenues will vary with the form of NOC/government participation.

- *Royalties*. Royalty arrangements also vary in content. At the most general level, we tend to distinguish between fixed percentage arrangements and those with sliding scales. A fixed percentage royalty does not take costs into consideration, but requires a contractor to pay a fixed rate (e.g., 12.5 percent) on all oil/gas produced. With a sliding scale, the size of the royalty changes in accordance with an agreed upon variable (e.g., level of field production; cumulative production; level of well production; location; production rate and price; R factor;[15] internal rate of return; or even particular field qualities such as water depth, oil gravity, and so on). Royalty arrangements also vary with respect to the point at which the royalties are determined (e.g., in/at the field, wellhead, block boundary, export terminal, and so on).
- *Taxes/Fees/Bonuses*. The host government also has access to a number of additional and more commonplace revenue instruments. The most common of these is the use of taxes (corporate income, dividend withholding, transfer, transit, environmental, and so on). Governments can also require bonus payments (e.g., signing, discovery, and production bonuses) when important milestones are met, or penalties in the wake of unexpected or unfortunate circumstances. Finally, the government can employ fees or levies. Among these are rental fees, where the IOC is required to pay a fixed rental payment, usually on an annual basis (e.g., for a fixed amount of the contract, or on the size of the field under its operation, or for a particular phase—for instance, exploration, production, and so on).
- *Investment Incentives*. The final set of instruments is associated with reducing the costs of production for IOCs. As part of the contract negotiations, the IOC and the government can agree over what is considered a legitimate cost of production, what the depreciation schedule should look like, and so forth. The list of possibilities is long, and costly: tax holidays; accelerated depreciation; tax credits; current expensing of exploration and/or development costs; exempting duties on imported equipment and capital goods; unlimited loss-carry forward; and so on. These sweeteners make it more attractive for IOCs to invest in the country—but the government loses revenue in the process.

One of the government's biggest challenges concerns the need to value a company's economic activities, so that this activity might be taxed accordingly. Because IOCs are international, they are accustomed to exploiting differences in national tax systems. Hence, the issue of transfer pricing is particularly vexing in the oil industry, and can cost the host country a great deal of money if proper controls are not put in place. Transfer pricing refers to the internal (non-market) pricing of goods or services when transferred within an organization, but among affiliates in different countries, with different tax regimes. Transfer pricing allows an organization to shift the site of its value creation to low-tax jurisdictions, minimizing their tax burden in higher-tax jurisdictions.

In the oil industry, examples of transfer pricing are infamous. IOCs rely on affiliate firms working in several different segments of the market (upstream, midstream, downstream), across many different tax jurisdictions. Hence, an IOC's affiliate might be exploring and producing oil in a Latin American country, on the basis of research and development work carried out by another of the IOC's affiliates in Europe. The crude oil produced might be shipped to a refinery in another country owned by yet another affiliate, with an affiliated shipping company registered under (still yet another) foreign flag. If the IOC can determine the price at which the oil is sold/bought between its various affiliates across these different tax jurisdictions, it can maximize its value creation in the jurisdiction with the lowest taxes.

In a 2000 speech to the Norwegian Ship Owners' Association, Michael Hudson described how Exxon's (then) treasurer, Jack Bennett, used transfer pricing to shuffle around the company's global earnings so that they would be taxed at the lowest possible rate. The oil companies, he noted:

> duly assigned all the profits from their billions of dollars of oil-well and refinery investment to their shipping affiliates, and reported these profits in the tax havens in which these "international" affiliates were registered. Of course, no taxes were owed on them. In this way the U.S. oil industry managed to avoid paying income taxes either in Europe or America. (Hudson 2000)

States use a plethora of tools to secure their share of the resource rent, and the particular mix of tools can vary across countries and time.[16] Because of this complicated collection of tools, it is impossible to generalize about the effect of the contract system on the size of the government take (indeed, there is little agreement as to how we should measure that take). Even though it is possible to distinguish between tax-based and production sharing-based systems, the ability of a state to secure a larger share of the resource rent depends on the particular way in which that a country assembles a host of menu options, and its skill in negotiating favorable outcomes with the IOC on each and every menu item.

While IOCs are able to influence decisions regarding a country's choice of contract and fiscal regime, and the particular tools within each regime, the sheer variety of options employed is evidence enough that states enjoy a great deal of latitude in deciding how to secure the government take.

Norway's Government Take

The remainder of this chapter considers how Norway secures its government take, and how its approach has changed over time. Our focus is mostly on domestic factors, but readers should be aware that the absolute size of the government's take (the size of the pie) also depends on the nature of the international context, for example the rise of OPEC; the influence of IOCs; the types and level of taxes employed by other states; and so on. In what follows, the section "Norwegian Licensing Arrangement" describes Norway's unique licensing system; the latter section, "Norway's Fiscal Toolbox," describes the changing set of fiscal tools used by the government to secure its petroleum rent.

Norwegian Licensing Arrangements

Before Norway could establish its own licensing arrangement for the blocks on the Norwegian Continental Shelf (NCS), it first had to ensure its legal right to the oil. When Phillips Petroleum first approached the government in 1962 to secure (what it hoped would be) exclusive rights to Norway's oil, legal ownership of the NCS's mineral rights was unclear. As a fishing nation, Norway had claimed fishing rights that extended just 12 nautical miles off the coast—but everything beyond that was considered international waters.

As mentioned in Chapter 2, the 1958 Geneva Convention on the Law of the Sea provided some guidance on this measure, but Norway had not yet ratified the convention in 1962, and the practice of offshore oil drilling was not very well established yet (consequently, the legal precedents were still under development).[17] After a round of hectic negotiations with both the United Kingdom and Denmark, the territorial boundaries of Norway's petroleum resource were largely settled, and announced by a cabinet decree in May of 1963.

With its ownership secured, the government employed a CA system, based on its previous experience with regulating the hydroelectric industry earlier in the same century (that is, *hjemfallsretten*). We would like to emphasize three central aspects to the Norwegian concession system. First, the Norwegian system is a negotiated one, in which political authorities employ the concessions to secure greater social and political objectives. It is not an auction, and it is not a state planning system. The system can secure these objectives because the government actually creates the JVs that develop the resource. Although individual companies (or groups of companies) apply to participate in production on the blocks, the Ministry of Petroleum and Energy (MPE) considers these applications and then negotiates with the applicants, to create a JV—that is, a group of companies, or license group. In short, the MPE had enormous influence because it decided the composition of these license groups, which company would be the operator in each group, and then distributed the work obligations within each license group.

This negotiated process was formally introduced in the second round of concessions, when the government wanted to reduce the number of blocks being offered (from 78 down to 14), while still including as many IOCs as possible in the process. But this decision reaped significant political rewards, as it allowed the Norwegian authorities to build licensing groups that could include nascent Norwegian firms, while encouraging IOCs to teach Norwegians the tricks of the trade, within the resulting JVs.

The creation of a license group might look like a government mandate, but it is the result of an interactive process, as depicted in Figure 5.3. The government begins the process by asking companies to nominate blocks for consideration (within a given geographic area). The Norwegian Petroleum Directorate (NPD) and the MPE then evaluate the nominated areas, and propose a formal announcement (which is then sent to other government entities, and to a public hearing, for comment).[18] Companies are then invited to apply for a production license. The NPD assesses these applications (in light of the applicant's geological understanding, technical competence, and financial capacities);[19] and other authorities assess the applicants' record with regard to their competence for managing operations in general and the health, safety,

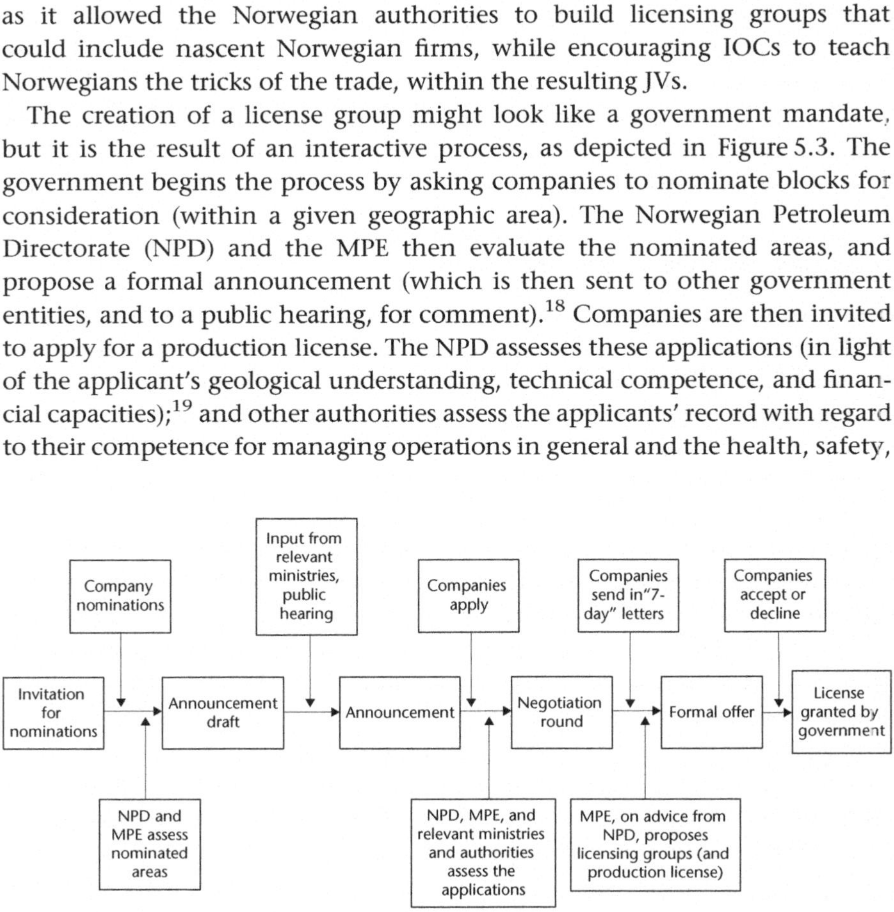

Figure 5.3 The concession granting process[a]

[a] Adapted from Riksrevisjonen (2010, 19). The original figure was from 2010; we have changed the institutional references (to bring them in line with contemporary institutions). While this figure describes a later process (with more steps), the main components remain throughout the period.

and environmental aspects (in particular). The MPE then proposes an area, work program, and participant share for each production license. This proposal is then negotiated with the companies involved, and the state's share of the bounty (the state's direct financial interest, or SDFI) is discussed with the Finance Department. The companies can accept, reject, or propose another arrangement (in what has become known as the "7-day" letter, because firms are required to respond in seven days). After the negotiations are complete, the MPE assembles a license group and selects its operator.[20]

This sort of negotiated outcome, where companies can influence policy, is an extension of the corporatist arrangements that are commonplace in Norway (as described in Chapter 3). Routines are developed in which all the relevant actors are expected to participate in (and influence) the making of policy. This provides policymakers with a great deal of flexibility, information, but also influence. Although the labor movement's influence is less explicit here (than in an incomes policy setting, for example), Labor is able to influence the process at several points along the way: when the proposals are being assessed by the different ministries (e.g., in the Ministry of Labor and Social Affairs), in the public hearings, and through Labor's substantial political influence (on government).

In relying on a concessions system, Norway chose a familiar path—one that had proven relatively successful in its past. In this system, foreign owners are allowed to own the oil at the wellhead, but this control is limited to an agreed-upon period of time. As in the hydroelectric sector, control of the underlying resource eventually returns to the state.[21] Norway did not feel it needed to experiment with alternative contract forms, as its administrators were already familiar with the existing system, and it seemed to work well: it maintained political control, generated impressive revenues, and yet still attracted foreign capital and expertise.

A second important component of the Norwegian concessions system is the role it played in building up Norwegian competence and expertise. As this element concerns building local content, more than securing state revenues, we discuss it in more detail in Chapter 8. For now, we simply want to point out that the government's authority to create license groups (which included Norwegian firms) was leveraged with other requirements that helped expand Norwegian participation (e.g., use of carried interests, a sliding scale, and technology and goodwill agreements).

Finally, it is important to emphasize that the Norwegian concessionary system changed frequently over time, making it difficult to summarize. In fact, one can distinguish between four distinct regimes, over time:

- Before 1972, the government bent over backwards to attract IOCs and to provide them with competitive terms;

- From 1972 to 1981 the system was used to expand Norwegian participation, by securing dominant shares in allocated blocks, and by ensuring that its interests (e.g., in Statoil) were carried;
- In the late 1970s, the system was used to expand the use of local providers/content (employing goodwill and technology agreements); and
- Since 1985, license allocations have become less explicitly political. They are now aimed at developing a more level playing field for commercial actors and better utilization of more mature fields.

For our purposes, the earliest developments are perhaps the most interesting, in that they show how the state was able to use the concessionary system to ensure a transfer of competence and resources to Norwegian firms and authorities. But we will close this section by looking at more recent changes to Norway's licensing regime.

DEVELOPMENTS OVER TIME

At the outset, the government seemed content to attract as many IOCs as possible, and get them to work in a responsible manner. Jens Evensen summarized the first concession round with respect to four objectives: they wanted to ensure that companies would conduct thorough explorations; that Norway would be able to secure a reasonable economic return; that the extraction of the resource would be carried out in a responsible manner, under public control; and that the activity would not be detrimental to other sectors (Midttun 2003, 14). In this first allocation round (1965), about 15 percent of the NCS was made available. This was the largest allocation round, in terms of raw acreage, in Norwegian history. In the end, 78 blocks were awarded to nine groups (including Norwegian partners).[22]

After oil was discovered, the concession rounds were used to increase government participation. We have already mentioned the use of carried interest clauses to help float Statoil and the nascent Norwegian oil industry. In 1974, the government introduced a second important tool for expanding state participation: a sliding scale—where state participation increased with the size of a field's production. By using this tool, Norway's level of state participation approached 80 percent in some of these concessions![23] The effects of these measures are evident in Table 5.1, where we can see that the Norwegian share of licenses, and the share of licenses that are operated by a Norwegian company, quickly increased over time.

In the early years, concession rounds were announced at irregular intervals, and additional awards were announced outside the regular (concessions round) process. These extraordinary allocations (the italicized rows in Table 5.1) were often used to award the most promising blocks (e.g., Statfjord in 1973 and Gullfaks in 1978), and Statoil was given a large share of these—50 and 85 percent, respectively (Skule and Grytli 1997, 31). Until the fifth concession round (1980–82), all activities were limited to blocks below 62 degrees north—as the government did not want to rush father north, where the challenges of E&P (owing to climate and weather) were formidable.

At the turn of the millennium, two important modifications took place. First, a majority in parliament agreed on the need to make it easier for smaller companies, and group application of licenses (*St. meld. nr. 39 (1999–2000)*). A couple of years later—in the wake of the Barents Sea project in 1997, and the North Sea

Table 5.1 Norwegian shares in license allocations, 1965–98[a]

Year	# Blocks	Concession Round	Additional Award	Norwegian Share	Norwegian Operator Share
1965	78	1		8	0
1969–71	14	2		15	0
1973	*2*		*Statfjørd (037)*	*52*	*0*
1974–1978	22	3		58	63
1977	*2*		*Ula (19B)*	*50*	*0*
1978	*1*		*Gullfaks (050)*	*100*	*100*
1979	8	4		58	68
1980	12	5		66	92
1981	9	6		64	50
1982	*1*		*Oseberg (079)*	*100*	*100*
1982	5	7		60	80
1983	*3*		*Troll (085)*	*100*	*100*
1984	17	8		60	60
1985	13	9		60	55
1985	*1*		*Additional Award 112*	*67*	*0*
1985	9	10A		64	67
1986	9	10B		65	56
1986	*1*		*Additional Award 129*	*67*	*100*
1987	22	11		59	62
1988	16	12A		58	38
1989	13	12B		64	67
1991	36	13		66	64
1991	*1*		*Additional Award 185*	*69*	*100*
1992	*2*		*Additional Award 085B*	*69*	*100*
1993	31	14		68	100
1995	*1*		*Additional Award 018B*	*8*	*0*
1995	*1*		*Additional Award 050B*	*100*	*100*
1995	*1*		*Additional Award 114B*	*90*	*100*
1996	47	15		53	44
1997	*19*		*Barents Sea Project*	*56*	*69*
1998	*1*		*Additional Award 037B*	*52*	*100*
1998	*2*		*Additional Award 019C*	*50*	*0*
1998	*1*		*Additional Award 053*	*100*	*100*
1998	*2*		*Additional Awards 102B &103B*	*50*	*0*
1998	*3*		*Additional Award 114C*	*90*	*100*
1998	*1*		*Additional Award 128B*	*87*	*100*
1998	*1*		*Additional Award 237*	*70*	*100*

[a] Adapted from Oljedirektoratet (1998, 110). Rows in italics are extraordinary allocations ("Additional Awards").

allocations in 1999—the parliament agreed to introduce a new type of concession system (*Innst. S. nr. 87 (2002–2003)*). Norway continues to allocate exploration licenses, but these are not exclusive, and they do not confer the right to drill. The real money lies in production licenses, which are awarded to companies that apply (and qualify) in one of two different types of licensing rounds: the traditional concessions rounds (which occur every other year), or Awards in Predefined Areas (APA),[24] which are presented annually. In recent years, more blocks have been allocated through the APA scheme than through the traditional (numbered) concession rounds (Riksrevisjonen 2010, 11–12).

The first group is reserved for underdeveloped (immature) blocks with unknown geology, technical challenges, and inadequate infrastructure. These are located in large parts of the Barents Sea, and in the deep-water areas and northernmost regions of the Norwegian Sea. The APA scheme was introduced in 2003, and is designed to be used in more mature/developed areas, where the geology is known, where there are fewer technical challenges, and where there is already a well-established infrastructure. The APA licenses were introduced to increase the effectiveness and simplicity of the allocation process in more mature areas, in order to better exploit existing infrastructure, and to provide more stability/predictability for firms working in the area.

The resulting production licenses provide firms with a right to prospect, drill, and produce in the specified areas. The formal definition of a Norwegian production license is provided in Box 5.1. These licenses are usually awarded to groups of companies and tend to be awarded for five years, but with the right to extend them for 30 years over 50 percent of the original area—provided that the work program has been fulfilled.[25]

Norway's Fiscal Toolbox

The Norwegian government secures its resource rent through a variety of instruments, and these instruments have changed over time. This section describes those instruments, and some of their more unique features. To do this, we begin with an overview of the main sources of government revenue, and then turn to describe the nature of its two most important components.

Box 5.1 THE PRODUCTION LICENSE IN NORWAY

"A production license entails an exclusive right to exploration, exploration drilling and production of petroleum deposits in areas covered by the license. The licensee becomes the owner of the petroleum which is produced."

—Norwegian Petroleum Act of November 29, 1996, No. 72, Chapter 3, Section 3-3.

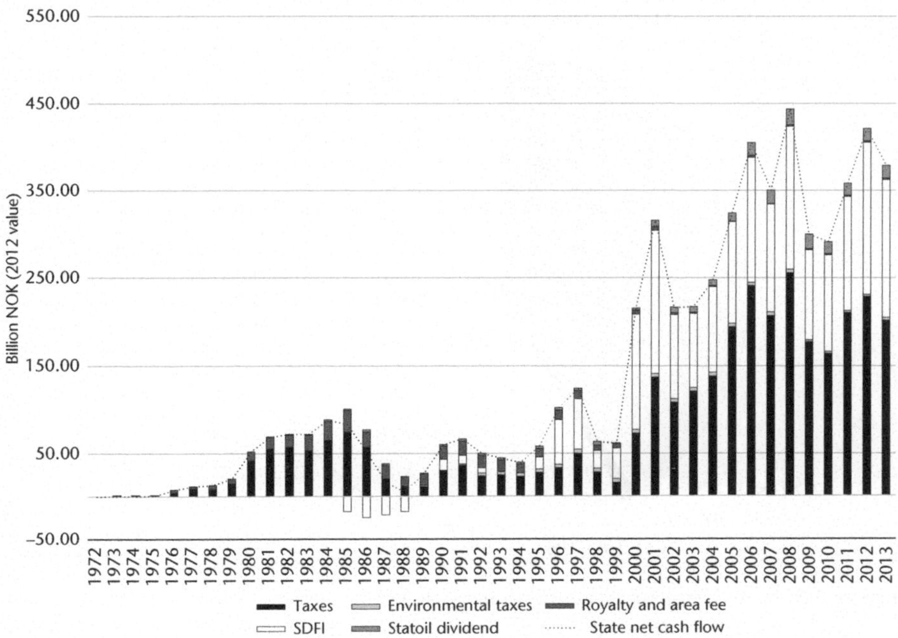

Figure 5.4 Net government cash flow from petroleum activities, 1972–2013[a]

[a] 1972–2011 is adapted from NPD (2013, Figure 3.4, 22); 2012–13 is from *Meld. St. 3* (*2013–2014*), 48.

If we look at the main sources of the government's cash flow from petroleum activities, as shown in Figure 5.4, we can see five important characteristics. First of all, these revenues have been largely growing over time, and are substantial. In 2013, they totaled almost 346 billion NOK (Deloitte 2014, 21 and 28). Second, taxes have always played the most important role in supplying the government with money from the petroleum sector, and this revenue source continues to dominate. Third, the SDFI has become the second most important revenue source, and has become almost as large as the tax revenues. Fourth, royalties—which are usually considered the central component in a CA system—have always played a minor role, and were phased out in the year 2000. Finally, the economic reward from Statoil, in the form of dividends, is hardly noticeable.

If we take this dynamic picture, and look at the most recent year (2013), as we do in Figure 5.5, we can get a better idea of the main sources of government revenue. In 2013, the dividend from Statoil provided the government with just 4 percent of its petroleum revenues. Most of the money is coming from taxes and fees from Statoil and other companies (58 percent), and from the SDFI (36 percent). For this reason, it behooves us to take a closer look at these two main revenue sources: the Norwegian tax system, as it has developed over

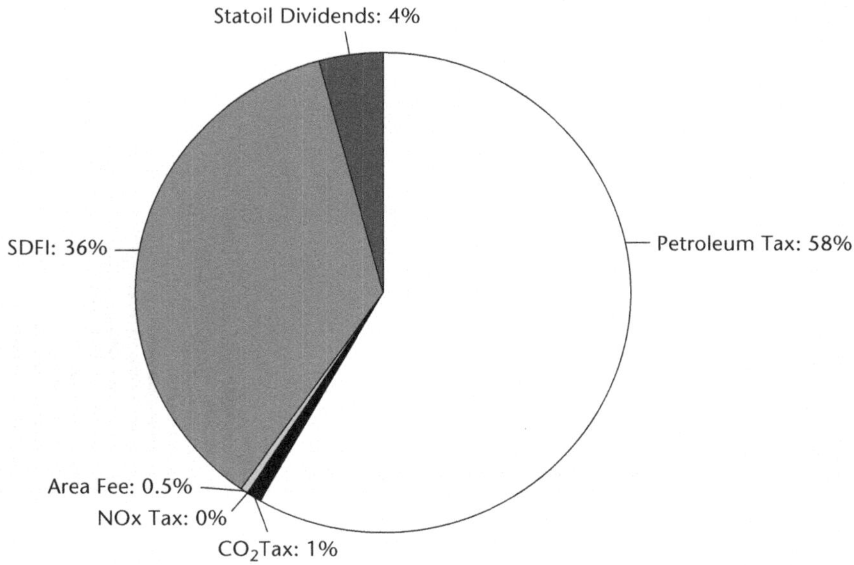

Figure 5.5 State petroleum revenue sources, 2013[a]

[a] Adapted from Deloitte (2014, 21 and 28). This constitutes the government's declared revenue stream. Total petroleum revenue in 2013 was 345.6 billion NOK.

time, and the SDFI. These two sources alone capture almost 95 percent of Norway's petroleum revenues.

THE TAX SYSTEM

From the very beginning, Norway sourced most of its petroleum revenues from taxes. With the first allocation round, in 1965, a "Tax Financial Committee" was established to work out tax issues (as some of the IOCs worried that the proposed government take might be exorbitant). One of the most important considerations, at this early moment in time, was to ensure that the IOCs would find it attractive to explore, thoroughly, the possibility of oil and gas on the NCS. In commenting on this committee's proposal for a new tax law, parliament's Finance Committee concluded:

> The committee has considered it an important objective to prepare the ground in such a manner that the oil companies would not, as a result of taxation provisions, refrain from undertaking the necessary investments that are required to initiate exploration of the subsurface in the Norwegian Continental Shelf.
>
> (Al-Kasim 2006a, 169–70)

Table 5.2 Important tax changes

1965	Act on Taxation of Subsea Petroleum Resources, adopted June 11 • Tax relief in the form of reduced rates, to compensate for uncertainty (and competition by neighboring states)
1972	1965 Tax Act abolished • Industry to pay taxes like everybody else • Up to 1975, taxes on general income, net worth, area fees, royalty, production bonus systems, net profit sharing licenses
1975	New Petroleum Taxation Act No. 35 (June 13) • Tax rates increase significantly • New special tax added • Norm price system introduced • Companies very upset about change in rules
1991	CO_2 tax introduced • Aims to reduce CO_2 emissions • Levied at a rate per standard cubic meter (Sm^3) of gas burned or directly released and per liter of petroleum burned
2005	Tax reimbursement system introduced • Reimbursement for exploration costs to entry barriers for new actors and to encourage economically viable exploration activity
2007	NOx tax introduced • Pursuant to Gothenburg Protocol of 1999, Norway is obliged to reduce annual emissions of nitrogen oxides (NOx)

Even before the concessions were granted, then, the parliament had passed exceptional measures which allowed reduced taxes for oil companies, and even a reduced royalty level—from 12.5 percent to 10 percent (Ryggvik 2010, 17).

Once oil was discovered, the other shoe dropped. The government was less concerned about scaring off potential investors, and more interested in protecting the government's resource rent. As mentioned in Chapter 4, the most important change came in 1974–75, when the government introduced an extraordinary petroleum tax of 25 percent of net income, against strong opposition from the IOCs (see Table 5.2). Still, the bill passed through a unanimous parliament (Skule and Grytli 1997, 29).

This new tax law provided the basic framework for taxing the sector in the years to follow. Oil companies were to pay an ordinary company tax at the same rate as all other (e.g., onshore) companies (about 28 percent), with amortization over six years for investment in production facilities, pipelines, and transportation. Then, on top of the normal income tax, a special tax of about 50 percent was placed on oil companies, to be calculated on the basis of net revenue. This tax is aimed at securing the resource rent.

This same basic system is in place today, as described in Box 5.2. This tax system is quite sophisticated, in that it is able to adapt to different stages in the development of a project, and it awards the government a substantial amount of money. Although the tax rates have changed slightly over the years, the basic framework remains. Today's tax rules are based on the ordinary

Box 5.2 NORWAY'S CURRENT PETROLEUM TAX SYSTEM[a]

Sales income (valued at norm prices)

- Operating costs (inclusive of exploration costs and indirect taxes)
- Depreciation (16.7 percent for six years)
- Interest costs
- Losses carried forward from previous years

= Ordinary tax base taxed at 28 percent.

- Uplift (including unused uplift carried forward from previous years)

= Special tax base taxed at 50 percent.

[a] Adapted from Ministry of Finance (2013). In 2014, the ordinary tax base was changed from 28 percent to 27 percent, while the special tax base was raised from 50 percent to 51 percent. After May 5, 2013, the uplift was reduced from 30 percent to 22 percent. See Deloitte (2014, 16).

corporation tax (27 percent) and the addition of a special petroleum tax (51 percent). Both taxes are based on the company's net profits, and all expenses relevant for the activities on the NCS are tax deductible. Investments are favored by a high depreciation rate. The uplift allowance allows a company to deduct 30 percent more than it invests, against the special tax.[26] In short, the system favors marginally profitable projects because the uplift allowance shelters profits from the full effect of the special petroleum tax (Radon 2005, 77).

In addition, the 1975 tax law introduced a new means for calculating the tax: it established a "norm price" for petroleum, to make it more difficult for IOCs to use transfer pricing to avoid paying their fair share of taxes in Norway. An institution called the Petroleum Price Board (*Petroleumsprisrådet*, or PPR), sets these norm prices for oil. The PPR meets quarterly to set the norm price for the previous quarter. As a rule of thumb, they set a norm price for each day, for each type of raw oil. They retrieve pricing information from a variety of different sources, including the oil companies, and they meet with those companies before the final norm prices are set.[27]

This tax pressure varied over the years, in response to a changing international price for oil, and additional taxing instruments were introduced, as outlined in Table 5.2, including a CO_2 and NOx tax (introduced in 1991 and 2007, respectively). As we saw in Figures 5.4 and 5.5, these additional taxes and royalties add little to the government's cash flow. In 2005, a new reimbursement system was introduced to reduce investment risks. This system allowed investors to recover 78 percent of their investments, even if they had not paid any taxes (they were refunded in cash). The reform was aimed to encourage production of smaller fields (by smaller companies, with lower costs/overheads), and longer tail-end production.

STATE'S DIRECT FINANCIAL INTEREST (SDFI)

The second most important source of government revenue from the petroleum sector is the SDFI. The SDFI is a system by which the state owns a share of many production licenses, fields, pipelines, and land-based facilities. That ownership interest is set when the production licenses are awarded, and it varies in size from field to field. For example, the SDFI has a 5 percent interest in license 018, and a 50 percent interest in license 169C.[28] As a partner in the license, the SDFI covers its share of costs/investments, and is entitled to its share of the income generated by the license.

As we mentioned in Chapter 4, Petoro is responsible for managing the state's share of these licenses and partnerships. Since it was established in 2001, the number of licenses managed by Petoro has grown from 80 to 179 (on January 1, 2014), along with 15 pipelines and land-based installations. As shown in Table 5.3, Petoro reported that SDFI generated 194,308 million NOK in 2013. Most of this money goes directly to the Norwegian treasury (outside the regular taxation system, described in the previous section, "The Tax System").

We would like to close by noting the difficulty of tracking down these diverse sources of revenue, as they are collected by different arms of the Norwegian government and paid for by a number of different commercial actors. Norway's system is complex, and it has developed over a long period of time, with the help of a large number of experienced administrators. Luckily, Norway was the first OECD country to implement the Extractive Industries Transparency Initiative (EITI) criteria, so its books are more transparent than most. But discrepancies occur, even in Norway,[29] and need to be tracked down

Table 5.3 Key figures, SDFI[a]

General	
Net income for the year	132,817 million NOK
Net transfer to the state's accounts	124,825 million NOK
Average realized oil price	647 NOK per barrel
Average realized gas price	2,31 NOK per Sm^3
Total production	153 million barrels of oil equivalents
Revenues	
Revenue, crude oil and NGL	92,614 million NOK
Revenue, gas	90,441 million NOK
Transport- and processing revenue	10,421 million NOK
Other revenue	205 million NOK
Net profit arrangements	627 million NOK
Sum revenue	194,308 million NOK

[a] Adapted from Deloitte (2014, 17). Data for 2013.

to minimize any incentive for corruption. The Norwegian system is complex, not just because so many institutions are involved, but because the nature of the JVs makes it important to establish which companies are paying (e.g., licensee or operator), and why.

License holders, license operators, Petoro, and Statoil are each making contributions to Norway's government take, via different channels. On the receiving end, there are four distinct institutions in the Norwegian government (the tax administration, the NPD, the customs' office, and the central bank), responsible for handling these revenue inflows. All of the state's revenue from this petroleum activity (petroleum tax, fees, the dividend from Statoil, and cash flows from SDFI) is transferred to a special fund, the Government Pension Fund, Global—which we will describe in more detail in Chapter 7.

Conclusion

One of the first, and most important tasks of a government with petroleum is to establish a reliable commercial framework for developing operational expertise and for harvesting those resources. This framework includes an underlying contractual system, or licensing agreement, and a collection of fiscal tools for securing the government's take. In deciding on the appropriate contract and fiscal tools, countries aim to create a system that is attractive enough to foreign capital and expertise, but strong enough to protect and deliver the people's resource rent.

Like many countries, Norway chose a familiar contractual system. Norway's CA system was originally designed to manage foreign interest in her hydroelectric potential, and the administrative framework was already established when oil was discovered on the NCS. When adapted for the petroleum sector, Norway's concessionary system became the central tool for managing the sector, as it could be used to expand Norwegian competencies, Norwegian influence and the government take.

In examining this system, it is evident how incredibly complex these commercial frameworks can become. The challenge of monitoring such a complex system is significant, even for a wealthy and developed country such as Norway. Luckily, there is growing awareness of the need to help journalists, voters, and the public become informed about the details of these commercial frameworks. We discuss some of these challenges in more detail in Chapter 10. In this chapter, we have relied on the work carried out by EITI, which provides a global standard to promote the open and accountable management of natural resources. This standard helps to inform public debate, and enhance trust—allowing the public to keep the industry and political authorities in check.[30]

6

Macroeconomic Balance

Up to this point, we have been considering the different institutions and policy choices that can help states succeed in managing their natural resources. We have described what policymakers can do to address the political challenges associated with the Resource Curse and the fiscal challenges related to maximizing government take. These are the essential first steps for every state that is blessed/cursed with natural resource wealth; efficient management is required by those who hope to transform resource wealth into national wealth.

Unfortunately, successful management of the resource is not enough. Worse, if policymakers prove to be successful resource managers, their troubles have only begun. This chapter and Chapter 7 consider the costs of success, and how such success can have a detrimental effect on a country's international competitiveness. In short, these chapters consider how to avoid the dreaded Dutch Disease.

As with the previous chapters, we begin with a general introduction to the subject matter, and what states can do to immunize themselves from Dutch Disease. This introduction describes several important symptoms and cures—one of the most important of which is determining the appropriate pace of resource development. As there is a tendency among economists to make the Dutch Disease more complicated than is necessary,[1] we employ clear and simple examples. By most accounts, Norway has managed to avoid the Dutch Disease, even as Norwegian policymakers remain on a constant vigil. It has achieved this by maintaining its macroeconomic balance. For this reason, we use the second half of this chapter to examine what Norway has done to counteract the threat of Dutch Disease, and how it has frequently changed its approach in response to evolving conditions on the ground.

Dutch Disease

The term Dutch Disease was coined by *The Economist* magazine in 1977, to describe what happened to the Dutch economy after it had discovered the biggest gas field in Europe (at the time). In the wake of this gas bonanza, Dutch workers (outside the gas industry) began losing their jobs, which was mostly the result of a real appreciation. This takes the form of a nominal currency appreciation (if the country has a floating exchange rate) or money inflows and inflation (if the country has a fixed exchange rate). A real appreciation is disconcerting because it makes it more difficult for the rest of the country's economy to compete internationally.

Dutch Disease is a popular and contemporary term for an economic phenomenon that has long been observed and understood. One useful way to think about the problem was provided by a former president of Mexico (1976–82), José López Portillo, who is said to have described it with a bodily analogy: "The capacity for monetary digestion is like that of a human body. You can't eat more than you can digest or you become ill. It's the same way with an economy" (quoted in Yergin 2009, 649). Seen in this light, Dutch Disease is the monetary equivalent of gluttony: the economy consumes more than it can possibly process. Overcoming Dutch Disease, then, is about matching consumption and production in a way that doesn't overstretch the domestic economy.

Imagine a gold rush town at the turn of the twentieth century. After discovering gold nearby, miners would flock into the local town to fill up with supplies and to celebrate (and cash-in) their discovery. When word of the discovery spreads, outsiders are also drawn to the town (and environs), seeking their own fortunes. In their wake, local merchants find more people with the means to purchase their (relatively limited) goods and services. This results in inflated prices. For example, the cost of whiskey and barbering became infamously expensive in these gold rush towns. Rather than travel to nearby towns for a drink and a shave (where the prices would surely be lower), miners preferred to stay near their staked claims and pay the inflated (local) prices.[2]

One finds a similar predicament in petroleum-rich economies. All of a sudden, the country is swimming in money, and consumer demand increases quickly. The booming oil industry draws workers to it, driving up their wages (and those in the surrounding labor market). With higher wages, workers are able to buy more things, for example, a house (or a haircut). As the housing stock is fixed (in the short term, at least) the inflow of money allows workers to bid up prices for existing houses. When that happens, houses become more expensive for consumers of all types, whether they work in the oil industry or not. Hence, wages in other sectors (say, those of school teachers) are ratcheted upwards, so that these workers can also afford houses; prices inflate from the bidding frenzy.

It is important to recognize the difference between the exposed (tradables) and sheltered (non-tradables) sectors. If you are working in an export-oriented sector (e.g., making TVs or farming rice), then the price of the goods/services you produce is basically set by the world market (i.e., by your competitors in the global market). If your prices go up, consumers will simply (and easily) switch to a cheaper (international) alternative. On the other hand, if you are working in the sheltered sector (e.g., as a lawyer, government bureaucrat, or barber), then you are not directly competing with international suppliers—your wages can go up, but consumers will have no alternative but to pay the higher price (as there are no easy international substitutes). Hence, when a country's economy begins to inflate, it forces additional costs onto producers; while sheltered producers can transfer their increased costs onto the customer, exposed producers cannot—if they try, they lose market share to their international competitors. If nothing is done, these export-oriented companies will be forced out of business. This, in a nutshell, is the essence of Dutch Disease.

Why does it matter if the exposed sector becomes less competitive? There are many reasons, but we will focus on two of the most important. First, a strong economy is a broad-based economy. No country wants to become overdependent on any one industry and reliance on the petroleum trade is especially risky, as its prices are notoriously volatile. A strong tradables sector can provide an additional leg, on which the economy can stand. Second, the exposed sector is the main channel through which an economy absorbs "best practices" from abroad. By competing in international markets, producers learn how to maintain long-run productivity growth (and with it, economic growth).

Ours is a rather generous description of Dutch Disease, in that it assumes that policymakers are affected even if they pursue good, solid management policies. Clearly, this is not always the case, and one variant of the Dutch Disease argument focuses on the effects of poor policy decisions. This (less-forgiving) description considers the resource rent in terms of a gift or a prize. A state's use of that rent can be seen as something akin to the way lottery winners often squander their unexpected (lottery) windfall. After all, it is not uncommon to hear about lottery winners quitting their jobs, buying houses for family members, and radically changing their lifestyles before ending up broke, destitute, and lonely. Perhaps we should not be surprised to find policymakers making the same sort of emotional, impulsive spending decisions when suddenly flush with oil money.

Inoculation

How then, can states inoculate themselves against Dutch Disease? To understand what is possible, we need to take a closer look at the different mechanisms by which this real appreciation occurs. It is then possible to devise instruments to affect each type of mechanism (as some of these will be more available than others in varying political contexts).

PACE OF EXTRACTION

The easiest way to minimize the threat of a real appreciation is to slow down the pace of development, so that the economy has time to adapt to the rise in domestic purchasing power. If an economy is healthy, and it enjoys sufficient time to adjust, domestic entrepreneurs can satisfy the rising demand that is fueling the appreciation. Indeed, this is the sort of organic economic growth that policymakers hope to encourage (even if it is often remarkably difficult to bring about).

There are two reasons why it is difficult to slow the pace of development. First, there are many important interests (including IOCs and government officials) who are chomping at the bit; they want to extract the resource as quickly as possible, and are under significant pressure (from lenders, shareholders, and constituents) to bring the petroleum to market quickly. Indeed, the longer a country has been extracting petroleum, the more difficult it is to slow down the pace—as the interests that benefit most from extraction have become more firmly entrenched. As we will see later in this section, there may also be good accounting/financial reasons for doing so. Second, it is not always possible to stimulate this local (domestic) supply. Sometimes, the requisite skills/capital requirements are too demanding for local entrepreneurs. Hence pitching demand at the right speed does not automatically translate to the growth of domestic suppliers (with all due respect to Jean-Baptiste Say!). Sometimes other problems with the domestic economy make it difficult for start-up firms to satisfy the demand.

These days, it is more common to set the pace of extraction according to other needs, with little thought given to Dutch Disease. When policymakers approach resource economists to ask about the most efficient pace of extraction (that is, does it make sense to keep the resource in the ground for future use, or should we drain the resource quickly?), the response often comes in terms of Hotelling's Rule.

Harold Hotelling (1931) argued that in a competitive market there was an equilibrium price path, in which the price of oil would rise at the prevailing rate of interest for capital invested in projects with similar degrees of risk. Hotelling's Rule would have a country slow the pace of extraction when prices are high, and increase production when the price falls below the prevailing rate of interest.[3] For Hotelling, then, the laws of supply and demand (not political, environmental, or surrounding macroeconomic factors) should determine the speed of extracting a natural resource; the rate of extraction should coincide with the exponential price path of the compound rate of interest.

What does this mean? First of all, Hotelling's main motivation seems to have been to keep regulators out of the oil industry, by using market incentives to motivate the pace of extraction.[4] Whatever the economic utility of Hotelling's Rule for linking the rate of extraction to market forces, one needs to be aware that there are other (social, political, environmental, and macroeconomic) consequences that also need to be considered when determining the pace of extraction.

This brings us to the second issue: how to manage this type of income? We cannot say it enough: petroleum is special in that it is a non-renewable resource. This implies that policymakers should manage petroleum resources like an asset (that will deplete), not a steady source of income. Consequently, tapping the resource should be understood in terms of draining the capital base, rather than using a source of income. In this light, Hotelling's Rule is interesting because it suggests that a country's optimal strategy might be to quickly convert most of its natural resource stock into financial assets. These assets could then be invested in a diversified portfolio, where the resulting return (interest on the financial asset) becomes a source of income for the state. By investing in a diversified portfolio, rather than relying solely on oil, a country can shield its economy from the notoriously unstable price movements that characterize the petroleum industry. In short, countries might extract the resource as quickly as possible, convert it into a diversified portfolio of financial assets, and then live off the (relatively stable, at least diversified) income/return that results.

It is in this context that we can understand the rising popularity of sovereign wealth funds, or petroleum funds. This issue is discussed in Chapter 7. For now, our interest is in pointing out that it is possible to mimic the effect of a slower pace of development by pumping up the resource quickly, putting the resulting income into an offshore fund, and repatriating that money very slowly in an economically and politically feasible manner (the environmental feasibility of this strategy is far from certain).

Our point is not to suggest that countries should extract their resources at breakneck speed and build up a petroleum fund. There can be many good reasons for keeping the resource in the ground (e.g., environmental, exploiting the OBM, taking time to learn the tricks of the trade, and so on), and empirical analyses have demonstrated that Hotelling's Rule does not hold in reality; the prices of most nonrenewable resource have actually fallen over time (Kronenberg 2006). Worse, if there is not a good, reliable mechanism for slowing the entry of money into the domestic economy, the money from such a fund could still flood the domestic economy (building inflationary pressure). Finally, many countries may have significant development projects that need financing, and it is possible to inject the petroleum money carefully, and selectively, into the economy in a way that will generate broader economic benefit, without the inflationary consequences. It would be a shame if needed money was squirreled away in an international investment fund, when it could be better spent to help the country develop and prosper.

SPENDING EFFECT

At the heart of Dutch Disease lies a spending effect: the economic windfall stimulates spending in excess of what the domestic market can handle. This spending effect flows through two distinct channels.

- **The private channel.** The effect from private consumption is the most obvious, and already mentioned. As a result of the influx of money, demand suddenly rises in both the tradables and non-tradables sectors. Because the prices on tradable goods and services are limited by a ceiling set by the international market, increased consumer power translates into greater demand for imports. This makes it more difficult for domestic producers to compete and they may well go out of business. As a result, labor and capital will move out of the sector to find better rewards (see the section titled "Factor Movement Effect"). The non-tradables sector doesn't work under the same price ceiling. When prices go up in this sector, it is able to attract more resources (capital and labor) from the surrounding economy. The result is an expanding, yet unproductive and expensive sheltered sector, alongside a shrinking and overpriced exposed sector (not counting the petroleum industry).
- **The public channel.** A spending effect also flows through the public channel. With access to increased tax and royalty receipts, public sector spending is bound to increase. Increased government spending can, of course, be a good thing—if the money is spent in a way that doesn't exacerbate the private spending effect already described. Unfortunately, fiscal policies in these countries are often procyclical (see Arezki et al. 2011). Economists tend to be skeptical of government spending (in general), and this skepticism is more pronounced in resource-rich contexts. This is because an economic windfall can encourage hasty and poor government decisions. Governments might be tempted (and able) to borrow money to finance larger development projects; and poorly designed investment projects will not develop the country's productive base. Indeed, in these contexts, a flood of public money could lead to politically motivated (but economically suboptimal) investments, inflation, and even asset bubbles.

Both private and public spending needs to be channeled in ways that can minimize the threat of currency appreciation. In each case, the government needs to find a way to channel the spending into areas which will maximize the economic impact, while minimizing its inflationary consequences. For example, wages in the sheltered sector need to be kept down, so they don't drive an inflationary (wage) spiral that will undermine the competitiveness of the exposed sector. The government needs to maintain the price competitiveness of firms in the exposed sector, by helping to keep their prices in check (and by stopping the outflow of mobile factors and production inputs).

FACTOR MOVEMENT EFFECT

The second central component of Dutch Disease concerns factor movement effects. The way to mitigate this effect can be clearly seen in the above-mentioned gold rush example, where increased demand for haircuts and whiskey could be met by enticing new barbers and barmaids to town (that is, in-migration). In the jargon of economics, this is a factor flow, and we expect that an economic boom in any given sector will draw factors of production (capital and labor) into it. Higher wages (for workers) and returns (for investors) will draw workers and capital into the booming petroleum industry, but it will also squeeze producers in the other sectors (both tradables and non-tradables), who will need to increase their wages/returns, or risk losing workers and investors. As noted earlier, the exposed sector cannot raise prices (because of international competition), so it is squeezed out of business by the increase in its costs (e.g., wages). The sheltered sector can simply increase its prices, to cover the rising costs. A similar sort of effect can occur when enormous production projects get under way, and crowd out the local market for production inputs (e.g., concrete, steel, asphalt, pipe, wiring, and so on).

Some of this increased demand can be met with foreign capital, workers, inputs, goods, and services, in the same way that entrepreneurial barbers and bartenders flocked to the gold rush towns to satisfy the increased wealth and demand. Policymakers need to balance the need to develop domestic firms and expertise, while recognizing the inherent limitations to domestic supply—they can do this by encouraging foreign investment, immigrant labor, and the import of supplies that are in short supply, during a transitionary period, while local suppliers gear up to meet the challenge.

EXTERNAL PRICE ADJUSTMENTS

Because the petroleum produced in a country is usually sold for export, this opens up another set of adjustment mechanisms, related to the exchange rate. We have already noted the threat of currency appreciation when a country begins to process and export its resource wealth. This happens because the money coming into the economy will stimulate demand (which cannot be satisfied by immediate supply)—hence domestic prices will increase. Addressing this inflow of foreign exchange can be dealt with in three different ways: (1) it can be redirected outward again; (2) the domestic economy can be protected/sterilized from these inflows; or (3) the authorities can manipulate the exchange rate.

First of all, a country can encourage its citizens to engage in international economic activities, so that the domestic demand for foreign currency will increase (which will result in less pressure on the currency to appreciate). For example, residents might be encouraged to travel abroad (on trips), or immigrant workers might be invited to work in the country, while sending their wages home in the form of remittances. Alternatively, the country can take the export income generated by selling its oil and invest it abroard.[5] Although we are not condoning the activity, the same effect is achieved when corrupt leaders hide their stolen funds in secret foreign bank accounts.

The second alternative is to establish a petroleum fund that trawls international markets in search of better (and more steady) returns. We will discuss this topic in more detail in Chapter 7. For now, it is enough to point out that by using a petroleum investment fund, the petroleum revenues can be kept out of the domestic economy and then reintroduced later, in smaller doses, or in times of emergency.

Finally, policymakers have a third alternative for dealing with the external price adjustment: devaluation. This tool is described in Box 6.1 because it is more challenging to employ, but it cuts right to the chase. By devaluing the international value of their country's currency, policymakers are able to regain international price competitiveness with the tap of a computer keyboard. The external value of the currency declines, such that exports become cheaper in global markets, while imports become more expensive (relative to domestically produced alternatives). Both price movements encourage growth in the domestic economy. On their own, devaluations often lead an economy down a slippery slope—generating more inflation, and the need for subsequent devaluations. But combined with other policy tools, devaluations can be used to claw back some of the real appreciation, and secure a more competitive footing during difficult times, for a limited period of time.

ATTITUDINAL ADJUSTMENTS

This last adjustment mechanism is the least palatable, as it smacks of moralism. This sort of argument is often (if mistakenly) traced back to the French political theorist Jean Bodin, and asserts that easy access to riches is said to breed idleness and lessen the incentive to accumulate or innovate.[6] Curiously, economists often embrace this moralist argument (in spite of their positivist pride). Our favorite example is provided by the Nordic and Baltic Correspondent for the *Financial Times*, who laments in an online video, entitled "Norway's risky good life," that

> Norway's riches encourage its population to work less, and play more. Symptoms are easy to see on a Thursday or Friday afternoon, when Oslo dwellers flee for their country huts. (Milne 2013, 1.38 min.)

Box 6.1 DEVALUATIONS

There are basically two types of exchange rate regimes: fixed and floating. In the former, a country's central bank buys and sells international currency to maintain a fixed exchange rate. In the latter, the exchange rate value is set (in principle) by supply/demand forces. When the value of a currency goes up, we say it has appreciated (floating regime) or revalued (fixed regime). When it goes down, the currency has depreciated (floating) or devalued (fixed).

Devaluation, then, is a decline in a country's (fixed) exchange rate. Central banks deliberately refrain from supporting their currency (sometimes they are forced into this, owing to lack of funds), and the value of the currency falls to a level that the central bank and markets think is sustainable/beneficial. As a result, the domestic currency buys less foreign currency (it is less valuable).

Consider an example. On one day the Norwegian krone (kr. or NOK) buys US $0.14 (7kr/$). On the next day, officials decide to devalue the krone, so that 1 NOK buys US $0.10 (10kr/$). From one day to the next, 100 NOK buys fewer dollars (from $14 to $10).

If you are a Norwegian fish farmer trying to export your salmon to the United States, the devaluation makes your fish much cheaper, relative to before the devaluation, and to your competitors (e.g., Chilean or Canadian fish farmers) who haven't devalued. To buy your fish, a US fishmonger would go to her bank to buy kroner (in order to buy your fish). Before the devaluation, she could exchange $100 for 700 NOK; after the devaluation, the same $100 will get her 1,000 NOK. In short, she can buy much more of your salmon for the same amount of dollars. With the blink of a computer screen, your fish became substantially cheaper than the fish offered by your (non-Norwegian) competitors.

On the other hand, importers in Norway are hurt by the devaluation. Consider a small taco importer in Oslo, who buys all his salsa and taco shells from the United States. This importer suddenly finds that each of his hard-earned krone buys fewer dollars (and hence fewer taco shells and less salsa). The krone price of US tacos has increased because of the devaluation, so that it is now easier for a Norwegian producer of salsa and taco shells to compete.

In short, a devaluation makes a country's exports cheaper (relative to their international competition), and its imports more expensive (relative to domestic producers). This happens immediately, in response to the new exchange rate. For obvious reasons, this is beneficial for the country, as it makes it easier to export its goods/services in world markets, and domestic consumers are encouraged to buy domestic goods (as imports have become more expensive), stimulating employment at home. Both developments act like a vitamin injection that stimulates domestic economic growth.

The problem is that these "vitamin injections" can be addictive if they are not coupled with policies that can secure the beneficial effect over time. Devaluations often generate greater inflationary pressure in the domestic economy; this, in turn, necessitates further devaluations in the future. If these knock-on effects are not stopped, countries can find themselves trapped in a devaluation cycle. But if policymakers can capture the competitive effect, by introducing parallel policy measures (e.g., an incomes policy that limits wage growth in the sheltered sector), then there can be longer-term beneficial effects from the devaluation.

It is hard to take this sort of analysis seriously, but there is a legitimate concern that increased wealth and government spending can undermine market incentives for workers (that is, they become decommodified), if they would rather stay at home collecting welfare payments than work. Governments can address this potential challenge by encouraging the creation of better jobs, and designing welfare programs that encourage rapid return to the workforce. When firms invest in more productive technologies, the resulting increase in worker productivity can be rewarded with increased pay and/or leisure time.

Norway

Since developing its petroleum resource the Norwegian authorities have used a variety of policy instruments to minimize the threat of Dutch Disease. This section begins by showing that Norway has indeed avoided the worst symptoms of Dutch Disease, before outlining two of three major sets of policies that it has used to minimize its threat: adjusting the pace of development and adjusting price developments (both internally and externally). The third set of policies, using a stabilization/savings fund with a soft budgetary rule, is described in Chapter 7.

Is There Evidence of a Norwegian Disease?

Over the years, there have been several studies seeking evidence of Dutch Disease in Norway. Most of these have been unsuccessful, in the sense that they have not uncovered strong evidence of the disease. The effects in Norway have been muted, or limited to particular sectors of the economy, or at particular times in the development of a sector.[7] The most comprehensive recent study finds "no evidence of Dutch disease [in Norway]. Instead, we find evidence of a two-speed economy, with non-tradables growing at a much faster pace than tradables" (Bjørnland and Thorsrud 2013, abstract). While it is clear that policymakers are constantly on guard to protect against the threat of a real appreciation, the Norwegian economy has escaped its most dramatic and damaging effects.

A study such as Bjørnland and Thorsrud's (2013) is quite sophisticated, but we can arrive at the same general conclusion by examining a couple of headline Norwegian statistics. In Chapter 1, we saw that Norway reveals few signs of a sick economy; total government expenditures in Norway have remained remarkably flat over time, and the country has maintained very low rates of unemployment and income inequality. When we add to the mix a remarkably low and steady level of inflation, as illustrated in Figure 6.1, it becomes clear that the Norwegian economy is not set on a run-away course.

Figure 6.2 presents Norway's real effective exchange rate, from 1970 to 2015. In this figure we see that the country's real exchange rate has varied significantly

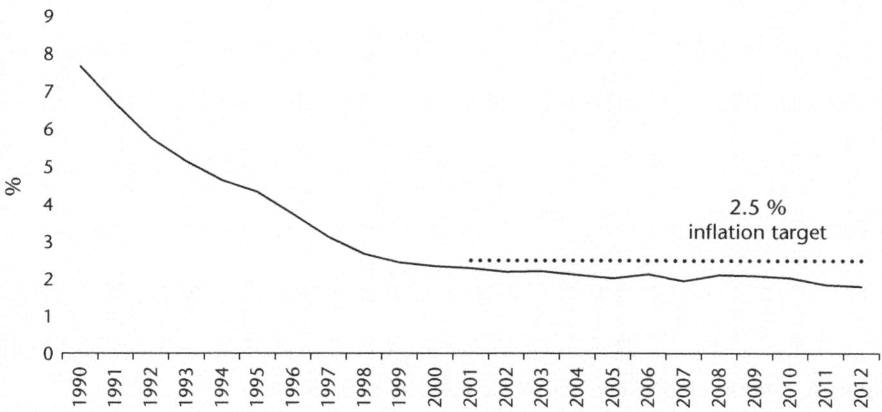

Figure 6.1 Norway's Consumer Price Index, 1980–2012[a]

[a] Adapted from Olsen (2013, 5). Ten-year moving average, as percentages, 1990–2012.

over time—but the variance has revolved around a fairly steady baseline (since oil was discovered). In short, the real exchange rate in 2013 was not very different from what it was in 1970 (about 95 on the Bank for International Settlements (BIS) narrow index). When we look closer, however, we see that the real exchange rate actually depreciated for most of the earlier period (from 1976 to about 2000)—when Norway was rapidly developing its resource—and then began to slowly appreciate (from 2000 to 2012). It now seems to be on a depreciating trend again.

As a small, open economy, there have always been threatening clouds on Norway's economic horizons. A real appreciation remains a constant threat, and it is possible to find traces of this in a handful of statistical artifacts (the most obvious is a menacing asset bubble in the property market). At any time, it seems, the Norwegian economy can tip into trouble—and this ever-present threat requires constant vigilance on the part of policymakers. We have not attempted our own study, but confer to the many thorough studies that have already been carried out, to demonstrate that the Norwegian economy has thus far avoided the worst symptoms of Dutch Disease.

Policy Cures

What have Norwegian policymakers done to avoid the Dutch Disease? The response has varied over the years, but we will focus our attention here on two types of responses: pace, and devaluation and incomes policies.

PACE

In Chapter 4 we discussed the importance of a slow pace of development, when Norwegian policymakers were leveraging the OBM to their advantage.

Figure 6.2 Norwegian krone, real exchange rate, 1970–2015[a]

[a] Adapted from BIS (2015). Narrow indices; CPI-based; Monthly averages. 2010=100. This is the BIS "effective exchange rate," where an increase implies an appreciation of the real exchange rate.

The reader will recall that it takes time for the state to build up the knowledge and competencies it needs when developing its resource and that a slower pace of development strengthens the state's relative bargaining position vis-à-vis IOCs. The same sort of argument can be seen with regard to building up the local economy to meet increased demand.

A concern for Dutch Disease was already evident in the 1973–74 parliamentary debates. Several lead-up studies had concluded that controlling the production volume was the best way to handle the economic and social consequences in a small-scale system like Norway's, and the government's white paper (*St. meld. nr. 25 (1973–74)*, 6–7) recognized that controlling the cost and pace of adjustment to the new oil wealth was the most important issue they faced:

> From a desire for long-term resource utilization, and after an evaluation of the total social costs, the government has decided that Norway should maintain a *moderate tempo* in developing its petroleum resource.
>
> (*St. meld. nr. 25 (1973–74)*, 15, emphasis added)

Predictably, the public response to this report was divided. Norwegian industry wanted to push ahead at full steam, and was quite skeptical of these warnings. On the other hand, "Almost all political institutions to the left of centre were however supportive of the need for moderation in the tempo of petroleum activities" (Al-Kasim 2006a, 39). This political consensus on the left resulted in a moderate pace of development during the early years.[8]

In 1982, a government committee was established "to give an account of all the conditions that are especially relevant concerning the future development of petroleum activities" (NOU 1983/27, 8). Under the leadership of Hermod Skånland (later governor of the central bank, Norges Bank), the Pace Committee

(Tempoutvalget)[9] laid the foundations for many of Norway's subsequent macroeconomic policies regarding petroleum management, including the establishment of a petroleum fund and the need for a budgetary rule to restrict spending.

The Pace Committee was so named because it suggested that the best way for Norway to minimize the threat of overheating was not to develop a petroleum fund, but to simply reduce the country's pace of petroleum extraction. This, they argued, was a much more viable solution, politically, as they were doubtful that politicians could keep their hands out of any investment fund.

Since the Pace Committee report, and especially after the establishment of the GPF in 1990, it became increasingly difficult for the Norwegian authorities to slow the pace of development. In stark contrast to the advice provided by the Pace Committee, Norway's petroleum reserves have been emptied with breakneck haste. "Rather than being consistently the most careful country, in 2009, Norway comes close to holding the world record for high-speed production, by comparison with the country's total reserves" (Ryggvik 2010, 84). The rapid growth and enormous size of the GPFG is a consequence of this high-pace strategy.

One reason for this change in pace (and attitude) is probably that the Pace Committee underestimated the growing political influence of the oil industry over time. In the early years, it was much easier to control the pace of development; but as more and more Norwegians made their living from the oil industry, it became increasingly difficult for political authorities to resist these powerful interests.

DEVALUATION AND INCOMES POLICIES

Throughout most of the early expansion of Norway's petroleum industry, the country's competitiveness was maintained by adjusting external and internal prices, by means of an active exchange rate and incomes policy. This was a remarkably challenging period of time to be managing a small, open economy (not to mention one swimming in oil!). As mentioned in Chapter 2, the Bretton Woods system was beginning to unravel, and the world's major currencies began floating vis-à-vis one another (after the US dollar was delinked from gold in 1971). At the very time that Norway was beginning to produce oil offshore, the government was frantically experimenting with a number of exchange rate regimes, trying to stabilize Norway's trading relationships with Europe (remember, Norway was considering membership in the EEC at this time), while a growing share of its income was being generated in dollars (the oil market is dollar-based).

During this time, Norway maintained its competitiveness by reforming its exchange rate regime and devaluing its currency several times. This activity is listed in Table 6.1. If we juxtapose the devaluations listed in Table 6.1 with the

Table 6.1 Norwegian exchange rate regimes[a]

Date	Event
August 21, 1971	**NOK allowed to float**
December 21, 1971	**Smithsonian Agreement (1% devaluation)**
May 1, 1972	**Joins the "snake in the tunnel"**
April 4, 1973	**Joins the "snake"**
November 15, 1973	5% revaluation
October 18, 1976	1% devaluation
April 4, 1977	3% devaluation
August 29, 1977	5% devaluation
February 13, 1978	8% devaluation
December 12, 1978	**Basket Index introduced**
August 2, 1982	3.5% devaluation
September 6, 1982	3% devaluation
July 2, 1984	2% devaluation
September 22, 1986	2% devaluation
May 11, 1986	10.2% devaluation
May 11, 1986	**Fixed regime established**
March 29, 2001	**Current floating regime (inflation target) established**

[a] Adapted from Moses (2000, 153). The "snake in the tunnel" refers to one of the first attempts at European monetary cooperation in the 1970s. It aimed to limit fluctuations between different European currencies with a set upper and lower margin (hence the tunnel). Regime changes are marked in bold. Jonathon W. Moses, *Open States in the Global Economy* (2000), Macmillan Press Ltd., reproduced with permission of Palgrave Macmillan.

real exchange rate developments tracked in Figure 6.2, we find that the krone's long decline in its real value was being driven by a number of important devaluations—for example, in 1976–78, and after 1986.

To ensure that these devaluations did not trigger wage inflation, the government encouraged a series of parallel incomes policies, as we first described in Chapter 3. Until 1978, the government was actively trying to coax out more attractive incomes policy outcomes, by encouraging coordinated incomes policies, mixed with important government sweeteners (Moses 2000, 128ff). This was the height of the Aukrust Model's influence and the time when the infamous "Kleppe-packages" were introduced:[10] wages in the exposed sector were constrained by a "leading branch" (*frontfag*) bargaining arrangement that ensured sheltered sector wages did not outpace wages in the exposed sector. After 1978, the government withdrew some of its fiscal sweeteners, but it still encouraged the labor market partners to work together. In effect, the state switched from a policy of providing financial support to encourage wage restraint, to relying more heavily on forced arbitration and legal constraints. In 1978, the parliament went so far as to introduce a wage and price freeze law, which lasted for 16 months, and the government went to great lengths to try and limit the bargaining power of Norwegian oil workers, for fear that they could topple this difficult price-balancing act:

> The International Labour Organization (ILO) has repeatedly criticized the Norwegian government for limiting wage bargaining in the oil sector. However this has

> not been a controversial decision in Norway; rather, it is seen as a normal way of limiting bargaining power in a particular sector that otherwise might lead to more wage dispersion. (Cappelen and Mjøset 2013, 53)

As we saw in Chapter 3, the Norwegian economy was already highly organized in a way that helped the authorities limit wage growth in the sheltered sector, and hence maintain the competitiveness of the exposed sector. By employing a series of devaluations, and by constraining wage developments (via the Aukrust Model), the Norwegian economy was able to defer the real appreciation that usually comes with oil, until it adopted a floating exchange rate regime in 2001. Since then, the authorities have been unwilling to use depreciations/devaluations to maintain competitiveness, and the krone has slowly appreciated (see Figure 6.2).

Consequently, the authorities have had to develop new ways to limit domestic wage growth. As seen in Figure 6.3, the number of immigrants to Norway has increased radically since 2005, and the primary draw for this inflow is jobs (SSB 2015a). Before the recent fall in oil prices, almost 80,000 immigrants were in Norway, and most of them were workers. To avoid the wage inflation that fuels a real appreciation, these workers are being funneled into high-growth sheltered sectors, such as construction—where migrants made up 12 percent of the workforce in 2011—and human health and social security activities—where migrants made up 16 percent of the workforce in 2011 (SSB 2015b). Absent is the ability or willingness to adjust the country's exchange rate; it seems that the Norwegian authorities are trying to limit the

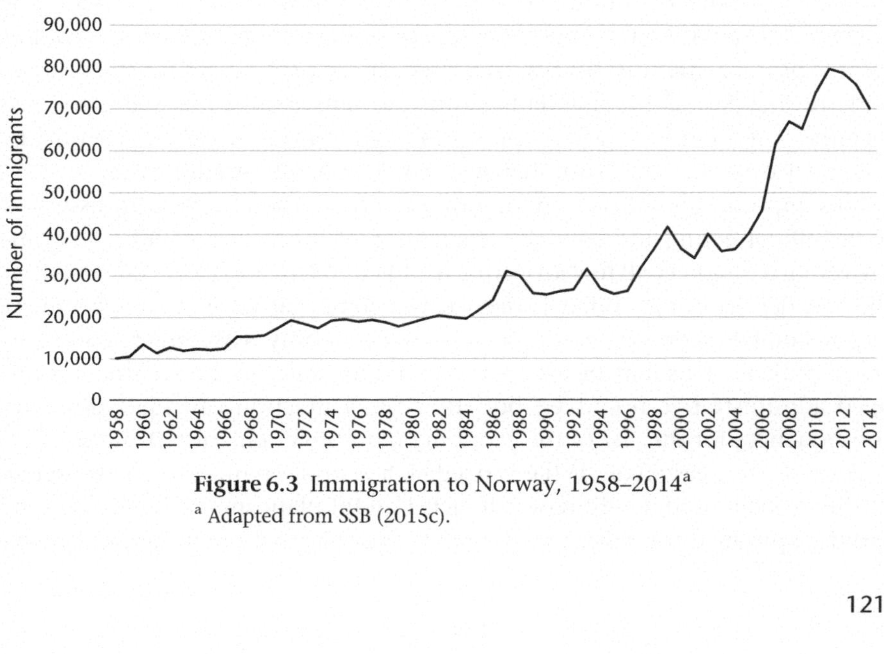

Figure 6.3 Immigration to Norway, 1958–2014[a]

[a] Adapted from SSB (2015c).

effect of wage inflation in the sheltered sector, in the tried and true manner employed by the gold rush towns: by importing workers.

There is one other important means by which Norway is able to constrain wage growth in its tradables sector. Recall (from Chapter 3) that much of Norwegian incomes policy is aimed at minimizing wage spreads: constraining wages at the top end, while pushing wages on the bottom upward. The political motivation for this is clear for any social democratic regime: the desire to minimize income differentials. But an important byproduct of wage equality is the ceiling placed on wage growth in the highest paying, most productive jobs—many of which are in the tradables sector. Consequently, the wages of Norway's most productive workers tend to be lower than market conditions could demand (and which international competitors are forced to pay). Hence, Norwegian employers in the most productive exposed sectors can benefit from this wage restraint at the high-skills end of the labor market. This complicated effect is evident in Figure 6.4, where we can see how Norwegian production levels in the engineering industry (in particular), but also in total manufacturing production, has outperformed the OECD average. Production is growing in this sector because Norwegian workers have maintained higher productivity levels than their competitors in Sweden, Germany, and the OECD (OECD 2015a); despite their higher unit labor costs (OECD 2015b).

Conclusion

We have considered a number of different ways in which Dutch Disease can affect the international competitiveness of resource-rich countries. In doing so, it is not our intent to suggest that there are any easy cures for the disease. Still, the first part of this chapter has aimed to demonstrate that states do have recourse to a number of policy options in order to minimize their exposure to Dutch Disease. But successful management is difficult, often illusive.

This important point is often forgotten at a time when markets are assumed to be self-correcting and eternally efficient. If left to its own devices, a booming resource market will fuel an outbreak of Dutch Disease, to the detriment of the broader economy. What is needed is a state that takes its management responsibilities seriously, and inoculates the economy with targeted and balanced policies. This may be asking too much for states that have underdeveloped administrative resources, but the cost of inaction can have dramatic consequences for the rest of the economy (and the population it supports).

There are few easy fixes. What is needed, first and foremost, is an awareness of the problem and a willingness to act. Careful planning and management can then produce outcomes that benefit the people, without undermining the

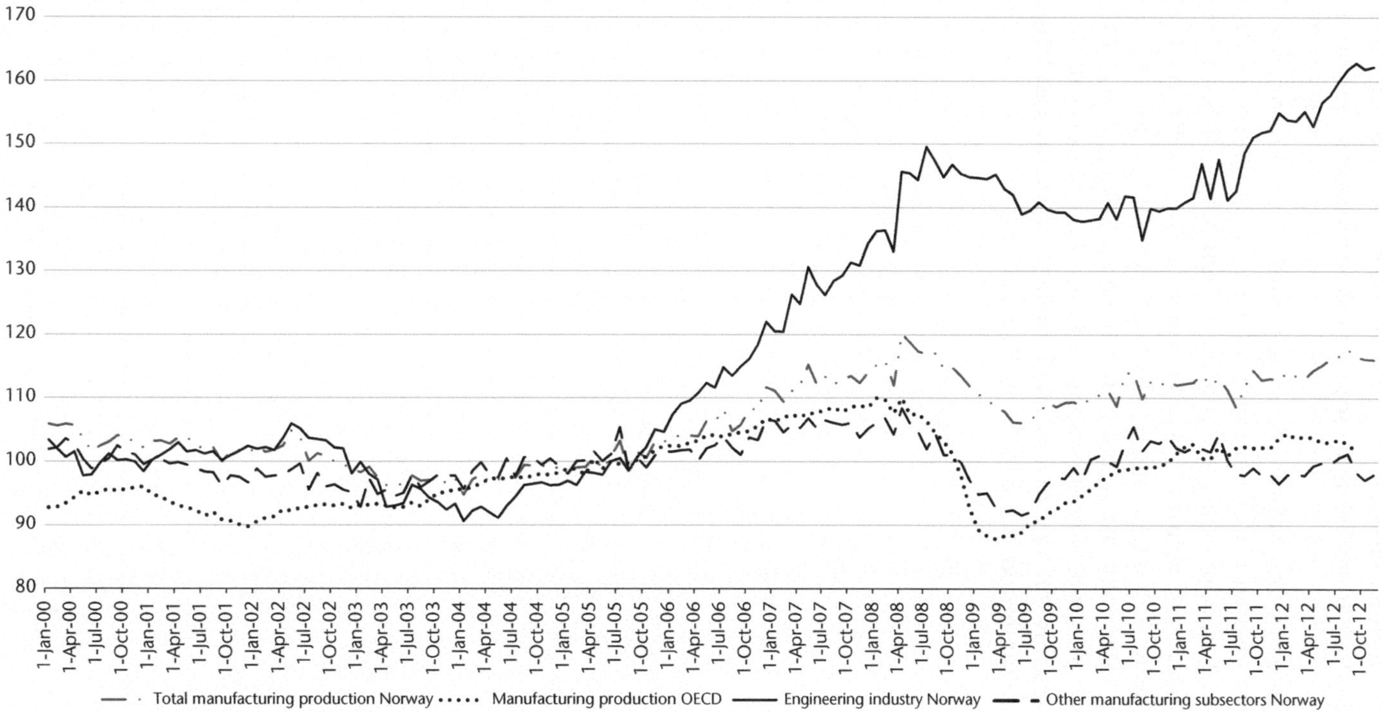

Figure 6.4 Production, by sector, 2000–12[a]

[a] Adapted from Olsen (2013, 7). Production index set at 100 for 2005. Monthly data are for January 2000–November 2012.

competitiveness of the rest of the economy. Finally, countries need to protect what few tools they have to sterilize and protect the domestic economy from a real appreciation.

As a small, open economy, Norway finds itself constantly adjusting to an international economy that places constraints on what it can (and can't) do. As the international economy changed, Norwegian authorities shifted from using exchange rate regime adjustments to increased factor mobility (mostly migration) as a means to mitigate inflationary pressures. All the time, it has developed a remarkable political consensus over the need to keep most of the petroleum incomes offshore, stored away for a time when the oil will disappear, and Norwegians will have to find a new niche to exploit in the global economy. This important tool is the focus of Chapter 7.

7

Wealth Management

In Chapter 6 we examined some of the most important tools Norway uses to secure a macroeconomic balance and to avoid economic overheating. By controlling the pace of extraction/development, and the pace at which wages and prices develop, Norway has managed to maintain international competitiveness and avoid catching the Dutch Disease. This chapter focuses on yet another way that Norway stays competitive: by managing its wealth effectively in a sovereign wealth fund.

There is much interest in Norway's spectacular Government Pension Fund, Global (GPFG). Since its inception, the fund has accumulated a great deal of wealth in a way that is both transparent and lucrative. It is also heralded as an important tool in the nation's struggle with Dutch Disease. But the real key to Norway's competitiveness lies in the unique way that this fund is integrated into the government's broader fiscal framework. Neither is the GPFG just about fighting Dutch Disease; it also plays an important role in addressing two of the most challenging aspects of an oil economy: that the underlying resource is exhaustible, and that its price is volatile.

This chapter begins with an important caveat before considering the different motivations for establishing an international investment fund for a country's petroleum revenues. We will refer to these funds as petroleum funds, but our argument extends to a wide variety of funds that rely on rents from non-renewable natural resources (e.g., from mining). The chapter then turns to the particular features of the Norwegian case. As with most of the other policy instruments described in this book, it is not the existence of an investment fund that explains Norway's success, but the way in which the fund is actually managed and integrated into the larger political machinery of governance.

Caveat Emptor

Investment funds for non-renewable resources have become quite popular since the turn of the millennium—in part, we suspect, because of the success

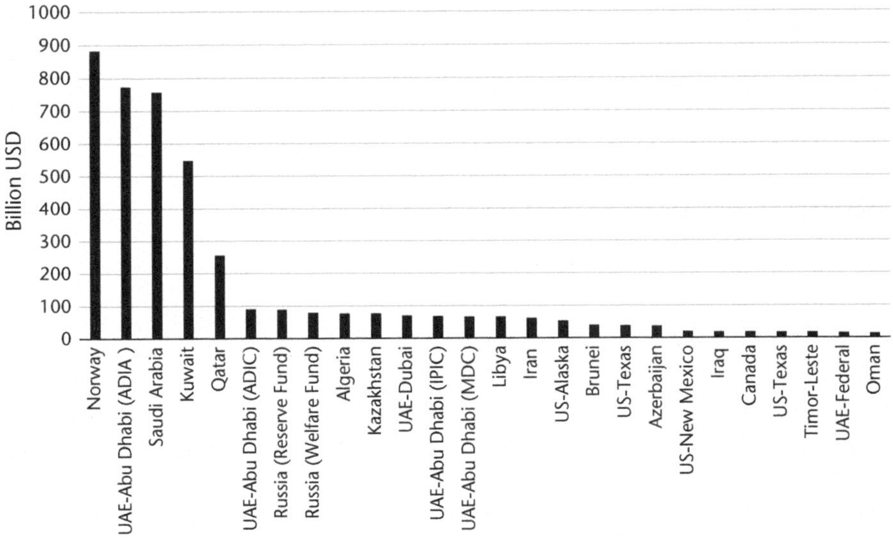

Figure 7.1 World's largest petroleum-based investment funds[a]

[a] Adapted from SWFI (2015) "Oil," "Oil & Gas," and "Oil & Other" based funds, above 10 billion USD. May 2, 2015.

of the Norwegian fund (Velculescu 2008). As shown in Figure 7.1, these sovereign wealth funds are remarkably large and increasingly popular among oil-producing states.

But sovereign wealth funds have also drawn a great deal of criticism (Ministry of Finance 2007). Investment markets are concerned about the lack of transparency and the uncertainty (surrounding the purpose of the investment activity) of such funds. Macroeconomists worry that these types of funds provide no guarantee that the government will maintain an appropriate fiscal stance (i.e., the existence of a fund is no substitute for sound fiscal and macroeconomic management). Development experts are concerned that a large fund can tempt corruption and fraud (administering a large amount of money endows patronage), and the existence of a fund can provide a false sense of security, so that politicians avoid taking difficult fiscal decisions (Stevens 2003, 30).

Curiously, the utility of resource-based investment funds finds little support in general economic theory (Davis et al. 2001a; 2003). For many economists, these funds appear as economic sideshows; any policy decision that is made in the context of an oil fund (e.g., how to spend the oil money over time) can be made in the absence of an oil fund. The trick, obviously, is to make the right (and often difficult) decisions. Worse, by shuffling money out of a public budget and into an investment fund, countries can erode governance, transparency; and accountability. For this reason, a handful of IMF economists have reviewed the harm that can be done through these types of funds, and concluded that:

> [t]he theoretical arguments for oil funds are unconvincing, and their operational implications can be negative. Oil funds can be poorly integrated with the budget, which can lead to a loss of overall fiscal control and create problems of expenditure coordination, such as duplication of expenditures or capital spending decisions made without considering their implications for future recurrent spending.
>
> (Davis et al. 2001b, 58)

To this list of concerns we would add another: such funds are often used to legitimate the transfer of precious investment capital from the developing world (where it is desperately needed) to rich countries—in the name of higher returns. For all these reasons, introducing an investment fund may not be the most appropriate course of action in many states, especially those that have only recently discovered natural resources.

When a country finds oil, and recognizes that the wealth generated by this oil can have a negative (as well as a positive) effect on the economy, its policymakers need a plan for getting the most out of the resource, with the least harm done to the surrounding economy (and community). As discussed in Chapter 6, perhaps the best plan is to moderate the pace of extraction, such that any resulting economic activity does not outpace the capacity of the state and the domestic market to accommodate. If new-found petroleum resources are allowed to trickle into the domestic economy slowly, they are more likely to flow to the right places.

After all, developing countries need investment capital—this is their defining characteristic. For government officials, the trick is to turn oil revenues into capital that can be accessed by, and employed in, the domestic economy. If too much money enters into the economy too quickly, the result will be inflation, appreciation, and a threat to any hope of sustaining a non-oil export economy. While the threat of overheating is very real, it should not be used to cut a developing economy off from the very capital streams it needs for subsequent growth. Rather, countries need a plan for leveraging their petroleum wealth as a source of investment for future prosperity. This is, at any rate, what the Norwegians did; officials spent the first two decades using oil revenues to pay off debts, build up local competencies, expand the welfare state, and improve the standard of living for Norwegians—before any money was put aside in a savings/investment fund.

Economic development is often handicapped by the lack of adequate health, education, and infrastructural investment. It is possible to invest in developing countries in ways that do not generate crippling inflation and appreciation. Doing so can produce an even greater return for future generations (than returns on international investments), if these investments lead to subsequent economic development, jobs, and prosperity at home. In light of their checkered reputation, then, it is important for states to proceed carefully when considering whether, and how, to establish a petroleum fund.

Banking on the Future

Oil in the ground is like money in the bank—except that it doesn't accrue interest (and is more difficult to steal). But once it is pumped to the surface, petroleum changes its nature in two important ways. First, when the resource remains in the ground, its value is shared by current and future generations. On the surface, however, and when sold in the market, that wealth is captured by the current generation, at the expense of future generations. Hence, governments need to consider the intergenerational effects of selling the country's birthright (in effect, oil constitutes the country's "family jewels"). These intergenerational costs only increase when one adds the environmental impact to this economic impact.

Far-sighted governments need to think about how petroleum wealth can hurt or deter national economic and political development. At the same time, any country that depends upon an exhaustible resource needs to plan for its eventual exhaustion. The more a country depends on petroleum resources, the more difficult it becomes for that country to transition to a post-petroleum economy (when the oil eventually runs out). Hence, governments need to develop a plan of action for dealing with three long-term consequences of petroleum wealth: (1) the challenge of intergenerational equity; (2) the need to protect the economy from becoming overreliant upon that resource; and (3) the day when the oil runs dry.

A price effect is the second way that oil wealth changes when it is pumped from the ground. The price for oil is notoriously volatile; it swings a great deal from high to low, to high again—in ways that are quite difficult to predict. As a result, petroleum constitutes a difficult source of revenue for governments that need to anticipate and plan for long-term developments (both economic and political). For example, when the price of oil is high, states come to depend upon deep streams of petroleum-based revenues, and expand their expenditures and activities accordingly. Worse, many states use this apparently steady source of income to secure loans for ambitious development plans. While these actions may make economic sense when oil is selling for $100 a barrel, they can threaten default when the price of oil falls to $40 (for example). When the price of oil falls (as it inevitably does), governments are forced to cut public expenditures, often at great political cost. Witness Venezuela in 2016.

A well-designed petroleum fund allows authorities to address one, the other, or both of these inherently challenging characteristics of oil. Such a fund can help states deal with the challenges of relying on an exhaustible resource and/or it can be used to isolate the government's revenue stream from the volatile price of oil. As hinted at in Chapter 6, petroleum funds are not the only way to achieve these objectives—but they offer useful tools, if designed correctly.

Let us first consider how a petroleum fund can address the problem of relying on an exhaustible resource. In Chapter 6, in the context of discussing Hotelling's Rule, we introduced the problem of relying on petroleum revenues alone—in that oil spent, is oil gone. In pumping oil from the ground and into markets, a country is simply draining its resource pool (and wealth), and leaving nothing in return. On the other hand, if the money from the sale of that resource is used to invest in a portfolio that can secure additional returns in the future, then a country's oil can be transformed from a pool of capital to a source of income.

Conceived in this way, a petroleum fund functions as a *savings* fund—where the government squirrels away its wealth with a simple deposit rule (e.g., as a fixed percentage of petroleum revenues, a percentage of total government revenue, and even a predetermined money contribution). If the fund is successful (or someone doesn't abscond with the money!), and its returns are plowed back into the fund when not needed, then the government's savings will grow until the fund is needed (e.g., when the oil runs out). This money could be used to fund future pension obligations, to retool the domestic economy to meet the challenges of a postpetroleum context, or even to invest in a more sustainable path of economic development. In creating a savings fund, governments recognize the need to plan for future generations and a time when the oil will run out.[1]

The second challenge of managing the wealth that comes from oil is a result of its volatile price; here a petroleum fund takes the form of a *stabilization* fund, and can be used to address two types of destabilizing threats. The first threat is a direct result of price volatility. To address this threat, the fund can be used to delink current government expenditures from the price of petroleum. In this way, government expenditures won't need to seesaw up and down along with the price of oil. This can be carried out by transferring petroleum revenues to a fund/account when the price of oil exceeds a pre-announced threshold level, then releasing assets from that account (to help finance government expenditures) when petroleum prices fall below that threshold level. A second source of destabilization comes in the form of Dutch Disease, where the very size of the resource bonanza can flood the money supply and appreciate the national currency. By investing petroleum revenues offshore, and by allowing them to trickle in at a rate that won't overheat the economy, petroleum funds can protect the domestic economy from succumbing to Dutch Disease.

Such is the general thinking about petroleum funds: they can be designed to address a country's need for stability and/or future savings. For them to work as designed, however, they also need to exhibit a number of basic underlying characteristics. For example, Daniel (2004, 9) notes the need for the funds to be accountable to elected representatives; for the funds to be subject to independent audits of transactions and activities; to employ a clear investment strategy with explicit benchmarking of desired returns in different market segments; and to appoint investment managers by means of a competitive process. We would add to this list the need to maintain transparent operations—that all reports and audits should be publicly accessible, and that the process by which investment decisions are made should be explicit and subject to debate. These are common-sense requirements, and should go without saying—but they are too often ignored or abused.

Creating a Savings and a Stabilization Fund

As we saw in Figure 7.1, Norway's GPFG is today the world's largest sovereign wealth fund, and it has been growing at a remarkable pace. This fund began life under a different name (the Government Petroleum Fund) and in the face of significant opposition. Indeed, it took over two decades after commercially viable oil was discovered before Norway's Petroleum Fund came to be.

While the size of the fund, and the motivation for its existence, demonstrates the authorities' desire to create a savings fund, the fund also plays an important stabilizing role. Actually, it is not so much the fund, as the unique way in which the fund has been integrated into the government's fiscal practice (by way of the "budgetary rule") that explains its utility as a stabilization fund.

The Government Petroleum Fund

As was briefly mentioned in Chapter 6, much of the intellectual groundwork for a petroleum fund was laid by the 1982 Pace Committee.[2] As part of its evaluation, the Pace Committee described the utility and attractiveness of a petroleum ("buffer") fund, and of investing the net revenues from the country's petroleum activity *outside* the Norwegian economy,[3] where it could accumulate for future generations.

In the end, the Committee did not think that such a buffer fund offered a very practical solution. Politicians simply couldn't keep their fingers out of the honey jar.

> Judging from the attitudes we know in both the political community and the general population at large, it is difficult to imagine that hundreds of billions could be invested abroad, while facing unmet needs at home. (NOU 1983/27, 90)

Because they believed that elected officials were not capable of saving over the long haul, the Committee held that the most realistic method for saving petroleum wealth, and distributing that wealth across generations, was to keep it buried. The Committee's logic was simple: unrecovered oil does not burn the same type of hole in the pockets of politicians.

Contemporary observers often forget that it took some time before Norway began to build out its fund. Although the idea was formally discussed in the 1982 Pace Committee report, it was not before 1990 that the Norwegian parliament adopted the Act on the Government Petroleum Fund (Act of 22 June 1990, No. 36). There are basically two main reasons why the fund's arrival was delayed—despite early interest. First, Norway needed to pay down its foreign debts and balance its budget. Simply put, there was no money left to save for a rainy day. Second, the Finance Department was concerned that the fund would undercut its position of power; it feared that the fund could be used to finance measures through channels that existed outside the normal budgetary channels—which the Finance Ministry controlled (Lie 2010, 345).

Once the Government Petroleum Fund was created, it took some time to grow; the first capital transfer didn't occur until 1996, and by 1998, the value of the fund was still relatively small—less than 175 billion kroner (see Figure 7.2). During this time, the government used its petroleum revenues to pay down its debts, while building out its welfare state and public infrastructure. Indeed, throughout the 1980s and 1990s, considerable sums were pumped into public budgets to maintain and improve these services (Ryggvik 2010, 88).

Thus, for the first 25 years of Norwegian oil production, Norway's income from oil was being consumed as rapidly as it was being produced. When we look at the entire period of oil incomes put together, about half of the money generated from oil has already been spent (Qvigstad 2011). What has changed is where the spending money comes from; while about 10 percent of the government revenues have always come from oil (since oil was produced), the government now gets its money from returns on investment (not from selling oil directly). It is only in the first two decades of the twenty-first century (when the price of oil was remarkably high) that petroleum incomes have grown so quickly that it has been possible to stash them away as a form of savings. This happy circumstance is the result of Norway having finally paid down its debts, but also because the country began to extract its resource at a much quicker pace.

The Budgetary Rule

Thus far, our description has focused on the important role that the Government Petroleum Fund plays as a store for government savings. But much of the fund's international reputation rests on its role in stabilizing the Norwegian economy. This reputation is somewhat unwarranted, in the sense that the key to stabilization is not so much the fund itself, but in the way that the fund's resources are channeled into the government's coffers. While most sovereign wealth funds receive their money as result of a government budget surplus, in Norway the oil money first goes to the fund, and a share of the return from that fund is then sent to the budget.

In 1982, when the Pace Committee was weighing the utility of a buffer fund, it also considered the need for a budgetary rule, or rule of thumb (*handlingsregel*), that could be used to siphon petroleum income into the state budget at a rate that corresponded to the real rate of return from the Petroleum Fund. While it took about seven years for the Petroleum Fund to come to fruition, the details of the budgetary rule were not nailed down until 2001. In that year, the Finance Ministry used a government white paper to parliament (*St. meld. nr. 29 (2000–2001)*) to announce two new policy rules in connection with its new floating exchange rate regime (see Table 6.1). The first of these was the introduction of an explicit inflation target (2.5 percent) for monetary policy (the result of which was evident in Figure 6.1). The second was the introduction of the budgetary rule. In short, Norway's budgetary rule was adopted after the country had been in the oil business for 30 years!

Norway's budgetary rule has two components, each of which are potentially contentious. The first concerns the amount of money that should be channeled back into the domestic economy (and how that amount should be determined). In Norway, a rule of thumb has been that the return should be determined by the expected (not the actual) real rate of return from the fund (*St. meld. nr. 29 (2000–2001),* 9). How this figure is then established becomes a matter of political interpretation.

When selling this idea to the Norwegian public, subsequent politicians and journalists have referred back to a well-known principle of resource management in Norway.[4] In these discussions, a 4 percent return was suggested as the modern equivalent of a traditional forest management technique (silviculture [*skogrøkten*])—as it takes generations of patient and conservative management to yield a viable forest, the rate for harvesting should not exceed the forest's rate of growth. The same principle was then applied to the fund.

The second issue related to the budgetary rule is: how should this money be spent? Here, too, there is much room for political interpretation. When the parliament agreed to the guidelines, however, it was emphasized that any increase in the use of petroleum incomes should be aimed at measures that can increase productivity, and with it the rate of growth, in the rest of the economy. A unanimous Finance Committee then pointed to the role that tax policy and investments in infrastructure, education, and research can play in getting more out of a well-functioning economy (*Meld. St. 2 (2014–2015),* 34).

As money from the budgetary rule is not earmarked for particular policies, it is impossible to know if the money is actually being used to increase productive activity in Norway. But this was the explicit goal behind the budgetary rule.

This budgetary rule is an important part of the reason for why Norway has been immune to Dutch Disease. There are two interesting characteristics worthy of note. First, as the 1982 Pace Committee recognized from the very start, it is remarkable that elected officials are willing to keep their hands out of the honey jar. Second, the agreement to focus on 4 percent is not set in law, but is more of an informal agreement made among the leadership of all the political parties.[5] Because it is more of an informal agreement, than a fixed rule, successive governments have enjoyed some flexibility in its application; spending from the budgetary rule is often used as an important countercyclical measure. Before the crash in oil prices in 2014, as the fund has grown in size, governments have been afraid to draw out the full 4 percent, for fear of creating too much inflationary pressure at home.

The budgetary rule works because consecutive Norwegian governments have maintained a very orthodox, even conservative, attitude with regard to fiscal policy. As part of that attitude, it is recognized that the accumulation of assets in the Petroleum Fund should reflect actual budget surpluses. To do this, the government aims to balance its (non-oil) revenues and expenditures. It then estimates what a 4 percent return on the fund (in that particular year) would look like, and readjusts the balance accordingly. All this assumes, of course, that the expected return from the fund surpasses the current budgetary rule percentage. For example, let's say that the budgetary rule produces an amount, X, for a given year—that the government can use to help finance its budget. In effect, the government can extend its expenditures by X, above the non-oil revenues it has secured. As we have just seen, this additional money is supposed to be spent in a way that will increase productivity in the rest of the economy. The remainder of the money (in the fund) is kept offshore, and any returns from its investments are then poured back into the fund. Hence, the Petroleum Fund grows in two ways: (1) by the state's oil revenues (as described in Chapter 5) which continue to pour in each year; and (2) from the returns on investments from past (and accumulated) revenues.

This setup is quite different from other states with petroleum funds in that the money generated from oil does not go into the government budget first, with the surplus (after government expenditures)—if any—subsequently deposited in the fund. In Norway, petroleum revenues go directly to the fund, and then 4 percent of the annual return is subsequently brought back into the country, to help balance the government budget.

To ensure that the reader understands the remarkable nature of this rule, let us elaborate. Norway is sitting on a large pool of oil and gas that is generating unimaginably large amounts of money. For fear of Dutch Disease, the parliament (and government) respects a budgetary rule that limits its access to this money. Each year, only a very small part of these revenues is allowed to enter the government's coffers. While the budgetary rule makes economic sense, it is also a remarkable political feat. Even though Norway is a very rich country, there is always pressure (often from very influential constituents) to provide more and better roads, teachers, health care, environmental protection, national defense, and so on. Non-Norwegians can look in astonishment at a democratic state this is able to defer important consumption decisions—decisions that have the support of significant political constituencies—out of respect for an informal budgetary rule.

Recent Developments

In 2006, the Government Petroleum Fund (GPF) was renamed the Government Pension Fund, Global (GPFG). This was an attempt by the government to generate more political support for the fund, and to highlight the fact that Norway (like most industrialized countries) faces a significant increase in future public pension costs. Even though the GPFG is now called a Pension Fund, it does not entail any formal pension liabilities. Indeed, there has never been an explicit political decision as to how the money in the fund might be used in the future.

As we saw in Figure 7.1, the GPFG is now worth about 900 billion USD (in 2015), or over 7,000 billion NOK. Although owned by the Norwegian people, the fund is managed by Norges Bank Investment Management (NBIM), at the country's central bank, on behalf of the Ministry of Finance. It is the Ministry of Finance (on advice from NBIM and discussions in parliament) that determines the fund's investment strategy. This strategy is based on a 2011 management mandate (recently updated in 2016) that defines the investment universe and the fund's strategic reference index (NBIM 2011; 2016a). Its relative size over time, in terms of market value and percentage of GDP, is shown in Figure 7.2.

In practice, the Ministry of Finance transfers its petroleum revenues to the GPFG. Since this was established, the government has transferred a total of 3,475 billion NOK to the fund (end of first quarter, 2016). In early 2016, the total market value of the fund was about twice that amount (7,079 billion NOK), with the remainder coming from returns on investment (2,591 billion NOK) and foreign exchange earnings (1,045 billion NOK) (Slyngstad 2016).

Before 1998, the fund was managed according to a strategy used by Norges Bank for its foreign exchange reserves (that is, the capital was placed in government and government-guaranteed bonds). This proved too conservative. Today, the GPFG invests in all sorts of non-Norwegian financial instruments, across the

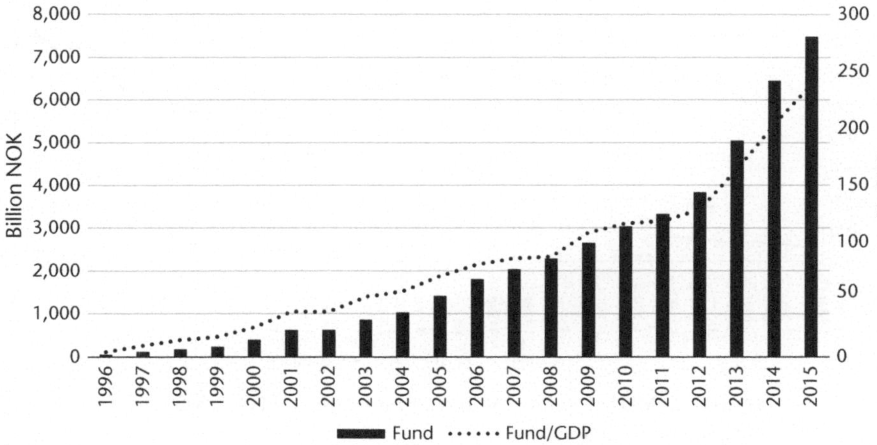

Figure 7.2 The Government Pension Fund, Global, 1996–2015[a]

[a] The figures for fund size 1996–2011 are adapted from NPD (2012); the fund size for 2011–14 are from Norwegian Government (2014c); the fund size for 2015 is from NPD (2016b); and the GDP figures are adapted from SSB (2016 [09189]). GDP figures capture the entire economy (not just the mainland GDP), in current prices.

globe, with an eye on trying to maximize the highest possible risk-adjusted return, by way of a diversified investment mix, in accordance with its mandate.

In order to invest in as many different countries, companies, and shares as possible (so as to obtain the highest possible return with only a moderate level of risk), the Finance Department has determined that about 60 percent of the GPFG should be invested in stocks, 35 percent in bonds, and 5 percent in real estate.[6] In 2011, the fund reduced its exposure in European markets and expanded its position in emerging markets. While the GPFG's investments in stocks and bonds are spread purposively thin (to avoid being overexposed to risk, or to be accused of influencing particular firms), the fund does own a large share of property projects in both London and Paris (often between 25 and 75 percent). The value of each of these investment types, and how they have changed over time, is shown in Figure 7.3.

In 2015, the GPFG owned shares in more than 9,000 companies, with investments in 75 different countries. This constitutes a remarkable 1.3 percent of the world's listed companies (NBIM 2016c). For a country with under 5 million inhabitants, this is an astonishingly large piece of the global economic pie: less than one-thousandth of the world's population owns 1.3 percent of the world's listed companies. Over time, a number of economic and political conditions have been placed on the way in which the GPFG is managed (see Box 7.1, but also Chapter 10), but the main objective remains to

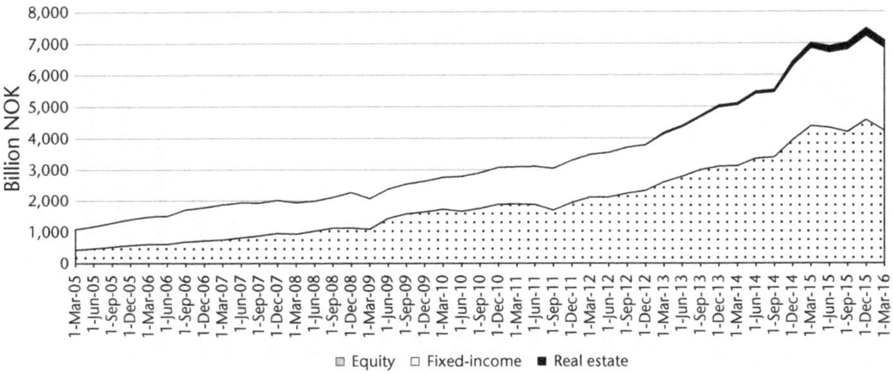

Figure 7.3 Market value, by investment type, GPFG, 2005–16[a]

[a] Adapted from NBIM (2016b).

Box 7.1 NBIM AND THE GPFG

Norges Bank Investment Management (NBIM) prides itself on its professional and transparent management of the Government Pension Fund, Global (GPFG). Online, there are a number of interesting links with videos and information to help us better see how the fund is changing over time. We recommend the following links:

On the fund in general:	http://www.nbim.no/en/the-fund/
NBIM's video on its activities:	https://www.youtube.com/watch?v=DKez6vEfR3Y
GPFG investment strategy:	https://www.nbim.no/en/investments/
How the fund votes its shares:	https://www.nbim.no/en/responsibility/our-voting-records/
List of countries excluded from the Fund	http://www.nbim.no/en/responsibility/exclusion-of-companies/
Ethical guidelines	http://etikkradet.no/en/guidelines/
Ethical Council's advice	http://etikkradet.no/en/tilradninger-og-dokumenter/
Investment principles	https://www.nbim.no/en/responsibility/responsible-investment/
2015 Annual (Responsibility) Report	https://www.nbim.no/en/transparency/reports/2015/responsible-investment-20152/

maximize the economic return. In doing this, the fund is prohibited from acting strategically, as it cannot obtain more than a 10 percent share of a single company that confers voting rights.

Conclusion

Norway's GPFG has grown to become the world's largest investment fund. As such, it is able to direct much critical attention to the firms and projects it decides to support (or not). We will discuss this influence, and other management issues related to the GPFG, in Chapter 10.

This chapter should also be read in connection with Chapter 6. The GPFG and its corresponding budgetary rule offer two additional tools that the Norwegian government uses to secure a macroeconomic balance and to maintain its international competitiveness. They can also provide a savings cushion to assist with the economic transition that will be needed when Norway runs out of gas and oil. These tools have worked well in Norway because they are teamed with a larger set of tools for securing macroeconomic stability. The burden placed on the fund and its budgetary rule have been made manageable because the authorities were careful to temper the pace of Norway's petroleum development and to manage Norway's price (and wage) developments—both internally and externally—in a way that protects a vibrant economy, beyond the oil sector.

Indeed, it is very easy to exaggerate the importance of this fund, especially in light of its impressive size. But as Skancke (2003, 318) notes:

> the total value of Norway's petroleum wealth (around 200 percent of GDP) corresponds only to around 6–7 percent of total national wealth. In comparison, the value of human capital is around 80 percent of total wealth. So reducing the size of Norway's labor force by 10 percent through longer vacations, shorter working hours, reduced retirement age, etc., will reduce its future consumption possibilities by the same amount as a drop in the value of Norway's oil wealth to zero. Clearly, maintaining the size of the labor force is the key economic policy challenge in the years ahead and the key lever to ensuring a sustainable fiscal policy.

Because less money is needed to help support Norway's non-oil economy, more money can be funneled into the GPFG. One reason for this willingness to store its wealth offshore is that many Norwegians and Norwegian firms have participated in, and already benefited from, this wealth creation. The following chapter examines the way in which the Norwegian authorities encouraged the creation of local content.

8

Securing Local Content

We expect that much of the new hydrocarbon production will occur in the developing world. Most of these countries will introduce local content requirements in an effort to create jobs, promote local enterprises, and secure new skills and technologies. This chapter describes what is meant by local content, and the policies that states can adopt to encourage its growth.

There are several reasons why more and more people are interested in local content policies (LCPs). The political imperative is clear: governments have a strong need and desire to secure economic opportunity for their resident populations. But there is also an economic imperative for creating local content, and it has two faces. First, and as we learned in Chapter 7, creating a resilient domestic support industry is an important weapon in the fight against Dutch Disease. Second, IOCs are using local content as a means to demonstrate their commitment to CSR, as we shall elaborate in Chapter 10. In short, local content has become an important political imperative and a strategic issue for the oil and gas industry.

LCPs are anything but new. The history of modern oil is the history of states attempting to secure a larger share of the petroleum pie; as early as 1944, Venezuela passed a hydrocarbons law that required IOCs to refine their oil in that country; and in 1953, Brazil's president, Getúlio Vargas, proclaimed that Petrobras (the Brazilian NOC), should use only Brazilian capital, workers, and know-how (Tordo et al. 2013, 18). Indeed, the rise of NOCs provides clear evidence for the desire of governments to increase the size of the local economic pie derived from petroleum.

The Norwegian experience is often heralded as a significant forerunner in this area, and many contemporary LCP strategies draw explicitly on Norwegian experience and examples. For this reason, the design of this chapter follows the design of those that have preceded it: we begin with a general overview of LCPs, and the role that they (can) play, before turning to document and explain the particular instruments employed by the Norwegian authorities.

Local Content Policies

In most countries, hosting a productive oil industry is not enough. For residents and authorities alike, the long-term goal must be to transfer productive capacity and wealth into greater value for the country—where value is measured in terms of domestic jobs, expertise, ownership, and the capacity for long-term sustainable economic growth. This transformation is directed by the country's industrial policies—and LCPs are one important aspect of a nation's industrial policy.

Economic and political development is about structural change; development means transferring resources from traditional activities into new, more productive activities. This is a fundamentally political process, as it is about addressing the just needs of residents (creating jobs, improving livelihoods, providing infrastructure, and so on) and the shortcomings of markets. Developing countries are littered with failed markets in need of redress; their markets for credit, labor, products, and knowledge are often inadequate, inefficient, or inept. To improve the workings of these markets, we need to study what states in similar conditions have done before.[1]

Every developing country, including those with petroleum, needs to design and implement industrial policies that can transform their domestic industries, diversify their economies, and extend their productive capacity (and the resulting tax base). In doing so, it is important to recognize that most jobs in the petroleum sector are not with the IOCs. It simply does not require that many people to bring oil/gas to the surface, once the well has been drilled. Most of the resulting jobs are found in the supply industry and the research and technological centers that feed those industries. To consider what is possible for states today, we need to know what the international community allows (that is, how that scope for autonomy is shrinking), and what states can do, within these political limits. These two considerations structure the first part of this chapter.

Scope for Autonomy

In Chapter 2 we learned how changes in the international context can constrain states that hope to manage their domestic economies. In the hope of creating level playing fields, avoiding corruption, and maximizing efficiency gains, states are told they must forego many of the policies that catapulted today's wealthiest states into their positions of power. These international constraints are especially pressing when states hope to procure a larger domestic share of the value extracted from their natural resources.

Our position is somewhat at variance with that of the industry's, where it is common to assert that all contracts should be awarded on the basis of international competitiveness for price, quality, and delivery. This standard is clearly beneficial for large foreign firms, and it can be a useful means to ensure that residents are getting the most for their money. But a fast and efficient return on investment is only one of several relevant criteria for policymakers. Many policymakers and residents are willing to forego some (short-term) efficiency gains in order to secure longer-term social and political objectives, such as protecting infant industries, balancing market power, and providing compensation for the adverse impacts of oil and gas investment on local communities and vulnerable groups.

The most pressing international constraints come in the form of obligations associated with the WTO, and the phenomenal power and concentration of the petroleum industry. As we saw in Chapter 2, a significant number of petroleum states are members or observers to the WTO, and this membership requires political sacrifice (in return for expected economic gain). While many petroleum-rich but economically poor states are not members of the WTO, their scope for policy autonomy is still affected by these constraints—as similar prohibitions are contained in several pan-regional and bilateral trade agreements (and these are expanding in number).

- **Trade-Related Investment Measures (TRIMs).** The TRIMs Agreement comes from Article III of the General Agreement on Tariffs and Trade (GATT) and (hence) pertains to the trade in goods (not in services). Remarkably, TRIMs are not actually defined in the Agreement, but it is clear that they are aimed at affecting investment measures that infringe on GATT Articles III and XI (that is, those that discriminate between imported and exported products and/or create import or export restrictions). TRIMs were subsumed into the WTO, and apply to all WTO members (that is, the Agreement is "multilateral," unlike the GPA); they explicitly prohibit: (1) local content measures that mandate particular levels of local purchases by an enterprise; and (2) trade-balancing requirements that restrict the volume or value of imports that an enterprise can purchase.[2] In short, TRIMs require a host country to treat foreign investors as they would domestic investors in similar circumstances. Clearly, this is the opposite of what LCPs are designed to accomplish, and makes it more difficult for countries to encourage, protect, and nurture local investors.
- **General Agreement on Trade in Services (GATS).** While TRIMs apply to investments related to trade in goods, the GATS applies to services. Although oil and gas have not been negotiated as a separate category under GATS, many petroleum and gas activities are included in other sectors, such as transportation, distribution, construction, consulting, and engineering. In effect, GATS extends the Most Favored Nation principle (which has long regulated the GATT's trade in goods) to the service

sector—but it adds two specific provisions related to market access and national treatment. With regard to the former, WTO members promise to treat service providers from all other WTO members in a way that does not reveal favoritism of any sort. The objective is to get rid of protectionist measures, such as limits on the number of foreign service providers, or limits on the value of foreign-provided service transactions; and limits on the number of employees or operations provided by a foreign service company (Warner 2011, 31). The GATS commitment to national treatment implies that a member state refrains from using discriminatory measures that would somehow benefit domestic services or service suppliers. In short, WTO members are obliged to treat foreign and domestic service suppliers in the same manner. As with TRIMs, these obligations hinder countries hoping to assist local suppliers.

On the other hand, it should be pointed out that the GATS does not contain a general prohibition on market-access barriers. Unlike other areas of GATT, GATS's national treatment obligation does not apply to all measures that affect trade in services. Hence, WTO members enjoy more regulatory autonomy with respect to the regulation of trade in services than they do with the regulation of trade in goods.

- **Agreement on Government Procurement (GPA).** As mentioned in Chapter 2, the GPA is different from the TRIMs or GATS in that it is a plurilateral, not a multilateral agreement. In short, the GPA opens up government transactions to international competition. The GPA explicitly prohibits so-called *offsets* (Article XVI), such as measures that encourage local development and improve the balance-of-payments accounts by means of regulations that affect domestic content, licensing of technology, investments, counter-trade, and so on (see Box 8.1). At the time of their accession, developing countries are allowed to negotiate conditions for the use of offsets (provided these are "used only for the qualification to participate in the procurement process and not as criteria for awarding contracts"), but most oil producing and developing states have been reluctant to sign the GPA.[3] This is not at all surprising, given that offset transactions play a central role in any government's LCPs.
- **Agreement on Subsidies and Countervailing Measures (ASCM).** This agreement takes aim at the use of subsidies as a protective measure. Under the ASCM, countries can use the WTO's Dispute Settlement Body (DSB) to have a subsidy removed. In particular, Article 3.2(b) of the ASCM prohibits "subsidies contingent, whether solely or as one of several other conditions, upon the use of domestic over imported goods," so-called "local content subsidies" (WTO 2016c). Indeed, a recent case involving the European Union and Canadian local content measures included reference

Box 8.1 OFFSET TRANSACTIONS

Offset transactions are contractual conditions that require the seller (usually a foreign supplier) to transfer additional economic benefits to the buyer (usually a host government or domestic company) as a condition for the sale of a base good or service.

Offsets can include:

- preferential hiring and training of nationals;
- preference lists for local sourcing;
- support for domestic suppliers to develop future competitiveness;
- support for the development of operational infrastructure to be made available for public use (e.g., roads, power, water supply, and so on);
- requirements to place a portion of contract value with domestic subcontractors;
- requirements to license technology to domestic manufacturers; and
- requirements for minimal inward investment (by international suppliers).

Offset transactions are common in the defense industry, and are often associated with corruption. When offsets include large amounts of money, suffer from a lack of transparency, and/or involve complex transactions, they become targets for corrupt officials and business people.

to this provision—making it clear that LCPs can also be affected by ASCMs (WTO 2016d, 50). In addition, certain R&D subsidies which were previously classified as non-actionable by the ASCM are now considered actionable.

Finally, it is important to note that the WTO has made provision for the "special and differential treatment" (SDT) of developing countries—allowing for exceptions to their general rules; but the impact of these remains limited. In particular, SDTs include:

- longer time periods for implementing agreements and commitments;
- measures to increase trading opportunities for these countries;
- provisions requiring all WTO members to safeguard the trade interests of developing countries;
- support to help developing countries build the infrastructure to undertake WTO work, handle disputes, and implement technical standards; and
- provisions related to least-developed country members (WTO 2016e).

These SDTs are sprinkled across the sundry WTO agreements. For our purposes, the most relevant include the 15 SDTs in the 1994 GATT agreements; the 13 SDTs found in the GATS agreement; and the 10 that exist in the GPA (WTO 2013, 5). Several of these SDTs allow flexibility to encourage foreign suppliers to assist in technology transfer and training through the use of offsets, or require WTO members to safeguard the interests of developing country members. But most SDTs for developing countries were allowed only during transition periods (when joining the WTO), and most of these have already expired. "In this regard, with the exception of Vietnam, all developing countries that produce oil and/or gas acceded to the WTO prior to 2006, which would suggest that their eligibility for transitional arrangements under GATT has expired" (Tordo et al. 2013, 28).

These growing international constraints are meant to secure efficient and fair market conditions. In practice they often serve an industry that is remarkably powerful. Although the major IOCs have lost power relative to states and NOCs, they remain powerful and rich. Alongside them has grown a service industry which is highly concentrated and global (in both outlook and reach). MNCs account for nearly 80 percent of the money spent on any particular project, and the top four to ten service providers in each segment account for 90 percent of the global market share. Market leaders in the service industry are active around the world: Baker Hughes operates in 90 different countries; Schlumberger in 80, Halliburton in 70; Saipem in 50; and (Norway's) Aker Kværner operates in 30 different countries (Sabahi 2015, 7).

These huge international companies have a significant comparative advantage, and they intend to keep it. States need to find a way to tap into this expertise and experience, but without becoming overwhelmed. In developing a strategy for expanding local content, it is important for officials to realize that no state is able to secure full control of the oil industry, and most developing countries are unable to meet the targets they set for themselves. In this light, it is important for states to develop realistic expectations of what they can hope to achieve, both in the short term and in the long run.

Local Content Policies in Practice

Within the narrow confines allowed by this new international context, what can states do to affect local content? This section tries to answer that difficult question by looking more closely at why LCPs are attractive, what is meant by local content, and how national objectives for securing local content can vary significantly. The section concludes with a simple typology for the different types of tools or instruments that states use to increase local content.

In light of our discussions in Chapter 3 about the utility of Developmental States and explicit industrial policies, it should be evident why countries want to employ LCPs. States hope to increase the value added (e.g., encourage diversification and new products/activities), minimize market failures/externalities (e.g., increase learning and knowledge links between local firms and IOCs), and further social objectives (e.g., increase employment, compensate for the adverse effects of oil and gas). They do this knowing that LCPs can expose their country to more corruption, and that this exposure needs to be minimized and monitored. The main arguments against LCPs are derived from problems when they are carried out poorly: poorly designed LCPs can lead to a misallocation of resources, which can generate unwanted inefficiencies. This is not so much an argument *against* LCPs, as it *for* creating well-designed LCPs.[4]

Simply defining local content can be a significant challenge. "Local" can refer to activities at various levels of aggregation: regional, national, subnational (regional), and even the surrounding neighborhood. For the most part, we will focus on national-level effects. Obviously, neighboring residents and communities are most detrimentally affected by the exploitation of natural resources, and these local communities are often the target of CSR efforts by firms (see Chapter 10). But we are interested in government policy, and this is usually penned at the national level. "Content" is similarly vague—as it might refer to workers, firms (their location, ownership, tax registration status), share of economic activity (percent of value of a given contract or market), or type of knowledge (e.g., encouraging the transfer of skills and technologies). Even defining a local firm is anything but straightforward, as we can see from Box 8.2.

Local content can be conceived in narrow or broad terms. LCPs aim to increase the total value added to a national economy through the localized production of select services and key materials, equipment, and goods related to target sectors of the economy, or they could be used to stimulate industrial development; promote and diversify the national economy; entice advanced technology and competence; and/or develop local capability to develop a competitive local supply base and technically competent workforce.

To better understand the utility and attractiveness of LCPs, we can return to a figure introduced in Chapter 5. In the original figure (5.2), the major components of a country's petroleum revenue were divided among costs, rent, and profit. Production costs were associated with the cost of exploration, development, and production, while rent and profit were seen as the basis for government take (taxes, duties, and state participation). Figure 8.1 adapts that original figure by subdividing the "cost" portion of the petroleum revenues into those that are locally allocated (this section is now hatched), and those that are not (horizontal stripes).

The reader will recall that the profit share in this diagram was set such that it is sufficient to keep companies interested in investing. As petroleum generates superprofits (rents), which are in excess of what is normal to entice investors to the resource, the government has to find a precarious balance: how much can the state take (as rent) without undermining the willingness of the IOC to remain and do business (at an acceptable rate of return)?

In Chapter 5 we were most interested in this division (between profit and rent). Now our focus is trained on the production costs, and who will be

Box 8.2 WHAT CONSTITUTES A LOCAL FIRM?

At first, the question seems clear: a local firm is contrasted against an IOC; it is locally based and owned. But what if an IOC establishes a local affiliate that is registered in the host country, and run by host nationals? Should that count as a local firm? How about a locally owned firm that registers offshore, to avoid paying taxes? Is that a local firm?

Common metrics vary by content. Consider the variety of metrics, whether the focus is on employment, goods, or service provision:

Employment

- Total headcount; headcount by job type; man-hours; wage value; wages by job type; and national wages and social security taxes paid, as percentage of total.

Goods

- Origin of company; nationality/ownership of company; origin of goods; and value of contract minus the proportion of imported components.

Services

- Total headcount of supplier; headcount under contract; headcount by job type; man-hours; wages; nationality/ownership of supplier; location of supplier; and registration of supplier.

Simply trying to determine the origin of the goods/services supplier can be tricky enough, and sundry measures include:

- address provided in vendor registration information;
- address on purchase order or invoice;
- share of equity owned by nationals, e.g., >50 percent;
- whether the supplier is incorporated in-country;
- whether the supplier is tax-registered in-country (e.g., for corporate or withholding taxes); and
- whether the supplier is a producer or provider of goods or services of "domestic origin."

Because of these complexities, we lack comparative indicators for local content; individual governments, lawmakers, and regulators tend to develop their own definitions. But it matters how we define and operationalize local content. Consider three different (entirely fictitious) firms supplying valves to the operator for an offshore oil platform, all three of which are registered in the host country:

1) *Exotic Valves* is a classic import agent, where the owner sits at a desk, with a phone, and arranges the import of reasonably priced valves from around the world. This company has only one (senior) employee, and she is a local. The company can generate much revenue, but has no domestic production.
2) *Valves R Us* is the world's largest producer of valves. The company has decided to invest in a very large manufacturing plant in the host country, and is a locally registered subsidiary. This company has a lot of employees, most of whom are host country nationals in production-based jobs. The senior management is foreign, however.

continued

Box 8.2 Continued

3) *Valve Nation* is a small, domestically owned and operated producer of valves. This firm employs fewer locals than *Valves R Us*, and generates less economic value, but the senior management are all nationals.

Depending on how a country defines local content, each of these companies can enjoy unique advantages, even though all three companies are registered in the home country (and qualify as local on that count). For example, if a country wants to prioritize local employment, then a contract with *Valves R Us* may be the most beneficial (even if it is less authentically local). If, by contrast, a country wants to maximize the number of nationals in senior management positions, then *Exotic Valves* or *Valve Nation* might offer the best bet. If a country wants to prioritize the total value of spending, then *Exotic Valves* might be the most beneficial. In short, deciding whether a firm is appropriately "local" can be very complicated.

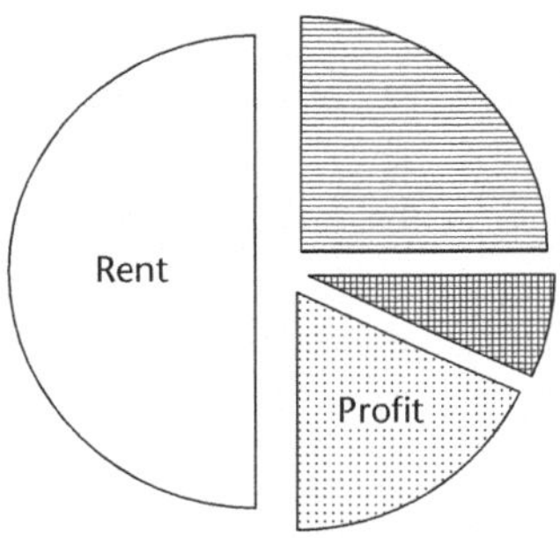

Figure 8.1 Maximizing benefits through local content

spending these. This share of revenues is more than just the cost of doing business; it is with "costs" that the opportunity for local content enters into play. Introducing "local" costs can change this figure in two important ways.

First, we can think of LCPs as an attempt by the state to secure a larger share of the costs in local hands and minds (that is, increase the size of the hatched area in Figure 8.1), without affecting the overall balance between profits, (total) costs, and rents. In effect, governments can influence the division of costs between local and foreign providers, with an eye on maximizing local competencies, employment, and value added. The increased share of local costs can have important social and economic consequences for the domestic economy.

Second, governments can change the cost calculation in ways that can affect the overall size of the pie, and the allocation within (that is, changing the relative size of the profits/rent/cost ratios). For example, in providing tax relief for exploration costs, or by encouraging the use of technologies that bring about fuller utilization of the fields, the entire pie grows (and with it, the component slices). In the early days of the petroleum industry, governments were satisfied with 25 percent or 30 percent oil recovery rates. Today, governments expect closer to 50 percent, and sometimes they are able to achieve as much as 80 percent. By adjusting the costs (in this case, making it cheaper for firms to develop and invest in new technologies), both the host state and the firms can benefit.

Alternatively, by increasing the cost portion of the revenue pie, the state can shrink the profit and rent portions. For example, if an LCP should increase the cost of production (e.g., the local firm is less efficient than the foreign firm), the money spent on local content will come at the expense of either profits or rents. Depending on how those rents are spent, the aggregate effect on the local population can be detrimental. Policymakers need to estimate the total gains from local content enhancement, and weigh these against the expected loss to government rent. The same sort of cost–benefit analysis needs to be conducted whenever the state introduces regulations or obligations that can affect the costs of production.

LCPs tend to fail when they are poorly designed, or when states are not explicit about what they hope to achieve. It is critical that policymakers establish targets that are within the reach or capabilities of local providers, and the needs of IOCs—or have a plan for increasing those local capabilities, and satisfying those needs. Policymakers need to reflect on what they hope to obtain, when they expect those results to have an impact, and what particular policies or instruments will be needed. In short, policymakers need to undertake a number of analyses to assess how their LCPs will map onto local conditions (see Table 8.1). Most important of all, policymakers need to be realistic, open, and careful.

Policymakers also have to be aware of industry trends and the concerns of their commercial partners. A remarkably large share of the industry experts surveyed, as was carried out by Hays (2013, 25), are most concerned about skills shortages (37.3 percent) and economic stability (25.3 percent). These two areas can (and should be) the focus of LCPs, as host states can encourage the relevant training of nationals, and secure the required economic stability with a broad-based industrial policy. These are long-term objectives that countries should embrace, regardless of petroleum.

LCPs will vary along a number of dimensions, depending on a country's particular needs, capabilities, and contexts. Some countries may wish to maximize local (national) employment creation; others will want to expand domestic ownership and entrepreneurial activity or encourage inward direct investment; and still others might want to focus on building up a domestic supply industry, providing local skills and technical competence. All of these things can fall under the rubric of LCPs.

Table 8.1 Mapping local capabilities[a]

Type of analysis	Description
Analyze national development priorities	Maximize the degree of overlap between local content ambitions and long-term development priorities.
Analyze local capabilities	Consider the scope of existing and potential capabilities. Map the preponderance of local expertise, resources, and capacity, and how these can be utilized by LCPs.
Map the supply chain	Identify opportunities for local participation, along with their strengths and weaknesses.
Assess the capability to meet international standards	Consider difficulties that local providers may have in meeting international health, safety, and environment (HSE) standards and/or experience with community engagement.
Cost–benefit analysis	Quantify and analyze the level of benefits, costs, and risk associated with various employment and supply options. Focus on long-term benefits.
Barriers and loopholes analysis	Locate and evaluate which barriers and/or loopholes, if any, can be used to protect and enhance local producers/suppliers.
Analyze financial support services	Access to adequate finance is usually a major hurdle for local firms. Analyze the availability of financial and business services/ support. Develop a long-term strategy to fill the gap.
Analyze technical capacity	Evaluate the availability of skilled personnel, and develop a long-term strategy to fill the gap.
Analyze infrastructure needs	Evaluate the suitability of local infrastructure (water, sewer, power, education, transport, pipelines, communications, and so on). Develop a long-term plan for supplying stable and cheap infrastructure support to help local producers.

[a] Adapted from IPIECA (2011, 5).

Policymakers also need to consider how and when LCPs will impact the local economy. Some LCPs are designed to deliver first-order economic impacts—they will directly stimulate local employees and contractors. Other LCPs deliver more indirect, second-order, effects—for example, through suppliers, subcontractors, and sub-subcontractors. Still other LCPs can be aimed at inducing third-order (multiplier) effects, from the spending effects of the income earned by nationals and resident workers.

To make matters even more complicated, the instruments used to secure these objectives (and their impact) can vary widely. The most common means of securing local content is by way of an NOC—states are able to use the licensing process to secure work for their NOCs, and NOCs are more likely to hire local employees, train local workers, and choose local contractors as their suppliers (while helping them become more efficient and experienced). In addition, states can employ a number of additional instruments that can be organized (broadly) in terms of sticks and carrots. It is important to point out that states have access to (and employ) all these different means to encourage local content.

Sticks are used to force firms into meeting explicit quotas, goals, and objectives. Almost every country has legislation that requires IOCs to use local

Box 8.3 ANGOLA'S LOCAL CONTENT LEGISLATION[a]

Article 27 of the Angolan Law No. 10 on Petroleum Activities of November 12, 2004 applies to all PSCs that are entered into by Sonangol (Angola's NOC): The article reads as follows:

27.1 Licensees, the National Concessionaire (i.e., Sonangol) and its associates, and any other entities which cooperate with them in carrying out petroleum operations shall:

(a) acquire materials, equipment, machinery and consumer goods of national production, of the same or approximately the same quality and which are available for sale and delivery in due time, at prices which are no more than 10 percent higher than the imported items including transportation and insurance costs and custom charge due;

(b) contract local services providers, to the extent to which the services they provide are similar to those available on the international market and their prices, when subject to the same tax charges, are no more than 10 percent higher than the prices charged by foreign contractors for similar services.

[a] Adapted from Sonangol (2004).

providers, so long as they are qualified and price-competitive. Some even add a buffer (e.g., you must prioritize a domestic provider, so long as their costs are not more than 10 percent more expensive). A good example can be found in Box 8.3, which shows how the Angolan authorities use the terms of their PSC to help secure local content. It is important to realize that these types of formal regulations are easy for IOCs to work around (the local company is said to be too expensive, incompetent, and so on), and this is one reason why actual local content levels often fall below the targeted levels.

In addition to these types of offsets, countries use a variety of procurement regulations, contracting strategies, vendor pre-qualification requirements, technical standards, bid documents, tender evaluation criteria, and contract conditions to enhance national competitiveness. They also can use trade barriers, such as import duties on goods and equipment, and/or restrictions on skilled immigrants (forcing firms to train and hire locals). Generally, these policies are aimed at encouraging the use of local products and services; training local personal; and transfering technology (often including training) from the IOC to the host country (or its NOC).

Carrots are more positive incentives to encourage the same sort of activity. The state can work hard to improve conditions for small and medium enterprises (SMEs) locally, and introduce other sort of incentives (e.g., leveraging tax and regulatory differences, subsidizing local inputs, and so on). Sometimes it is sufficient for a host country to simply signal their strong interest in building local competencies, as IOCs are anxious to demonstrate good will.

Many developing countries suffer from problems that are much larger than oil, and these problems make it difficult to build up local content. In such states, it is often difficult and/or expensive to start up new companies. To expand local content, the state will need to play a more active role in encouraging entrepreneurial growth.[5] In particular, states can work actively to improve the competitiveness of small and medium-sized enterprises (SMEs) by:

- *Reducing start-up costs.* States can reduce many of the bureaucratic barriers to entry, making it less costly and time-consuming to register a business.
- *Address pre-qualification challenges.* States can help local firms meet challenging international-quality, health and safety standards.
- *Lower credit costs.* These countries often have underdeveloped financial markets, and SMEs lack sufficient equity. States can help secure necessary credit lines.
- *Lower regulatory compliance costs.* States can reduce fees and reporting obligations related to legal, tax, health, safety, and environmental requirements, as these often are more difficult for small firms to meet. (On the other hand, familiarity with stringent local regulatory standards can provide local firms with a competitive advantage.)
- *Balance tax laws.* States can help smaller domestic firms by changing tax laws that benefit larger more than smaller firms.
- *Improve labor productivity.* States can provide incentives (e.g., tax incentives) so firms will invest in better-skilled workers and technologies.
- *Improve the human and physical infrastructure and utilities.* Governments can provide stable and cheap access to important infrastructure items (e.g., clean water, energy, transportation, information, education).
- *Unbundle contracts.* The state can make it easier for local (smaller) firms to participate in public procurement rounds by breaking up the contracts into smaller components. This decreases the necessary capital costs and capacity requirements—allowing smaller firms a chance to compete.
- *Develop research links.* The state can facilitate contacts between local entrepreneurs and the local university/research community to help firms secure a technological advantage.
- *Balance market power.* States need to be aware of the market power wielded by large companies, and endeavor to level the playing field.

These are common-sense reforms that states can adopt to help local entrepreneurs enter into, and compete in, open procurement tenders. Even if the bid is competitive, however, there may be other things that affect the likelihood of a local or foreign firm getting the job. For example, foreign and domestic vendors may be exposed to very different tax regimes. A prospective foreign supplier may be subject to import tariffs and withholding taxes, which make it less price competitive. On the other hand, foreigners may have access to export subsidies or export credit guarantees from their own governments. Alternatively, a domestic supplier may be disadvantaged because of a local value added tax that is not applicable to the foreign supplier, or a local firm might benefit from access to subsidized sources of domestic credit (for example, from national development banks, or from subsidized energy costs). These advantages and disadvantages lie in the margins of many contracts and policymakers need to be aware of them. They can be certain that their counterpart IOCs are exploiting these advantages. As we saw in Chapter 5, with reference to transfer pricing, MNCs are expert at international regulatory arbitrage. It is not without reason that IOCs are staffed with large and expensive legal teams.

Norwegian Local Content

Although Norway is often heralded as an example of a country that has successfully implemented LCPs, the path it has chosen to encourage local content lies low in the terrain. This path was rather ad hoc, meandering, and strongly influenced by local and past conditions. In following this path, however, the authorities have boosted an indigenous supply industry that has become international in scope—so that Norwegian firms are no longer limited to the NCS.

Context Matters

In 1962, when Norway's oil adventure was about to start, the country had remarkably few relevant experiences, institutions, or pieces of legislation. The country lacked domestic firms that could explore, produce, transport, or process petroleum—and the requisite infrastructure to do so. Norway's public shortcomings were just as evident. The country's higher education and research institutions were not oriented toward the needs of the petroleum industry; there was little institutional capacity for regulating that industry; and Norway lacked a legal framework that was particularly relevant for petroleum exploration and production.

If it was largely unprepared, Norway did have two things working in its favor. First of all, its oil and gas were located in a very difficult and costly place, so that new methods and equipment would be required to bring it to market. IOCs in Norway very quickly discovered that conditions in the North Sea were more demanding than those that existed in the Gulf of Mexico, where much of the equipment and expertise was first developed.[6] The colder climate, higher seas, stronger winds, and deeper water depths all introduced challenges that made it impossible to draw from off-the-shelf equipment, skills, and knowledge. The North Sea required new thinking and new technology, and this provided Norwegian authorities with an incredible opportunity.

This leads us to a second piece of luck: Norway had experience in tangential industries that it could adapt relatively quickly to fill these needs. Because new methods and equipment would need to be developed, Norway simply had to show that its firms and research facilities were well-suited for the tasks. After all, Norwegian ships had been plying these difficult conditions for centuries, and Norwegian builders of ships, docks, harbors, bridges, and dams all required specialized competencies and training. Long-standing demand for these specialized skills had resulted in an education system that was already producing engineering degrees in a number of tangential areas. For example, there were already Norwegian shipbuilding and marine architectural companies intimately familiar with the demanding conditions that characterized the North Sea (and could slowly transition to the needs of an offshore petroleum industry). A century of building massive dams, and an even longer history with the demands of building bridges, harbors, and piers that could withstand North Sea conditions had produced a strong local competence in concrete structures and engineering.

What the Norwegian government did, then, was to facilitate a marriage of convenience between the demands of Norway's special circumstances and the specialized competencies that lay just beyond the reach of the global petroleum industry. This second part of the chapter looks at how this marriage came to fruition.

The main framework for this partnership was set by Norway's unique petroleum administration, and its strong political components (see Chapter 4). From the very start, the Ten Oil Commandments made it clear that Norwegian institutions and the Norwegian people would be in charge of subsequent development. After all, the very first commandment held that "[n]ational supervision, [and] control of all activity on the NCS must be ensured." But the third, sixth, and eighth commandments also provide a political foundation for greater local content (see Box 4.2).

Likewise, Norway's tripartite administrative model (with regulatory, operational, and policy responsibilities divided and balanced against one another) encouraged the development of technical competence in Norway, and facilitated the expansion of Norwegian commercial interests. If we take the original representation of this Tripartite Model (Figure 4.2) and add to it some of the specific responsibilities associated with each corner of the triangle, we can better see how each part of the model contributes to expanding Norwegian content (see Figure 8.2). As we shall in the subsection that follows, allocation rounds were used to prioritize Norwegian firms and develop Norwegian competencies (the responsibility of the MPE); many Norwegians developed their expertise or secured their first contracts by working with the NOC (Statoil);

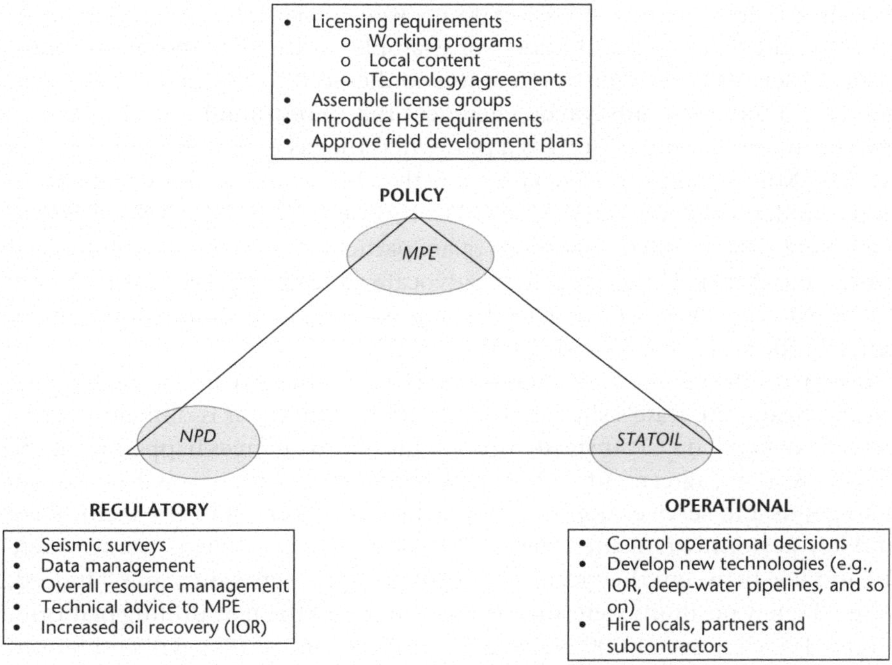

Figure 8.2 Local content in the Norwegian model

and the regulatory authority, NPD, introduced and organized new technical projects that would depend upon Norwegian research competencies (such as increased oil recovery (IOR) projects).

It is beyond the scope of this book to go into the details of Norwegian LCP, but the main outlines of that policy can be divided up into three distinct periods. For most of Norway's oil history, the government has not played an active role in encouraging local content. In the first period (1963–70) and the more recent period (1986–today), the government has largely remained on the sidelines, trying to create a level playing field upon which Norwegian firms could compete. It is only in the intervening decade and a half (1970–86) we find a more neomercantilist policy, when the government actively encouraged and facilitated the expansion of local content. While most of our focus will be on this interim period, it is important to remember that this period was, in many respects, a policy outlier over most of Norway's petroleum history.

PERIOD I: A LIBERAL START (1963–70)

As we saw in Chapter 5, the Norwegian authorities employed a hands-off strategy in the first couple of allocation rounds. It was generally assumed that the capital demands and risk exposure were too large for Norwegian firms, and that they should stand aside. For example, in the first allocation round (1965), there were no explicit references to the need to rely upon Norwegian goods and services—only vague references that "contributing to the Norwegian economy" would be seen as a plus when the licenses were allocated.[7] As we saw, Norwegian firms were given mostly table scraps from the first allocation round. Indeed, the Norwegian Oil Consortium (NOCO) and Norsk Hydro were very disappointed that Norwegian participation was not emphasized more, and Norsk Hydro began to advocate (already in 1965) for a more "national oil policy" in order to develop Norwegian technological competence in the sector (Skule and Grytli 1997, 22).

Even with the second allocation round (1969), where we find a greater push for increasing the state's share of the economic activity, it is difficult to see a detailed or explicit LCP. Instead, we see some minor changes happening at the level of working agreements, where new requirements might privilege Norwegian goods and service suppliers. In practice, however, all of the technology and key personnel remained foreign (see Box 8.4). Norwegians workers were far and few between; they received mostly low-pay, low-status jobs.[8] Indeed, when 35 new positions were advertised for work on the new drilling platform, Ocean Traveler, over 1,000 Norwegian workers applied (Ryggvik and Smith-Solbakken 1997, 27). It also bears remembering that the first Norwegian company to secure a contract in the sector was Christiania Dampkjøkken, a catering company from Oslo (Ryggvik 2010, 56).

As the Ekofisk example illustrates, this first period was one in which the government relied heavily on foreign expertise and equipment, which was imported en masse (mostly from the Gulf of Mexico) to the North Sea. The

Box 8.4 EKOFISK[a]

Norwegian content was remarkably small in the first large production installation on the NCS: the Ekofisk field. As the main operator, most of the field development was planned in Phillips Petroleum's offices in Oklahoma. As was its custom, Phillips assigned large consulting tasks to American engineering companies, who were heavily involved in every aspect of the field's development. Phillips' purchasing office was in London; most of its technology and equipment was imported from the Gulf of Mexico; and the company relied heavily on tried and true (off-the-shelf) technologies. Every single large contact signed by the Ekofisk group from the fall of 1970 to the end of 1972 went to foreign firms, mostly American. The only exception was for a concrete storage tank, where the tender went to a new Norwegian company: Norwegian Contractors. With help from this enormous tank project, the Norwegian share of the Ekofisk still only reached 17 percent.

[a] Adapted from Skule and Grytli (1997, 38ff) and Kvendseth (1988).

IOCs could draw from a tribe of nomadic workers that traveled with the rigs, and a network of suppliers that were already familiar with one another. This is the close-knit community that Norway wanted to join.

PERIOD II: THE RISE OF LCPS (1970–86)

The apex of Norwegian local content provisions occurred between 1970 and 1986. Here we find four main tools for securing local content:

A. local content legislation;
B. the negotiated nature of the licensing agreements;
C. the large share of activity than went to Statoil (and its reliance on Norwegian suppliers and subcontractors); and
D. explicit technology agreements.

The impetus for this new emphasis on LCPs was a crisis in the shipping industry in the wake of the 1973 OPEC oil boycott; many companies canceled their order for tankers, and Norwegian shipbuilders suffered greatly. The Norwegian government felt an obligation to keep Norwegians employed and wanted to encourage the Norwegian shipbuilding industry to adapt in order to service the growing petroleum industry.

A. Local Content Legislation

At the start of the country's oil adventure, Norway avoided formal or contractual commitments to secure Norwegian goods and services. License holders were encouraged to employ Norwegian suppliers when they were competitive (in terms of quality of service, time of delivery, and price)—but they weren't obliged to do so. In 1969, this principle was included in the licensing requirements; in 1971 it was elaborated upon in a government white paper,[9] and by 1972—after two large Norwegian firms (Aker and Kværner) protested over how little they were called on to help in the Ekofisk project (Ryggvik 2014)—the principle was formally introduced as Article 54 of a Royal Decree from December 8, 1972, which read:

> Licensees shall use Norwegian goods and services in petroleum operations to the extent that these are competitive in terms of quality, service, delivery time and price.[10]

That same year (1972), the Ministry of Industry established a Goods and Services Office that monitored the contracting and procurement process to ensure that qualified Norwegian firms were on the various tenders' lists, and that the bids included a sufficiently large Norwegian content (measured in personnel and monetary terms). In 1976, the government required that operators had to report all planned bidding invitations, and who eventually won the contract. After 1978 these reports were analyzed thoroughly in an attempt to see whether Norwegian firms were invited to provide bids. Then operators were required to explain which supplier they chose, and why a Norwegian alternative was not chosen (where this was the case). As a result of this new focus, Norwegian industry became more relevant and integrated (Noreng 1980; 2004). By the second half of the 1970s, the MPE was able to document remarkable increases in the Norwegian net share of deliveries to petroleum operations: by 28 percent in 1975, by 42 percent in 1976, by 50 percent in 1977, and by 62 percent in 1978 (*St. meld. nr. 53 (1979–80)*, 27).

This same report goes to great lengths, however, to note that these requirements are not meant to protect Norwegian industry from competition. They were meant as a means to secure opportunities for Norwegian industry—to participate on an equal footing. This is entirely consistent with Norway's corporatist traditions, which recognizes the need to maintain international competitiveness in order to survive in a global economy.

B. Licensing Agreements

As we noted in Chapter 5, the Norwegian authorities were able to exert a great deal of influence during the negotiation phase when assembling sundry licensing groups. They could pair up different companies as partners in a licensing group, and weigh the voting shares such that the state controlled a majority within each group. This allowed the authorities to prioritize Statoil and other Norwegian firms. Indeed, it was through these negotiations, and the resulting licensing groups, that Statoil was able to grow so quickly and become so powerful. Statoil then used its position to train a very skilled and experienced workforce,[11] build up experience, and prioritize Norwegian suppliers.

By the end of the 1970s, as part of the detailed negotiations between IOCs and the government over the nature of license groups, the Norwegian authorities found a nuanced tool for increasing local content; it was not a law or a formal regulation, but a flexible solution that was employed at every step of the way, along traditional corporatist lines (where interest groups have an important role in the formulation and implementation of policies). During each allocation round, the Norwegian authorities would negotiate with the IOCs to secure agreements that would team up Norwegian companies with more capable IOCs, so that the Norwegian firms could learn from their more experienced international partners.[12] By linking Norwegian firms with IOCs, and encouraging cooperation in individual firms and projects, the authorities facilitated a great deal of technological and knowledge transfer, to the benefit of Norwegian firms.

C. Statoil

Statoil blossomed in this protected environment, where it enjoyed many privileges (see Box 8.5). From the beginning of the third allocation round

Box 8.5 STATOIL PRIVILEGES

Operatorship

- Granted operator status after a transition/learning period or alongside "technical operators."
  - e.g., Statfjord—becomes operator from Mobil after ten years of production.

License shares

- Introduced in the third allocation round (1974–78).
  - At least 50 percent share in all licenses.
  - Extraordinary allocation of Statfjord blocks.
- Statoil privileges dropped with the fifteenth allocation round (1996).

Carrying

- Introduced in the third allocation round.
  - Carrying until commercialization.
- Dropped from all new licensees with the eleventh allocation round (1987).
  - Dropped for all production licenses from 1991.

Sliding scale

- Introduced in the third allocation round.
  - Scale initially went up to 66–75 percent, but increased to 80 percent in the fourth allocation round (1979).
  - Right for Statoil to increase participating interest to a predefined level (decided at the time of award).
- Adjustment of sliding scale practice from the twelfth allocation round (1988).
  - Dropped for all production licenses (except 3–4 defined fields in production) in 1994.

Voting rights

- From the third allocation round to 1985.
  - 50 percent of rights (enough to commit the license).
- From the tenth allocation round (1985).
  - Statoil loses its ability to vote on the government's shares in a license group; in effect, losing its veto power over field decisions.

(1974), a number of measures were put into place to prioritize Statoil's interests. These measures included arrangements that transferred ownership over to Statoil after a given period of time; secured a (government) majority of shares and voting rights within licensing groups—to influence outcomes; carried Statoil interests (as described in Chapter 4, above); and introduced a sliding scale for increasing state control (along with the size of the field under production).

To illustrate the sizeable influence of Statoil, and its allied (state) partners, Figure 8.3, depicts the share of state influence in Norway's offshore license groups, operationalized as Statoil alone, and together with other state interests

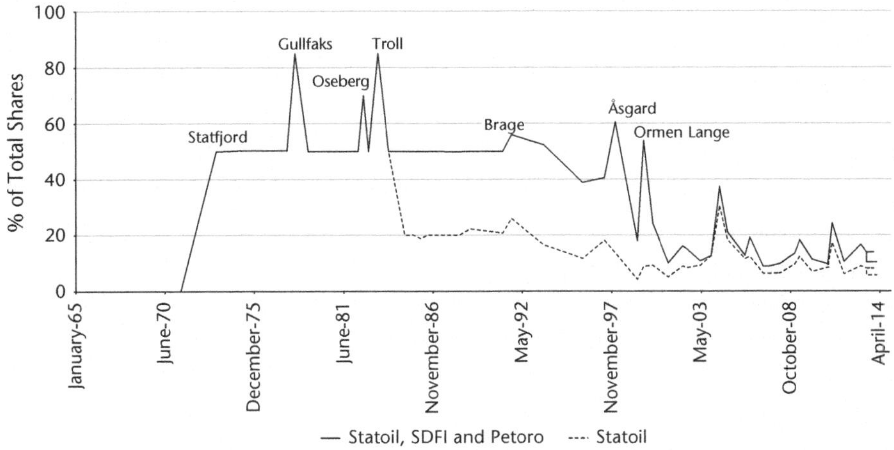

Figure 8.3 State share of licenses on the NCS, 1965–2014[a]
[a] Adapted from NPD (2016a).

(SDFI and Petoro). Here we can see that Statoil received 50 percent of the Statfjord field in 1973, and subsequent allocation rounds prioritized Norwegian interests, such that the Norwegian state's share of the licenses allocated remained above 50 percent until the early 1990s. Not only were Statoil and other Norwegian companies (e.g., Norsk Hydro, Saga) getting most of the licenses, but they were also being allocated fields with the greatest potential. This is arguably the most important means for expanding Norwegian local content on the NCS.[13]

Being an NOC, Statoil recognized that it had social as well as economic obligations.[14] This argument is often used to explain why NOCs can be less efficient, compared to their private competitors. But the social gains to the country can be significant, if such obligations are delivered correctly. For example, Statoil was willing to invest in high-risk technology projects in order to gain research and learning experience. Examples include risky investment projects in pipeline development (Statpipe); improved oil recovery (Gullfaks, Statfjord fields); subsea installation projects (Gullfaks, Norne fields); and carbon capture (Sleipner field). Although the risks associated with these projects were larger, so too were the social rewards—as Norwegian firms gained experience and expertise at the cutting edge of the industry.

As the size of the government's rent increases with the size of a field's productive capacity, the government introduced a sliding scale to ensure these rents would stay with the Norwegian people. This was achieved by changing Statoil's participation share from a fixed and flat rate, to a sliding scale (from the third to the tenth allocation rounds). What this means is that

Table 8.2 Statoil participation, typical sliding scale[a]

Maximum License Producing Rate (thousands of barrels per day)	Statoil Working Interest (%)
0–75	50
75–100	52
100–125	54
125–150	56
150–170	58
170–190	60
190–200	61
200–210	62
210–230	63
230–250	64
250–270	65
270–290	66
290–310	67
310–330	68
330–350	69
350–370	70
370–390	72
390–410	74
410–430	76
430–450	78
450 and above	80

[a] Adapted from Kemp (1987, 40). Alexander G. Kemp, *Petroleum Rent Collection around the World* (1987). Reproduced with permission from the Institute for Research on Public Policy www.irpp.org.

Statoil's share of total production was allowed to increase along with the size of peak production from the field. The precise share was subject to negotiations, and not generally published. But typical shares, from the late 1970s and early 1980s, are given in Table 8.2. By the seventh, eighth, and ninth rounds, the sliding scales become truncated, before the process was dropped in the tenth allocation round (1985).[15]

To gain an idea of the importance of this sliding scale, consider the fact that peak production levels in the Statfjord field were estimated to be 700,000 bbl (barrels)/day; at Oseberg they were 500,000 bbl/day; and at Gullfaks they were also 500,000 bbl/day, as shown in Figure 8.4.

D. Technology Agreements

The fourth important part of this puzzle came into play at about the same time. In 1978, as part of the fourth allocation round, it was determined that at least 50 percent of research and development (R&D) efforts related to field development should occur in Norway. Applicants were required to sign R&D agreements (so-called technology agreements) with Norwegian institutions. These varied in form, from general cooperative agreements with Norwegian

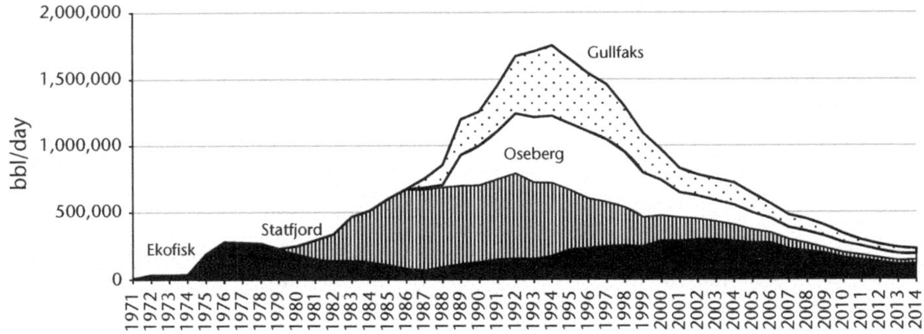

Figure 8.4 Oil production rates for major Norwegian fields, 1971–2013[a]

[a] Adapted from Norsk Petroleum (2016).

R&D institutions to the allocation of funds for specific R&D projects to be carried out by selected Norwegian institutions.

The logic was simple: as Norway was paying most of the bill (through tax deductions and direct funding as a major shareholder) it can expect that the money be spent to develop Norwegian expertise. The introduction of these technology agreements triggered an impressive development in petroleum-related technology in Norway. When the idea was introduced to the parliament, the government argued that this sort of R&D emphasis was to play a central role in a national policy that aimed to: (1) develop independent Norwegian enterprise to ensure that Norwegian resources would be developed in a socially responsible manner; (2) strengthen the competitiveness of Norwegian industry; and (3) ensure that petroleum operations would develop in a way that was compatible with Norway's stringent safety and environmental protection regulation, while shining light on the societal consequences of the operations (*St. meld. nr. 53 (1979–80)*, 67). In short, the needs for protecting workers' safety, the environment, the national economy, and society in general, would propel new research.[16]

As Wiig and Nås (1992) have noted, Norway was the only country in the world that employed these sorts of technology-transfer agreements—in effect trading oil for research. The technology agreements were anchored in law and integrated into the concession framework. These consisted of two main types: *50 percent agreements* mandated that operators conduct at least 50 percent of the research necessary to develop its field in Norway; while *offer agreements* required companies to cooperate with Norwegian R&D institutions within defined areas, for defined amounts, if they obtained a concession. In addition, Norway employed non-binding *goodwill agreements*, which were voluntary (but they surely counted when license allocations were being decided). In these goodwill agreements, companies declared their intent to conduct as much of their petroleum-related R&D in Norway as possible.[17]

Meanwhile, the government was expanding its petroleum-related educational offerings, to fuel the increased demand. NTH in Trondheim (now the Norwegian University of Science and Technology), the universities in Oslo and Bergen, and the District College in Rogaland all trained engineers and workers in a way that was relatively new for the industry. In contrast to the United States, where past experience was often prioritized over theoretical or academic competence, the Norwegian sector became better trained and educated at every stage of production. As a consequence, it could rise to the challenge and consider more radical and technically demanding solutions for banking hard-to-reach resources. This emphasis on technical competence and solutions was buttressed by a tax system (described in Chapter 5), which allowed the oil companies to write off much of the expensive R&D costs. High labor costs in Norway provided a similar incentive to switch to more technical (less labor-intensive) solutions.

These different tools came together to build up a new Norwegian-style approach and design to offshore drilling, which prioritized Norwegian competence and rewarded Norwegian experience. In a milieu that encouraged foreign companies to collaborate with local firms, IOCs shifted to a new, more modular system of construction which allowed smaller companies to participate and compete in the process. When the bar was first lowered, several Norwegian shipbuilders were enticed into the industry, building platforms, supply ships, and so on. At about the same time, Condeep platforms came into play, taking Norwegian concrete expertise and experience to new levels. As long as rigs were being constructed from steel, it would be hard for Norwegians to compete; but with the design of new PDQ platforms made of concrete,[18] Norwegian industry had a real chance to compete.

In fact, the development of these new concrete and modular systems provides a good example of how Norwegian regulators helped to spark the growth of domestic competencies. Originally, the offshore platforms were designed with the workers' quarters located directly atop the drilling rig (atop a "colossal primus," in the words of one observer).[19] The NPD opposed these plans early on, and suggested that the living quarters should be located on a completely different platform, to protect workers' safety. Although the operators complained that this was cost-prohibitive, they agreed to move the living quarter module to a different part of the platform (hence the development of more modular designs).

Although we will focus on these types of regulations in Chapter 9, it is important to note how they played an important role in favoring Norwegian workers and firms, who were most familiar with them:

> Measures whose initial purpose was to establish a high safety and environmental standard, to regulate the workplace or health issues, played perhaps the greatest role in giving an advantage to Norwegian firms and workers. Even the more conservative government which took over in 1982 demanded that foreign firms should accept the rules of the game within the established Norwegian tripartite relationship between trade unions, employers' associations and the authorities. The requirements that the oil fields' operational organisations had to be based in Norway, that all documents and contracts had to be set out in Norwegian, and that the working language on the platforms was to be Norwegian, could be justified on social and safety grounds. They nevertheless had the effect of giving companies who understood the Norwegian system an advantage. (Ryggvik 2010, 59)

This most active period combined several different tools to increase Norwegian content. The most significant effects came from licensing groups that were a product of the negotiated allocation processes. These groups often included Norwegian firms, especially Statoil—and these domestic agents became catalysts for expanding local competence and experience. The most unique of the Norwegian tools was the development of technological agreements that catapulted Norwegian firms to the industry's research frontier. When this period of active LCPs comes to an end, in the mid-1980s, Norwegian firms were competent, established, and ready to face increased international competition.

PERIOD III: RETURN TO LIBERAL POLICIES (1986–PRESENT)

In 1986, the price of oil began to drop precipitously. Given the high cost of production on the NCS, the Norwegian parliament felt it needed to act quickly in order to stave off the massive flight that was being signaled by the IOCs. The response was a substantial number of reforms aimed at keeping the IOCs in Norway. These reforms were two-pronged: on the one hand, the government signaled a significant reduction in the tax level, in exchange for a promise to maintain offshore activities.[20] On the other hand, the government began dismantling its local content regime. The motivation for doing so was partly a need to placate IOCs, partly a recognition that many of its tools would no longer be acceptable in a changing international context, and partly a desire to ensure that Norwegian firms maintained their international competitiveness.

Norway's retreat from LCPs was quick and effective. As early as 1986, the technology agreement requirements were dropped, in response to the falling oil prices. The carrying requirements followed suit on January 1, 1987. In preparing to join the EEA, the government began to cut back on a number of measures that the EU considered protectionist. For example, in 1992, §54 (from the 1972 Royal Decree) was dropped, and the Goods and Services Office that monitored Norwegian hiring was closed. In the following year, 1993, the sliding scale was removed. From 1994, the requirement that the state and

Statoil should participate in at least 50 percent of the activities was dropped, and the first allocation that did not include Statoil or the Norwegian state came in 1996.

By the middle of the 1990s, IOCs were no longer obliged to establish Norwegian subsidiaries, or explain why they hadn't chosen Norwegian partners/suppliers. Norwegian goods and service providers could no longer be prioritized, and being Norwegian was no longer a legitimate criterion for the allocation of licenses. Even the government's demand that the oil/gas be landed in Norway was dropped.

By this time, however, Norwegian firms had established themselves enough to compete on the international scene. At the turn of the millennium, the Norwegian supply and service industry was increasingly international, as shown in Figure 8.5. Since 2011, these companies generated more sales in international markets than they did at home, with the total sales from Norwegian suppliers reaching 524 billion NOK in 2013 (Rystad Energy 2014, 7). By 2014, over 1,250 petroleum service companies making a significant income were registered in Norway (Rystad Energy 2014, 4). These firms are the direct benefactors of the Norwegian LCPs in the 1970s and 1980s.

As can be seen in Figure 8.6, the two largest foreign markets for Norwegian petroleum-related service firms are in South Korea (where Norwegian suppliers are filling orders for platforms being built in South Korea) and Brazil (where Norwegian experience with underwater production is attracting Brazilian

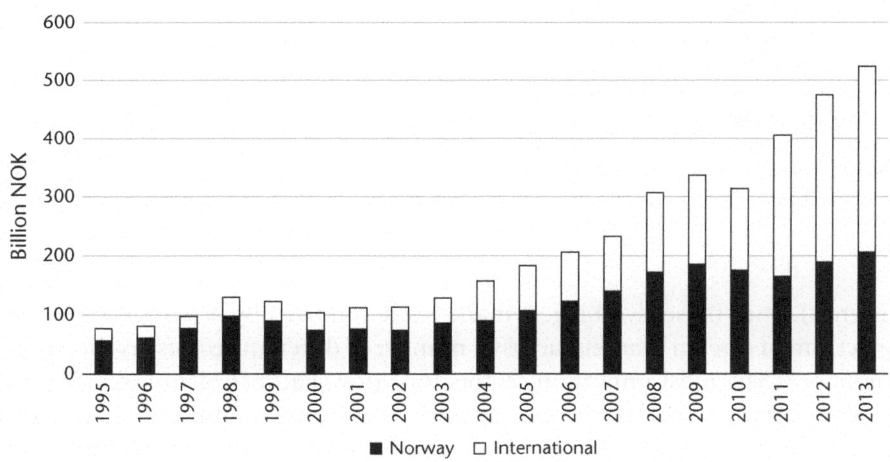

Figure 8.5 Total sales, Norway and international, 1995–2013[a]

[a] Adapted from Rystad Energy (2012, 18; 2014, 7). Norwegian petroleum service providers. Billions of NOK. The approach for calculating these figures seems to have changed between the two reports. In the 2012 report, Rystad Energy reports the transition year (2011) figures to be 152 in Norway; 209 for International, with a total of 361 billion NOK. The 2011 figures used here are from the 2014 report.

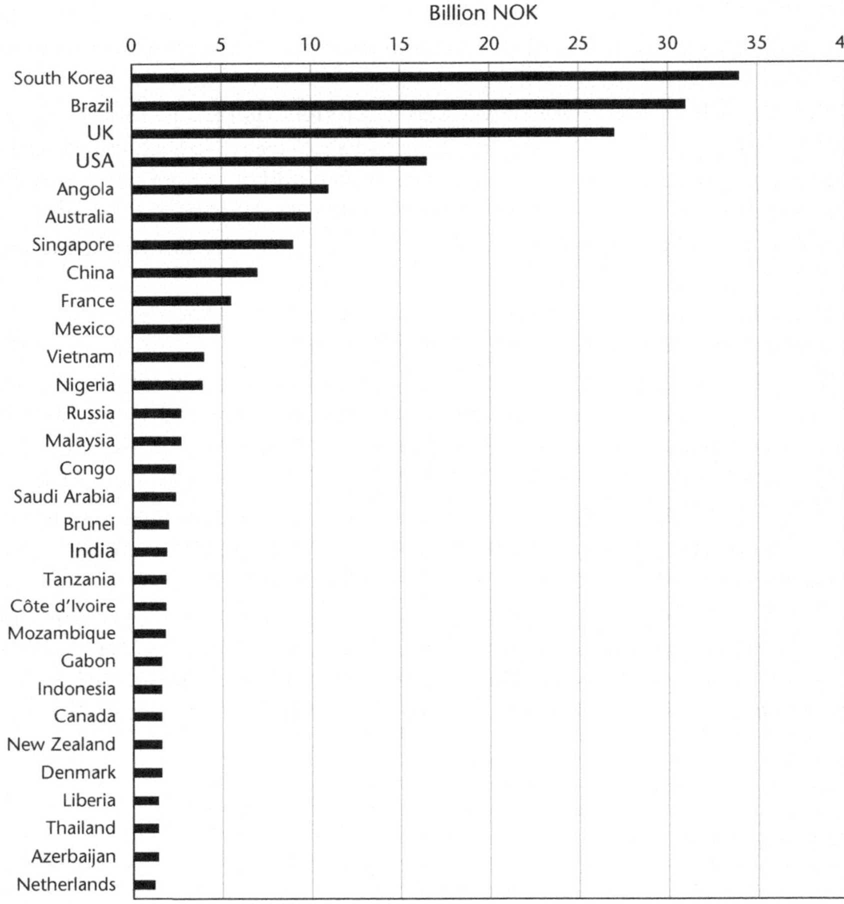

Figure 8.6 Thirty largest markets[a]

[a] Adapted from Rystad Energy (2014, 25). International sales by Norwegian petroleum service providers in 2013. Billions of NOK.

interest). Still, these two biggest markets are followed by a remarkably wide spectrum of foreign markets, across a number of different sectors, as shown in Figure 8.7. The most important sectors for Norwegian petroleum service providers working overseas are topside and processing equipment, along with rig and drilling services, but the Norwegian industry is in no way limited to these two areas (e.g., subsea equipment, transport, and operational services are important sectors). As in geographic terms, Norwegian industry is spread widely across a number of sectors.

This diversity is less evident when we focus in on the 20 largest petroleum-related service firms in Norway. These can be grouped into two main sectors,

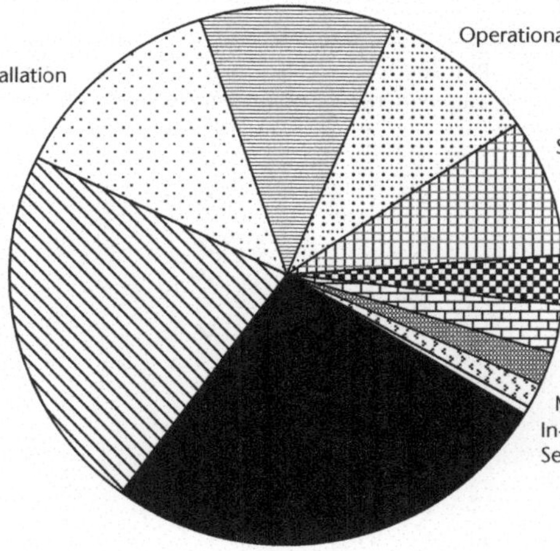

Figure 8.7 International sales, by sector[a]

[a] Adapted from Rystad Energy (2014, 30). Norwegian Petroleum Service Providers in 2013. G&G refers to geological and geophysical activities.

as is the case in Table 8.3: either rig- and shipowners, or offshore maritime and equipment services (in 2012, there was a shipyard among the top 20 firms (Kværner), but it has since dropped in the rankings). Remarkably, just six of the top 20 firms are subsidiaries of IOCs. These are mostly homegrown companies, doing well in international markets.

Conclusion

As we have already seen with regard to other areas of petroleum management, countries will differ in terms of the policy tools they choose (as their specific economic and social objectives will differ). For this reason, there is no one-size-fits-all standard package for securing local content. Instruments that work in some countries and in some period of time may not work at other times, or in other countries.

Countries rely on LCPs to create jobs, promote enterprise development, and accelerate the transfer of skills and technologies. In effect, these policies are aimed at strengthening the productive capacity of the host country: to increase economic growth, international competitiveness, and the country's standard of living. In the 1970s and early 1980s, Norway was able to employ a number of different policy tools to build up local competence and expertise.

Table 8.3 Top-20 Norwegian petroleum service providers, by sector and foreign subsidiary[a]

	Rig- & Shipowners	Offshore & Maritime Equipment/Services	Has Foreign Parent Company
Aker Solutions		✓	
Archer		✓	
BW Offshore	✓		
CGG-Veritas		✓	✓
DNV		✓	
DOF	✓		
Dresser Rand		✓	✓
Farstad Shipping	✓		
FMC Technologies		✓	✓
Fred Olsen Energy	✓		
Kongsberg Group		✓	
National Oilwell Varco		✓	✓
Ocean Rig	✓		
Odfjell Drilling	✓		
PGS		✓	
Polarcus		✓	
Rolls-Royce Marine		✓	✓
Schlumberger		✓	✓
Seadrill	✓		
Solstad	✓		

[a] Adapted from Rystad Energy (2014, 33). Firms listed alphabetically, 2013 data. The 20 largest firms were determined by the companies' international sales figures.

The result is evident today in a remarkably strong, competitive, and internationally oriented Norwegian petroleum industry.

Norway's success is largely owing to good management techniques and strategic thinking. Norway's good management practices facilitated the development of local content—finding new niches and opportunities for domestic industries to exploit and grow. Given Norway's difficult (sometimes extreme) working conditions, new technological solutions needed to be developed—and Norwegian firms were encouraged to fill the void. They could do this at a time when the international economic context allowed states a freer hand to prioritize national firms and workers. By the mid-1980s, that window was shutting for Norway, and it was forced to jettison its most effective measures. This suited Norway fine, as the country had already secured enough time to build an infant industry and prepare it for international competition. Today's developing states will find it much more difficult to secure this space for policy autonomy.

As we saw briefly, one of the tools employed by the Norwegian authorities to build up local content was an extensive regulatory regime offshore. We turn now to the role that safety, health, and environmental regulations play in resource management.

9

Regulating the Workplace and the Environment

As oil and gas E&P extends into new corners of the globe, the industry finds itself working in an incredible mix of geological, climatic, and environmental settings: from the frigid waters of the North Sea to humid and tropical conditions around the equator. Petroleum resources lie under countries that are rich and poor, democratic and authoritarian, young and old. It should not be a surprise, then, that firms also face an equally broad spectrum of workforce and environment regulations. This chapter describes that regulatory variance. In so doing, we explain some of the biggest challenges facing states that wish to protect their workers and environment from a powerful, dangerous, and dirty industry.

We begin the chapter by considering the domestic demand for regulation, and the role that it plays in modern markets. This regulatory demand has evolved over time, such that countries (and firms) have moved from prescriptive regulations to embrace more systematic approaches to minimizing risk. These new approaches rely heavily on a growing industry of standards providers and analysts. The first part of this chapter introduces this network of regulatory providers.

The second part of the chapter describes the regulatory path chosen by Norwegian policymakers. Although it took some time to get there, the Norwegian path has led to an impressive and unique regulatory edifice, which combines worker participation and a great deal of self-regulation. The Norwegian story shows how important it is to secure an atmosphere of trust between the relevant actors (state, workers, industry), and to use that trust to build flexible, yet robust, regulatory regimes.

Regulatory Regimes

For many of us, regulation sounds like a dirty word—something that governments place in the way of efficient markets. While there can be no question that regulations can be cumbersome, they are an essential tool for governing an industry that is as powerful, dirty, and as potentially dangerous as petroleum. Regulations are also costly. Introducing environmental, health, and safety regulations inflates the "cost" slice of the revenue pie (see Figure 8.1) in a way that is very similar to what we saw with local content measures: the costs of E&P expand at the expense of the other components slices (profits and rents) in the total revenue pie. But the costs of not regulating, should this result in disaster, can be astronomical.

There are several reasons why market regulations are necessary. The most common of these include conditions of imperfect competition, the existence of negative externalities, inadequate information, and the need for social protections. While each of these market shortcomings can prove an important motivation when designing a regulatory framework, the first two are especially pressing when regulating the market for petroleum.

A Dirty, Dangerous Industry

It is hard to imagine an industry that is better suited for regulation. There are many reasons for this. First of all, it is a complex industry. As the editors of a recent volume on *Risk Governance of Offshore Oil and Gas Operations* note:

> These offshore activities involve sophisticated analytical methods, heavy engineering, large-scale investment, and complex projects, and they must be managed appropriately to ensure that benefits are gained without incurring major accidents and other unacceptable harms to the public, the workers involved, and the human and natural environments. This requires partnership between public regulators and industry, the involvement of labor and other stakeholders, a supporting role for researchers, mutual trust that best practices will be used and continuously improved, and much more. (Lindøe et al. 2014, xv)

Second, bringing petroleum to market is a dirty, dangerous business. Workers are exposed to significant risks at every stage of development; they work with extremely flammable/explosive materials, often highly pressurized, and under challenging working conditions—often times far from the regulatory gaze of officials in the nation's capital. The US Occupational Safety and Health Administration (OSHA) figures that on-the-job fatalities in oil and gas extraction were seven times greater than the rate for all other US industries between 2003 and 2010, and that these fatality rates are nothing new (OSHA n.d.). Globally, a survey of 49 influential IOCs, working in 107 different countries, revealed that 80 of their workers died in 2013, down from 88 in 2012 (OGP 2014, vii). While fatality rates tend to be declining in recent years (both in the United States and globally), too many workers continue to die each year from a variety of different accident types.

These workers need protection from a demanding industry, sometimes unscrupulous employers, and a workplace that can be extremely dangerous. To secure a better workplace (in general), the International Labor Organization (ILO) promotes a Decent Work agenda, based on four main components. Countries should: (1) defend a wide spectrum of rights at work (e.g., minimum wages, healthy working environments, volume and form of work, freedom of association, and so on); (2) secure stable employment for their residents (as this has an obvious effect on social security); (3) extend social protections, including access to health care and income security (particularly in cases of old age, unemployment, sickness, invalidity, work injury, maternity, or loss of a main income earner); and (4) encourage social dialogue, including the need for negotiation, consultation, or simply an exchange of information and views between employers, workers, and governments (ILO 2012 and 2015a; and Ghai 2006). In the petroleum industry, worker safety is increasingly tied to the use of appropriate technology and workplace cultures—so governments need to focus their attention on these factors.

In addition, the industry works under a constant threat of environmental disaster, as wells can blow out, storms hit hard, and tankers run aground. Countries need to protect themselves from these threats. After all, the oil industry's history is littered with dirty, dangerous episodes—and many of these names remain lodged in the public's collective memory—for example, Deepwater Horizon, Ixtoc, *Amoco Cadiz, Exxon Valdez* (for a ranking, see Moss 2010).[1]

As with safety and workplace regulations, national regulators have to develop regulatory systems to protect the environment. In general, there are three types of environmental risk: (1) operational emissions; (2) unplanned emissions; and (3) long-term climate effects. Operational emissions concern sea emissions, which consist of cleaned water that originates from the reservoirs (produced water) and drilled rock mass (drill cuttings) that originates from drilling activity, and air emissions, such as the exhaust from the energy production needed to operate the facilities.[2]

The second type of environmental risk, unplanned emissions, is the one that makes the headlines; unplanned emissions are the result of accidents (e.g., tankers sinking, pipes breaking, and so on) that can have more detrimental environmental costs. Regulations are used to minimize the threat of these and to facilitate rapid and effective responses, once the disaster takes place.

The third risk concerns climate change and global warming. Here the petroleum industry finds itself standing on the wrong side of history. Not only were the industry's most vocal members slow to admit publicly that climate change existed, and lobbied strongly against preventive measures, but their level of commitment to pursuing alternative energy remains uninspiringly low. In short, gas and oil account for a significant share of global warming emissions, and yet the IOCs are proceeding with a business model that generally ignores this. All this is happening at a time when the international community has agreed to limit the use of oil and gas reserves, in an attempt to stave off global warming (see Chapter 10 for more on this point).

Citizens and workers need protection from such a hazardous industry, but they cannot expect the oil companies to ensure this on their own. We are not suggesting that the oil industry embraces environmental disaster, global warming, and worker fatalities—the industry is highly motivated to minimize these risks. The problem is that an unregulated market provides little incentive for them to do so. Each firm finds itself locked in a collective action dilemma; in pursuing a strategy that appears optimal for the individual firm, they are contributing to an outcome that is collectively suboptimal. Worse, in the absence of clearly defined and assigned costs and responsibilities, each firm has an incentive to free ride.

In a competitive (non-regulated) context, firms are under greater pressures to reduce their costs of production (including the costs of protecting workers and the environment), and each has an incentive to cut corners. Regulation provides a means to overcome this collective action dilemma; it provides a level floor, upon which every firm can compete fairly.[3] Government regulations secure the necessary protections in a way that takes threats to the environment and workforce exploitation out of the competitive dynamic. These regulatory costs should be the same for every firm, so that no individual firm has an incentive to cut corners (and risk safety, health, or the environment).

Regulatory Approaches

Governments have long recognized the dilemma, and have worked hard to improve the way in which they regulate the industry. As the nature of production has changed, along with the sophistication of international markets, approaches to regulation have also changed. The most significant changes came in the waning decades of the twentieth century, in response to a number of serious industrial accidents that clearly demonstrated the inadequacies of existing regulatory approaches.

Originally, health, safety, and environment (HSE) challenges were considered to be mostly technical in nature, and hence required technical solutions.[4] The resulting regulatory regime was reactive and prescriptive: safety norms were made explicit and firms were expected to follow them; regulators approached the industry with a checklist in hand, permitting companies that complied with the regulations listed. In short, governments and firms created technical "fixes" that were monitored by a regime of technical inspections.

This regulatory regime came under scrutiny after a series of high-profile industrial accidents, lasting from the mid-1970s to the end of the 1980s—for example, the Seveso industrial accident in Italy (1976); the partial nuclear meltdown at Three Mile Island in the United States (1979); the capsizing of the Alexander Kielland platform off Norway's coast (1980); the Bhopal gas tragedy in India (1984); the catastrophic accident at the Chernobyl nuclear power plant in the Ukraine (1986); and the explosion on the Piper Alpha platform in UK waters (1988). These accidents occurred in a wider spectrum of industries (not just petroleum). But when several related industries experienced catastrophic environmental and human disasters, one right after the other, officials (in both the private and the public sector) became aware of the need for radical changes in the way we regulate industrial production.

As a result, a new regulatory approach turned the industry's focus away from technical checklists and toward more systemic and organizational features; human, behavioral, and organization/cultural issues came to play a central role.[5] Industrial organizations and their production techniques were becoming more complex, and were changing quickly over time. In response, a new regulatory regime needed to be more proactive; there was a growing recognition of the utility of a standards-based and self-regulatory approach to HSE regulation. This new emphasis on industrial autonomy and self-regulation fits hand-in-glove with the general shift in ideology and power that we witnessed in other areas of the industry at the end of the twentieth century.

The result was a number of more systematic HSE approaches that rely heavily on independent standards and third-party verification procedures. In the 1990s, standards-based approaches to Occupational Safety and Health Management Systems (OSH MS) were developed that drew on Total Quality Management frameworks (specifically, Quality Management Systems according to the ISO 9000 standard). This led, in turn, to the development of OHS MS standards (the most common of which is OHASAS 18001), and the ILO introduced their OSH MS Guidelines in 2001 (ILO 2001 and 2011). Box 9.1 provides an overview of the most influential standards, guidelines, and principles.

This effort to systematize the analysis of risk factors has spawned an industry of professional standards providers, analysts, and controllers. The result is a new hierarchy of regulatory control, where regulatory forces (and logics) can be found at three distinct levels: national, industry, and firm. Figure 9.1 illustrates this, schematically. National policymakers generate the laws, regulations, and guidelines that steer more general developments (nearer to the ground). These regulations often refer to particular standards, provided by a wide range of standards providers (international, industrial, national)—for example, the management standards and guidelines presented in Box 9.1. Finally, the field operator (often working alongside other firms in a shared license) will have to

Box 9.1 SAMPLE OF INFLUENTIAL STANDARDS, GUIDELINES, AND PRINCIPLES

IFC Policy and Performance Standards (2012)

- Generated by the World Bank's International Financial Corporation (IFC). Now the global benchmark for managing environmental and social risk by financial institutions.

Occupational Health and Safety Assessment Series (OHSAS 18001)

- A leading international occupational health and safety management system specification, created by the world's leading national standards bodies, certification bodies, and specialist consultancies. Provides the minimum requirements for occupational health and safety management best practices.

ILO Guidelines on Occupational Safety and Health Management Systems (ILO-OSH 2001)

- Generated by the International Labor Organization (ILO) to help organizations, national institutions, employers, workers, and other social partners establish, implement, and improve occupational safety and health management systems.

ISO family of Quality Management Systems (ISO 9000)

- Generated by the International Organization for Standardization (ISO) to address various aspects of quality management, including some of the ISO's best-known standards. They are not specific to any one industry, but are used for quality management and quality assurance purposes.

Equator Principles

- Adopted by banks that account for over 85 percent of global project financing. Commits institutions to apply IFC standards in project loans they provide.

World Bank Group Environmental, Health, and Safety Guidelines (EHS Guidelines)

- Provides performance guidelines on a variety of issues (including environment, occupational health and safety, community health and safety and construction and decommissioning), and specific Industry Sector EHS Guidelines for onshore and offshore oil and gas development, liquid natural gas (LNG) facilities, gas distribution systems, crude oil and petroleum product terminals, natural gas processing, and petroleum refining, among others.

UN Principles for Responsible Investment

- Six general principles for responsible investment, based on environmental, social, and corporate governance (ESG) issues.

negotiate site-specific specifications to synchronize different firm-level standards requirements and their fit with local conditions.

Figure 9.1 helps us to see the importance of the middle-level standards industry in this new regulatory regime. On the one hand, regulators find it convenient to rely on international management standards to ensure that occupational health and safety needs are being protected, and to facilitate cooperation within firm partnerships or licensing groups.[6] Projects that rely

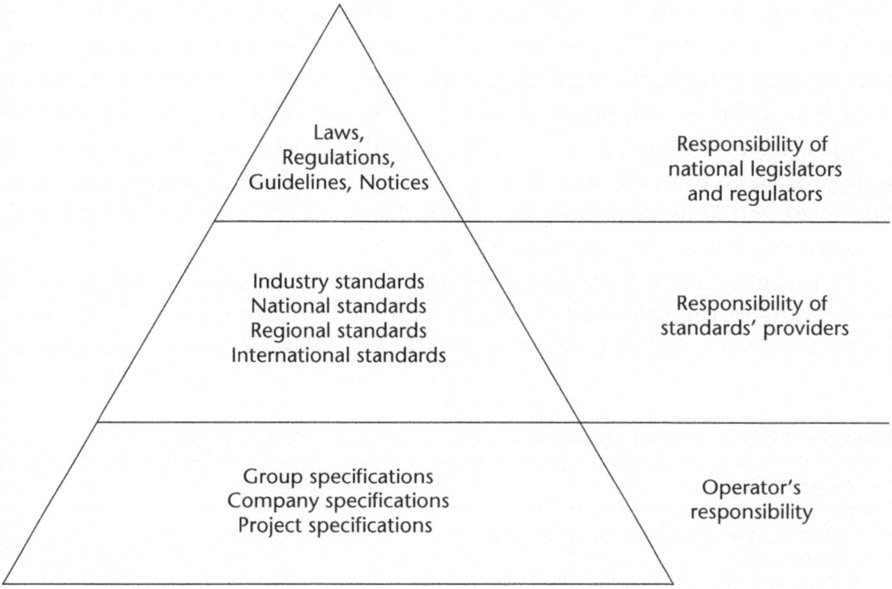

Figure 9.1 Hierarchy of regulatory control[a]

[a] Adapted from OGP (2010, 3).

on external project finance often face a requirement to comply with World Bank, IFC, or Equator Principles requirements, and countries may require that voluntary codes of corporate contract are followed (e.g., UN Principles for Responsible Investment or those provided by the IFC and/or national export credit agencies).

At the same time, regulators employ more particular (fine-grained) standards when defining their specific schemes, regulations, and guidelines. Given the complexity of the industry, and its propensity to change, policymakers find it increasing difficult to create a checklist of dos and don'ts. Neither can they afford the labor needed to go out into the field and control whether the checklist is being followed. Instead, regulators latch onto a number of different standards, from a remarkable array of standards providers. They then require license holders to meet those standards, and hire third-party certification agencies (e.g., DNV, Lloyds, BVQI, and so on) to conduct dedicated audits that confirm they have "justified confidence" in the management system, and that the organization complies with all relevant requirements.

A 2010 survey of 14 national regulatory documents by the International Association of Oil and Gas Producers (OGP) found that regulators relied heavily on standards. Most regulators used a wide variety of individual (and unique) standards. In particular, the survey found 1,140 references to different, individual standard titles. Of these, 989 (or 87 percent) were referenced by just one regulator (that is, only 13 percent of the standards were used by two or more regulators). Table 9.1 provides a sample of those standards used by more than one regulator in the OGP survey. These standards came from 60 different international, national, regional, and industry standard development organizations, from around the world. Globally, the most common (44 percent) standards came from the industry (those from the American Petroleum Institute (API) dominate), followed by national standards (35 percent) and international standards (21 percent) (OGP 2010, 1).

At the same time, the practice of relying on these standards is embedded in three layers of international developments: (1) a growing network of international treaties; (2) an expanding number of bilateral investment treaties

Table 9.1 Most commonly shared international standards (illustrative sample)[a]

#	Standard Title	Standard Provider
5	API Spec 6A Wellhead and Christmas Tree Equipment	American Petroleum Institute
5	IMO Mobile Offshore Drilling Units (MODU) Code	International Maritime Organization
4	API RP 14B Design, Installation, Repair and Operation of Subsurface Safety Valve Systems	American Petroleum Institute
4	API Spec 17J Unbounded Flexible Pipe	American Petroleum Institute
4	IEC 60331-11 Fire Resisting Characteristics of Electric Cables	International Electrotechnical Commission
4	API RP 14C Analysis, Design, Installation, and Testing of Basic Surface Safety Systems for Offshore Production Platforms	American Petroleum Institute
4	API PR 521 Guide for Pressure-Relieving and Depressuring Systems	American Petroleum Institute
4	API Spec 14A Subsurface Safety Valves	American Petroleum Institute
4	NFPA 13 Installation of Sprinkler Systems	National Fire Protection Association (US)
4	ANSI/ASME VIII-1 Pressure Vessels	American National Standards Institute/ American Society of Mechanical Engineers
4	NACE MR 01-75 Standard Material Requirements, Metals for Sulfide Stress Cracking and Stress Corrosion Cracking Resistance in Sour Oilfield Environments	National Association of Corrosion Engineers (US)
3	DNV OS-F101 Submarine Pipeline Systems	Det Norske Veritas (Norway)
3	EN 1838 Emergency Lighting	European Norm
3	IEC 61892-7 Fixed and Mobile Offshore Units—Electrical Installations	International Electrotechnical Commission
3	ISO 10418 Analysis, Design, Installation and Testing of Basic Surfaces Safety Systems	International Organization for Standardization
3	NORSOK Z-013 Risk and Emergency Preparedness Analysis	Norwegian Competitive Position on the Continental Shelf
3	UKOOA Guidelines for The Management of Competence and Training in Emergency Response	UK Offshore Operator Association, now UK Oil & Gas

[a] Adapted from OGP (2010, 123). Standards referenced by three or more regulators. Only two standards were referenced by five regulators in their survey (API Spec 6A and IMO MODU Code); nine standards were shared by four regulators; and 25 standards were used by three regulators. We have only included a sampling of those three-regulator standards, to illustrate the variety of providers employed.

(BITs); and (3) the spread of international contracts that include stabilization clauses and independent arbitration venues.

Most obvious of these is the growing number of international treaties and agreements that protect workers and the environment and limit the power of unscrupulous policymakers and employers. There are several international and regional (e.g., EU) conventions that deal in whole or in part with the sundry environmental and social issues associated with petroleum activities. These include the law of the sea (UNCLOS 1982), international conventions on marine waste disposal (IMO 1972), the rights of indigenous peoples (ILO 1989), biodiversity (UNCBD 1992), and a variety of labor and human rights topics—the latter of which we will discuss in more detail in Chapter 10.

The most notable is the growing body of international agreements regarding different aspects of the environment, especially in the wake of the 1972 UN conference on the Human Environment in Stockholm, and the 1992 UN Framework Convention on Climate Change in Rio (UNFCCC 1992), with its 1997 Kyoto Protocol. Since 1992, the environmental obligations of signatory states have been extended through over 20 multilateral environmental treaties and the emergence of customary international law relating to the environment. Decisions by the International Court of Justice (ICJ) and the European Court of Human Rights (ECHR) have confirmed that states are responsible for protecting against the harmful effects of pollution (Cotula 2008, 168; Rivkin et al. 2015, 133; and IPIECA 2007).

International working standards are provided by the ILO in the form of conventions (legally binding international treaties, which are ratified—or not—by member states) and recommendations (non-binding guidelines). Of these, the Governing Body of the ILO points to eight particular conventions, covering the most fundamental principles and rights at work: freedom of association (Nos. 87 and 98); forced labor (Nos. 29 and 105); discrimination (Nos. 100 and 111); and child labor (Nos. 138 and 182). These ILO Conventions are listed in Box 9.2. Of course, the conventions are not ratified by all countries;[7] and ratified conventions are not always respected. But they provide an important baseline for countries when they decide how to regulate and protect their workers.

These international treaties provide a general framework that facilitates the spread of international regulations that can protect workers and the environment. Two concomitant international developments tend to pull in the opposite direction.

As we described in Chapter 2, the rise of Bilateral Investment Treaties (BITs) can impose significant constraints on national sovereignty (and its capacity to defend workers and the environment) when they include Investor State Dispute Settlement (ISDS) clauses. Since the 1990s, the share of concluded BITs with environmental language has risen precipitously; by 2011, 62 percent of

Box 9.2 THE ILO'S FUNDAMENTAL CONVENTIONS[a]

No.	Title (Year)
29	Forced Labor Convention (1930)
87	Freedom of Association & Protection of the Right to Organize Convention (1948)
98	Right to Organize & Collective Bargaining Convention (1949)
100	Equal Remuneration Convention (1951)
105	Abolition of Forced Labor Convention (1957)
111	Discrimination (Employment and Occupation) Convention (1958)
138	Minimum Age Convention (1973)
182	Worst Forms of Child Labor Convention (1999)

[a] Adapted from ILO (2015b).

all BITs publicly available included some form of environmental language (Rivkin et al. 2015, 148; Gordon and Pohl 2011). In this context, mentioning environmental language in a treaty does not necessarily imply environmental protection. But the introduction of HSE regulations can be challenged in third-party arbitrational venues (that include ISDS), if IOCs see them to be infringing upon their investors' rights. They can use this legal foothold to recover money they have already invested, or demand compensation for profits that were allegedly lost (or which they expect to lose in the future). Research conducted by John Ruggie and others showed that "provisions protecting foreign investors' interests and the rulings of arbitration panels [have] become increasingly expansive over time, particularly where host governments lacked bargaining power" (Ruggie 2013, 86).

A similar development can be seen in the rise of stabilization clauses that are included in contracts where the host state commits to maintaining the original regulatory framework that governs the investment (hence "stabilization"). Stabilization clauses were popular in the 1970s and 1980s, were less popular in the early 1990s (with the rise of BITs that secured the same ends), but are now increasingly popular again. They are especially common in natural resource/energy projects, where there are high and fixed costs, which require large capital injections early on, and where the investments can be sunk for very long periods of time. They have become the preferred means of correcting OBM imbalances. The earliest stabilization clauses focused on protecting investors from the threat of nationalization, and relied heavily on "intangibility clauses" (that is, that both contracting parties need to provide consent to modify the contract). Over time, their scope has broadened to include changes in the regulatory framework, fiscal regime, tariff structure, and so on.[8]

These stabilization clauses are embedded in host government agreements, which are a result of negotiations that turn on the relative power of the state vis-à-vis the IOC. As we saw in previous chapters, the host government can improve its bargaining position by delaying the negotiations over such clauses (or their extent). Indeed, there is a growing realization that broad stabilization commitments tend to shift negotiating power from the host state to the investor, and make it much more difficult for a government to introduce an autonomous regulatory policy:

> As a result, a host state that adopts regulations raising environmental standards and seeks to apply such standards to the ongoing investment projects would have to restore the economic equilibrium of the contract, or to compensate investors for the economic impact of such regulations. This legal liability may make it more difficult for host states, particularly poorer ones, to adopt legislation or ratify treaties raising environmental standards. (Cotula 2008, 169)

Before the evolution of contemporary regulatory regimes, the state was able to use domestic regulations as a means to prioritize local suppliers and national concerns. This meant that IOCs faced a wide spectrum of different regulatory regimes, making it more difficult and costly to adapt to local regulatory regimes. That older system was clearly not up to the task of regulating a fast-paced, technically driven, and sometimes dangerous industrial development.

The adoption of more systematic and standards-based regulatory regimes, and the changing international context in which they are embedded, has made it more difficult for states to affect the nature of their domestic regulatory regime. As a result, much of the competence and work associated with regulation offshore have shifted to international consulting bodies and standards providers.

At the same time, the growing reliance on BITs and stabilization clauses reflects a shift in the balance of power from governments to firms. These developments can constrain government efforts to protect workers, human rights, and the environment. This is why international advocacy groups, such as Amnesty International (2006), have been critical of stabilization clauses in that they often exempt foreign investors and their projects from complying with human rights' laws.

Norwegian HSE

Norwegian HSE regulations have undergone the same sort of transformation as described: the authorities have jettisoned prescriptive lists of regulations for a more systematic management approach. But the resulting Norwegian regime is unique in at least two important ways. First of all, the Norwegian system—more than most—shifts the burden of regulatory responsibility onto the industry and firm. Although the state maintains a strong supervisory role, it is up to the license holders to ensure that everything is being done to protect their workers and the environment. In many contexts, this sort of self-regulatory regime could be a recipe for increased disaster. The reason it works in Norway is a result of the country's second unique component: a heavy reliance on worker input and engagement. Norway's corporatist institutions and traditions secure a legal and regulatory context that severely limits the industry's ability to abuse the self-regulatory regime.

This section of the chapter provides a brief introduction to this change in regulatory regimes, and the disasters that propelled the change. It then describes the current regulatory landscape, and the most prominent institutions in that landscape.

Texas North

In 1967, Norway was the first country in the world to introduce safety regulation for offshore activity,[9] when it began to apply its Workers' Protection Legislation (*Arbeidervernloven*) offshore (Hanisch 1992; Svensen et al. 2014). From the very start, Norwegian authorities were concerned about the environmental and safety hazards associated with extracting oil under difficult offshore conditions; recall that safety and environmental concerns were evident in at least three of the ten commandments (see Box 4.2).[10]

This concern was driven in part by the challenges of drilling for oil offshore in the North Sea's unforgiving conditions. But it was also in response to the HSE culture that accompanied the American firms that came to explore offshore. To Norwegian eyes, offshore working conditions on US-based rigs were Neanderthal; the workers that followed these rigs had been trained in the southern United States (Gulf of Mexico), in non-union (so-called "right to work") states where workers tend to get the short end of any stick. IOCs were accustomed to treating their workers like disposal factors of production, and the Norwegian deckhands were no different (Karlsen 1982, 17). The industry fought hard to keep Norway's onshore unions at bay, they pushed their own workers hard, and they paid little attention to the niceties of onshore Norwegian labor law. As a consequence, the rate of accidents was high, there was little by way of safety regulations and preventative measures, and the acceptable grounds for termination were vast, and often arbitrary. In this culture, workers took pride in more macho attitudes, and accidents were seen as a normal price of doing business (Smith-Solbakken 1997). The price was high, and growing; in 1978, for example, 26 workers died offshore (Jensen et al. 1979, 74).

But the 1967 law, in itself, was insufficient—as the state at first lacked an effective enforcement apparatus. A number of different state directorates and organs were responsible for varying health and security concerns, while the Norwegian Maritime Directorate (*Sjøfartsdirektoratet*) was the coordinating institution. As a result, there was some confusion as to which rules would apply, as oil companies received different, and often contradictory, directives. As we saw in Chapter 4, it was this need to gather regulatory competence in one institution that was used to justify the launch of the NPD in 1972.

This early regulation regime was rule-based; the state designed regulations that companies were obliged to follow. The government drew up lists of very distinctive requirements, and the industry was expected to follow them in an exacting way. The thinking was clear; if safety measures were spelled out in the acts, offenders could be easily punished. As new and more complex technical developments drove the industry forward, however, it became increasingly difficult for the state to keep up with introducing and monitoring relevant regulations. Regulations always seemed to be one step behind the technology front. When things went wrong, as they inevitably did, the industry was able to hide behind the fact that the state's regulations were inadequate.

Between 1967 and 1978, 82 people died working on the NCS (Jensen et al. 1979, 74). The Ekofisk field alone cost 45 lives between 1971 and 1977–16 of which died in helicopter crashes. But most of these casualties occurred by way of small accidents, where one or two workers lost their lives. In 1980, however, these fatality statistics grew exponentially worse, when the Alexander L. Kielland platform capsized, taking 123 people with it (see Box 9.3).

A number of serious accidents pushed safety concerns to the top of the political agenda. People began to wonder if the NPD was not suffering under a conflict of interest—where the needs of the industry were being prioritized over the need to protect workers and the environment.[11] Eventually, as we have already seen (Figure 4.3), regulatory responsibilities were slowly sliced off from the NPD and allocated to the Norwegian Environment Agency (in 1974), the Norwegian Coastal Administration (in 1974), and eventually (in 2004) to Norway's Petroleum Safety Authority (PSA(N)).

More specifically, there was a realization that effective HSE regulation required the inclusion of firms and workers in the regulatory process. The result was the sort of classic tripartite approach that we associate with the Norwegian model (see Chapter 3), where the state, the industry, and workers' organizations cooperate to discuss, develop, and implement HSE regulations. Each of these actors is central, and their frequent exchange builds trust and competence. Their relationships and duties are depicted in Figure 9.2. In short, tripartite exchanges facilitate decision-making at a very high level of detail, by those who are ultimately responsible. This is one of the most important and unique aspects of the Norwegian regulatory regime.

The important role that industry and IOC standards plays in the regulatory regime has been discussed in the section "Regulatory Regimes." This section will focus on the more unique aspect of the Norwegian regulatory regime: the role of workers and the function-based system of laws, regulations, and protocols.

Box 9.3 MAJOR NORTH SEA DISASTERS

Ekofisk Alpha fire (November 1, 1975)

- The jacket structure on Ekofisk Alpha experienced a riser burst, followed by fire and an explosion.
  - During evacuation, one lifeboat with six people aboard was inadvertently dropped from a height, killing three and injuring the other three.
  - Remaining 65 crew members were evacuated safely.

Ekofisk Bravo blowout (April 22, 1977)

- Unignited blowout while doing a workover. This incident caused the worst oil spill in Norwegian history and exposed serious deficiencies in the country's defenses against such pollution. Nevertheless, the environmental impact was slight.
  - No injuries.
  - Platform successfully and calmly evacuated by four capsules in 15 minutes.
  - Largest oil spill in the North Sea: 12,700 m^3.

Statford A fire (February 28, 1978)

- A welder unintentionally started a fire on the shaft deck that killed five workers.

Alexander L. Kielland Platform capsized (March 27, 1980)

- Semi-submersible platform suffered structural failure causing the platform to list and then capsize. The rig turned turtle after just 20 minutes.
- 123 out of 212 died.

Workers Front and Center

Labor unions are one of the most important drivers of increased safety in Norway. Originally, labor's position offshore was significantly weaker than it was on the mainland, as much of the earlier work was carried out by US firms with an aggressively antiunion posture, or under the laxer regulatory regime covering sailors (rather than onshore workers). Indeed, until the late 1970s, there was an organizational struggle for offshore workers, and Norway's normal pattern of negotiation and rules of conduct were suspended on the NCS.[12]

The oil companies, drawing from their US experiences, preferred local company unions, unaffiliated to the main labor federations. In 1977, these company unions, organized by Phillips, Mobil, and Elf, joined together to create the Federation of Oil Workers' Trade Unions (*Oljearbeidernes Fellessammenslutning*, OFS), and Norway's main trade union federation, the LO, responded by organizing the Norwegian Oil and Petrochemical Trade Union (*Norsk Olje- og Petrokjemisk Fagforbund*, NOPEF). At the time, many of the workers offshore believed they could secure better terms outside of the LO, in

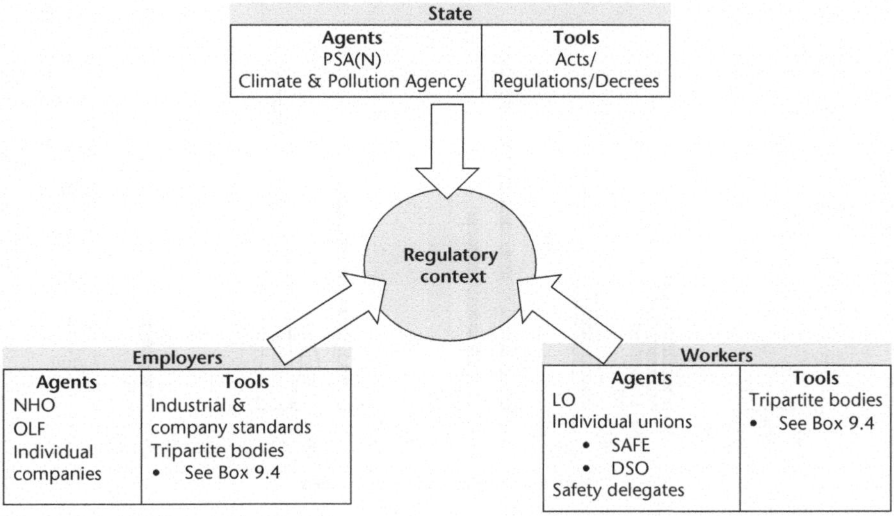

Figure 9.2 Tripartite HSE[a]

[a] DSO: The Collaborating Organizations (*de Samarbeidende Organisasjoner*); SAFE: Norwegian Union of Energy Workers (*Sammenslutningen av Fagorganiserte i Energisektoren*); OLF: The Oil Industry's National Federation (*Oljeindustriens Landsforening*); PSA(N): Petroleum Safety Authority (Norway).

large part because the LO was towing the government line on the need for wage restraint.

As production began to take off, and more workers from traditional (onshore) industries began to work in the industry, the organizational battles between unions, between unions and management, and between unions, management, and the state began to harden. Between 1978 and 1985, workers in the oil industry went on strike 26 times more often than other industrial workers in Norway (Remøe 1986; Ryggvik and Smith-Solbakken 1997, 232). Another couple of strike waves are evident in the early 1990s and at the turn of the millennium. This pattern of strike activity is evident in Figure 9.3.[13]

There are many explanations for the early rise in strike activity. High on the list was a concern over the growing number of fatalities and injuries offshore, and a (related) response to the authoritarian and high-handed working culture of the foreign oil companies. This was clearly dangerous work, and the existing regime seemed incapable of taming the industry.[14]

Over time, however, worker frustration was increasingly aimed at Norway's larger negotiated incomes policy arrangement, between labor, capital, and the state. While the LO and the Labor Party were calling for wage restraint (for fear of Dutch Disease), the IOCs were willing to pay higher wages in exchange for labor quiescence offshore.[15] For a while, workers appreciated this, and

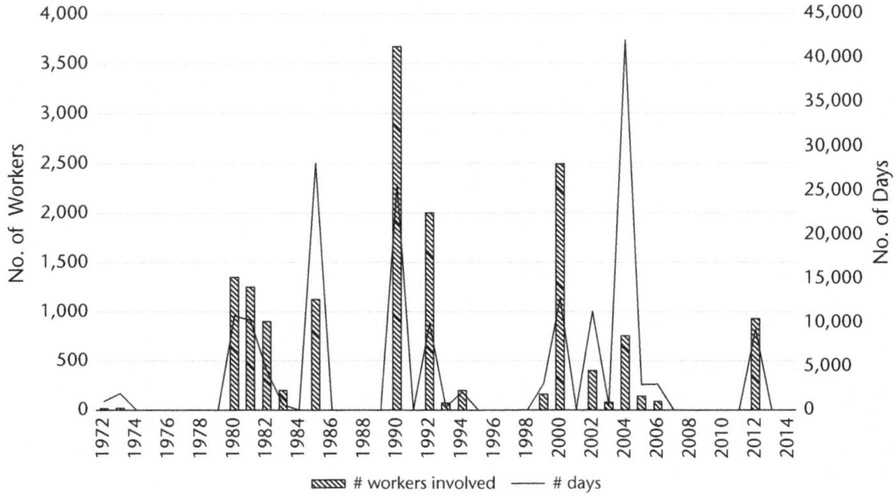

Figure 9.3 Labor conflicts in petroleum and mining, 1972–2014[a]

[a] 1972–1995: adapted from diverse issues of SSB's "Arbeidsmarkedsstatistikk"; 1996–97 data missing; 1998–2014 data adapted from SSB (2016 [07952]). See note in text for further caveats about this dataset.

prioritized their own bank accounts over solidarity with workers in the rest of the economy.

By the early 1980s the level of strike activity and wage increases were beginning to threaten the competitiveness of the larger Norwegian economy. In the wake of the 1981 election, a newly elected (Conservative) prime minster, Kåre Willoch, called all of the foreign oil companies into his office and told them that they needed to behave in the way that Norwegian firms were expected to behave (Willoch 1981). Now known as the "Willoch Declaration," the new prime minister declared that he was not going to let developments offshore torpedo the rest of the Norwegian economy; employers were expected to join the national employer organization and accept Norwegian labor unions as their respected partners in business, and begin to behave in a more responsible manner. In response, the oil companies immediately created the Norwegian Operators Employers' Association (*Norske Operatørselskapers Arbeidsgiverforening,* or NOAF),[16] which was affiliated with the NAF. The creation of NOAF helped to coordinate and collect the sundry company agreements into a single agreement, covering most of the industry. With time, NOAF/OLF, OFS, and NOPEF established normal cooperative arrangements, like those that predominated onshore.

As we learned in Chapter 3, industrial partners in Norway are accustomed to cooperating in order to secure collectively optimal outcomes while competing in global markets. With the trade unions now involved (and their safety delegates in particular), HSE policy offshore began to take on a different character. For example, when some disagreement developed between the partners as to the HSE conditions (or proposed changes), new institutions and fora were developed to encourage discussion and compromise.

By the mid-1980s, as part of a major reform of the offshore regulatory regime, a new, collaborative External Reference group for Regulations (ERR) was introduced, which would subsequently (2004) be extended by the Pollution Control Authority and the health authorities and renamed the "Regulatory Forum." In late 2000, a Safety Forum was also created. Each of these tripartite institutions—and others that were created to help educate, develop standards, improve practices, and so on (see Box 9.4)—drew representatives together from the state, the employers, and the workers to discuss, create, and implement HSE regulations.

These tripartite institutions/arenas play a critical role in improving and implementing the safety norms that are contained in regulations drafted by the PSA(N), NORSOK industry standards, and Working Together for Safety's "best practices." They also provide important feedback when the PSA(N) circulates new regulations for public hearing. But because these groups meet so regularly, most of the challenges are already resolved before drafts are circulated for comment. In short, these corporatist fora provide an opportunity to work out differences before they get too large.

Finally, the NPD launched a Monitoring Program—"Trends in risk level in the petroleum industry" (*Risikonivå i norsk petroleumsvirksomhet*, or RNNP)—to highlight major accident risk in relation to loss of life and occupational accidents. The RNNP draws on an even broader cooperative base, as it includes research institutions in addition to the usual tripartite members (industry, workers, government). The PSA(N) is responsible for day-to-day operations of the RNNP, and also organizes the network collaboration. The first report came in 2001 (RNNP, 2001), and there have been annual studies/reports on risk trends ever since. In 2007 the project was expanded to embrace the PSA(N)'s responsibility for land-based petroleum plants.

Legislative and Administrative Changes

During this early, tumultuous time of industrial growth, the government was also reforming its regulatory approach, and the new regulatory approach was both novel and far-reaching. It was also strongly influenced by the tripartite developments described in the preceding section.

Given the complexity of petroleum activities, and the fact that they cover many different types of activities (industrial, maritime, different aspects of working environment, and so on), there is no single act that covers the HSE aspects of petroleum activities, exclusively. Rather, HSE regulations are spread

Box 9.4 MAIN TRIPARTITE ARENAS FOR NORWEGIAN HSE[a]

Safety Forum (*Sikkerhetsforum*)

- Established in late 2000.
- Central arena of cooperation, to "initiate, discuss and follow up relevant safety, emergency preparedness and working environment issues in the petroleum industry."
- Has the broadest mandate of all arenas—its participants are high-level representatives of their organizations. Chaired by the Director General of PSA(N).

Regulatory Forum (*Regelverksforum*)

- Began in 1986 as the ERR.
- Purpose is to facilitate openness and information on documents that determine the outer framework for oil and gas operations. For example, they discuss the development of new HSE regulatory strategies.

Working Together for Safety (*Samarbeid for Sikkerhet,* or SfS)

- Established January 2001 by industry group.
- Parallels Safety Forum. PSA(N) is involved as an observer. Includes several technical subcommittees that mainly work on recommendations to the industry, based on "best practices."

Compliance in Rules and Regulations for the Petroleum Industry (*Regelverkskompetanse,* or RVK)

- Education program to give individuals and enterprises a better understanding of how important it is to comply with the rules and regulations.
- Input from government, industry, and labor unions.

Petroleum Board of Standard Norge (*Sektorstyre Petroleumsindustri*)

- Responsible for petroleum-related standardization work carried out in Standard Norge.
- Product of NORSOK (*Norsk sokkels konkurranseposisjon*): new standards introduced to reduce costs and improve competitiveness on the NCS.

[a] Adapted from PSA(N), SfS, RVK, and Standard Norge internet sites. See also Bang and Thuestad (2014).

out across a number of different acts, the most important of which are the Working Environment Act; the Petroleum Act; the Pollution Control Act; the Ship Safety and Security Act; and a broad range of health legislation. For the time being, our focus will be trained on the first three acts: the Working Environment Act, the Petroleum Act, and the Pollution Control Act.

But the acts themselves provide only a general, often vague, framework. Much of their importance lies in the way they delegate authority to the government to issue royal decrees, which detail the circumstances, conditions and rules applicable to protect workers and the environment. For that reason, it is useful to examine some of the more important decrees, along with the main institutions used to issue and monitor HSE legislation in the petroleum sector.

Finally, the number of relevant institutions, and their constant evolution, makes it difficult to follow developments over time. In the past 30 years, there have been many institutional reforms, with several institutions and acronyms trading places. Table 9.2 provides an overview of the most central institutions with regard to Norwegian HSE regulation.

As discussed in Chapter 4, the NPD was established by parliament in 1972, to administer Norway's growing oil sector, and to collect regulatory competence in a single institution. Originally, the NPD was run out of the Ministry of Industry (from 1972 to 1978). But after the Bravo accident (1977), responsibility for oil and gas matters was placed in a new Ministry of Petroleum and Energy (MPE), which was established in 1978. The NPD was then moved into the MPE, but it had to report to the Ministry of Local Government and Labor on working environment issues. In other words, the NPD would regulate general industrial matters and report to the MPE; but it would also regulate workplace environment issues, and report to the Ministry of Local Government and Labor.

In 1985, this division of regulatory responsibility came under increased scrutiny. A royal decree assigned coordinating regulatory authority to the NPD; it was to follow up on all aspects concerning HSE by coordinating the government agencies with independent regulatory authority. It was given responsibility for safety in all remaining areas. In 2004, as part of a comprehensive restructuring, the parliament transferred safety regulation to the PSA(N).

The PSA(N) is an independent government regulator with responsibility for safety, emergency preparedness, and the working environment in the Norwegian petroleum industry.

As we shall see, the PSA(N) does not approve plans or applications, as such—but it requires duty holders to apply for consent to commence and carry out their operations. Because of this, the agency is relatively small (about 170 employees), as most of its work is advisory-supervisory in nature, and does not involve "on-site" checks. By cooperating with other HSE regulators, and employing audits, the PSA(N) ensures that the petroleum industry and related activities are supervised in a coherent manner. It supplies information and advice to the players in the industry, and establishes appropriate collaboration with other HSE regulators (both nationally and internationally). Finally, the PSA(N) provides input to the supervising ministry on its respective matters.

Table 9.2 HSE institutions in Norway

	English	Norwegian	Supervising Body	Description
DN	Norwegian Directorate for Nature Management	*Direktoratet for naturforvaltning*	Defunct	Joins with Klif to become the NEA.
Klif	Norwegian Climate and Pollution Agency	*Klima-og forurensnings-direktoratet*	Defunct	Klif implemented government policy on pollution. In 2013, it was subsumed by the NEA.
NBHS	Norwegian Board of Health Supervision	*Helsetilsynet*	Ministry of Health and Care Services	An independent supervision authority, with responsibility for general supervision of child protection, health, and social services in the country.
NCA	Norwegian Coastal Administration	*Kystverket*	The Ministry of Fisheries and Coastal Affairs	Responsible for preparing the state's response to acute pollution, and for coordinating private, municipal, and state players in a national emergency preparedness system.
NDH	The Norwegian Directorate of Health	*Helsedirektoratet*	Ministry of Health and Care Services & Ministry of Labor and Social Affairs	The Directorate of Health aims to improve the health of the nation through integrated and targeted activities across services, sectors, and administrative levels. It functions as a subministerial agency, as a regulatory authority, and as an implementing authority in areas of health policy. (Prior to 2008, known as the Norwegian Social and Health Directorate (NSHD).)
NEA	Norwegian Environment Agency	*Miljødirektoratet*	Ministry of Climate and Environment	Provides advice on the development of climate and environmental policy, collates and communicates environmental information, and exercises regulatory authority. The agency manages and enforces several petroleum-related acts, including the Pollution Control Act, the Product Control Act, the Greenhouse Gas Emission Trading Act, and the Nature Diversity Act.
NFSA	Norwegian Food Safety Authority	*Mattilsynet*	Ministry of Agriculture and Food; Ministry of Fisheries and Coastal Affairs; & Ministry of Health and Care Services	Uses regulations and controls to ensure that food and drinking water are as safe and healthy as possible for consumers and to promote plant, fish, and animal health. Drafts and provides information on legislation; performs risk-based inspections; monitors food safety, as well as plant, fish, and animal health; and provides updates on developments in its field and plans for emergencies.

NLIA	Norwegian Labor Inspection Authority	*Abeidstilsynet*	Ministry of Labor and Social Affairs	The agency's mission is to oversee that the industry follows the Working Environment Act.
NMD	Norwegian Maritime Directorate	*Sjøfartsdirektoratet*	Ministry of Trade and Industry & Ministry of Environment	In cases concerning environmental issues that are connected to a specific ship, and the protection of the marine environment, the authority sorts under the Ministry of Environment and assists the PSA(N) in implementing the petroleum legislation on the NCS.
NRPA	Norwegian Radiation Protection Authority	*Statens Strålevern*	Ministry of Health and Care Services	The national authority with competence in the area of radiation protection and nuclear safety in Norway.
PSA(N)	Petroleum Safety Authority (Norway)	*Petroleumstilsynet (Ptil)*	Ministry of Labor and Social Affairs	Created in January 2004. Responsible for safety, emergency preparedness and the working environment in Norway's petroleum industry. Sets the terms for HSE and emergency preparedness in the petroleum sector, and follows up to ensure that the industry maintains high standards in this area. Oversees the Petroleum Act.
SFT	Norwegian Pollution Control Authority	*Statens forurensningstilsyn*	Defunct	In 2010, becomes Klif, now subsumed by the NEA.

WORK ENVIRONMENT ACT

The development of the tripartite institutions described in the section "Workers Front and Center" was not happening in a legislative vacuum. In 1977, Norway passed a new and radical Work Environment Act (WEA). This act included two important components with regard to HSE. First, the law required employers and employees to cooperate in collaborative organs and venues.[17] It was this law that gave impetus for greater worker involvement in the workplace and in forming the institutions that would deliver Norway's sundry HSE regulations. The idea was to involve workers to improve products, processes, and working conditions, as part of an effort to improve the working environment and expand workplace democracy. In contrast to earlier HSE regulations, which were the result of discussions among the state and various employers, the WEA broadened those discussions to include workers.

The WEA required that all HSE work needed to be informed by extensive dialogue and cooperation between employers and employees. Employees are to participate both as individuals, but also collectively through safety deputies (*verneombud*) and their shop stewards (*tillitsvalgte*). Larger firms were required to establish Occupational Environment Committees (*Arbeidsmiljøutvalg*), where employers, employees, and occupational health and safety personnel would be represented (Andersen et al. 2009, 165). In short, workers gained a formal say in HSE conditions.[18]

The second component required employers to reconceptualize the relationship between their workers and technology. Whereas in the past, employees were expected to adjust to the needs of new and existing technology, the WEA required that technology be adapted to the needs of employees. Following the logic of Maslow's hierarchy of needs, employers were expected to create workplaces where employees could achieve higher levels of individual potential or self-actualization (Ryggvik 2010, 73). This component encouraged firms to adapt technologies that would reduce the threat of HSE accidents, with important spillovers effects for local content (as described in Chapter 8).

With backing in the WEA, and the pressure being exerted from the Willoch Declaration, Norway's traditional (corporatist) means of conducting industrial policy began to kick in: the labor market partners were to have a major influence over the preparation, and implementation of HSE legislation.[19]

PETROLEUM ACT

The second important legislative component of Norwegian HSE policy comes in the form of a Petroleum Act and its supporting regulations. After the WEA, a new regulatory framework begins to shift responsibility for HSE protection away from the state and over to the producers (increasingly fortified with input from workers). This shift was deemed necessary to ensure that new

HSE measures would be introduced quickly, in response to a rapidly changing technology front.

This authority is anchored in several regulatory instruments—most importantly, an act, a royal decree, and a number of more focused regulations, guidelines, and standards. While the regulations are a part of Norwegian legislation (and hence legally binding), guidelines and standards (both international, national, and industry-based) are used to help the industry adapt quickly to changing circumstances.

The central piece of legislation is the Petroleum Act (first passed in 1985; since revised). This act provides the basic and general safety requirements for petroleum operations in Norway, and stipulates that "the petroleum activities shall be conducted in such manner as to enable a high level of safety to be maintained and further developed in accordance with the technological development" (§ 9–1).

The current HSE regulations were introduced in 2001, and consist of one bridging framework regulation (in the form of a royal decree) and four supporting regulations. The original Framework Regulation (*Rammeforskriften*)[20] was issued jointly by the PSA(N), the Norwegian Pollution Control Authority (SFT), and the Norwegian Social and Health Directorate (now the Norwegian Directorate of Health, or NDH). As the name suggests, this regulation provides a broader framework for offshore safety regulations. "It defines the common scope of application for all the regulations, their common purpose and definitions, who is to be responsible for complying with all the regulations, and the common main principles for health, safety and environment, including what is labeled 'health, safety and environment culture'" (Kaasen 2014, 115). Together, these regulations make it explicit that the PSA(N) has regulatory responsibility for safety, emergency preparedness, and the working environment in the petroleum activities,[21] even if it frequently works together with other governmental bodies and organizations to secure these objectives.

As explained in the Framework Guidelines, the system is designed to provide flexibility for the involved parties (operators, licensees, personnel, the managers onshore) in order to achieve the desired safety level. As with the Petroleum Act, the focus is on what needs to be achieved, rather than a list of dos and don'ts; there are no references made to standards in the regulations themselves—as that would create an administrative burden on the PSA(N) when operators or others wanted to change things (OGP 2010, 35).

Further details are set out in four additional regulations, covering specific areas of petroleum activities, as illustrated in Figure 9.4. Each of these four subordinate regulations has been issued by their competent directorates, where enforcement capacity lies,[22] and each covers a separate aspect of safety

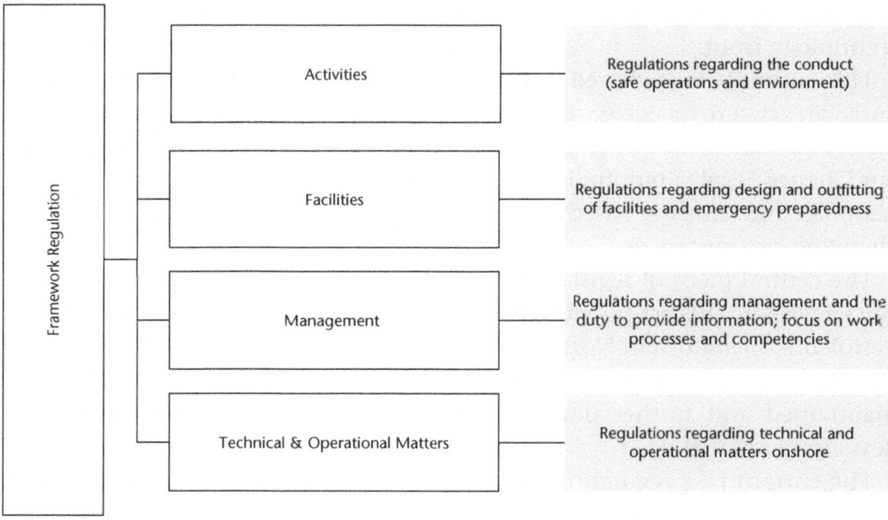

Figure 9.4 Regulatory framework for Norwegian HSE[a]
[a] Adapted from PSA(N) (2016a).

issues: activities, facilities, and management, as well as technical and operational matters.

The result is a management system that prioritizes three unique elements:

- **Internal control.** The principle of internal control preceded the legislation (it was already introduced in 1976; guidelines were then developed in 1978).[23] But it received a stronger legal footing when the concept of internal control was formally introduced in 1981 and became a central part of the 1985 Petroleum Act. The fundamental principle here is that it is the license holders who are responsible for compliance with the rules and regulations in their activities. They are therefore also responsible for their own internal auditing and control. New target-oriented objectives were introduced and companies are expected to develop their own internal safety systems to secure these targets (hence, internal control). Although these systems are checked by the authorities, it is the firm's responsibility to ensure that safety is secured. The PSA(N)'s role is to ensure that the companies have adequate HSE management systems—not to have full oversight over what was happening on each and every offshore installation. The PSA(N) intervenes, if (and when) necessary.
- **Operator responsibility.** It is the operator and license holders that are responsible for finding solutions. Firms have an explicit duty to establish, follow up, and further develop management systems that will protect workers and the environment. This responsibility is made explicit in

both the Petroleum Act (§10–6) and the Framework Agreement (§7). The operator is required to take on explicit duties with regard to delivering safety (a so-called "see-to-it" duty [*påse*]), and a follow-up duty to see that others performing work for them are also following the regulations.

- **Creating an appropriate HSE culture.** In §15 of the Framework Regulations (and in a subsequent white paper, *St. meld. nr. 7 (2001–2002)*), a strong emphasis is placed on creating an appropriate HSE culture. Employers are expected to nurture an HSE culture where safety and the environment are protected and respected.[24]

This new regulatory system had widespread consequences. With this shift in responsibility (from the state to firms) came a dramatic reduction in the number of regulations on the NCS. In 1985, there were 22 explicit regulations in effect; a decade later (in 1996), that number had dropped to 14 regulations (Andersen et al. 2009, 20). With the widespread introduction of tripartite solutions, in 2002, the number of relevant regulations dropped to just five.

Now freed from the task of monitoring whether this (shrinking) list of regulations is being followed, the authorities have more time to focus on their other tasks: to specify the objectives and overall standards for work on HSE protection; and to monitor activities and ensure that the license holders are conducting their operations in accordance with their own codes and standards.

Companies, on the other hand, have focused their energies on developing internal safety systems, with their corresponding standards. License holders are required to satisfy basic preconditions—for example, demonstrating knowledge of the requirements and introducing suitable management systems to define necessary methods, to achieve and comply with the requirements. Firms working on the NCS tend to employ a variety of international, national, and industrial standards (e.g., NORSOK, ISO, API, and so on), and are required to document systems of internal government, functional requirements, technical requirements, and work processes.[25] Non-compliance with health and safety legislation is penalized by daily fines. In the event of serious or repeated violations of the regulations, the government can revoke a license granted under the Petroleum Act.

Finally, one unanticipated consequence of the new regulatory regime was an added incentive for local content, as hinted at in Chapter 8. The 1985 Petroleum Act included a provision specifying that the industry had to lead their activities from Norway.[26]

> Expertise and knowledge about Norwegian regulation and regulatory principles thereby acquired increased significance, and the subject became an important area of collaboration in terms both of regulating and supervising the activities as well as the important tripartite partnership which developed after the 1985 paradigm shift. (Bang and Thuestad 2014, 248–9)

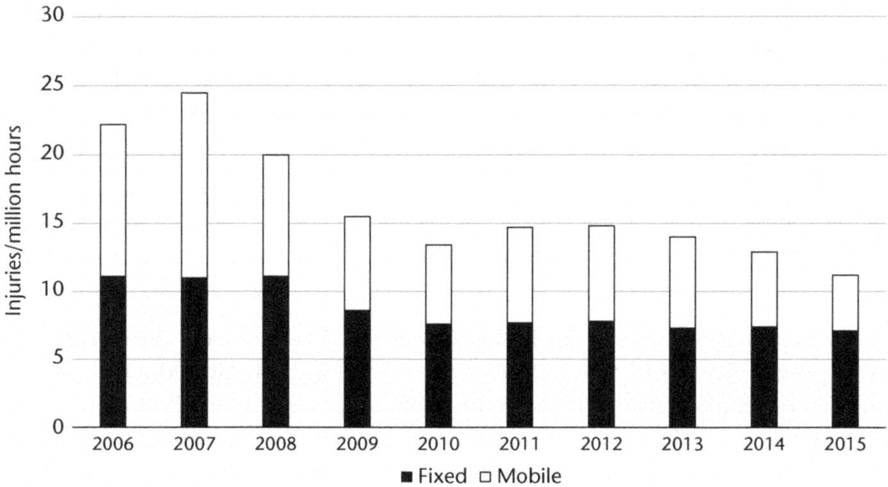

Figure 9.5 Personal injuries on the NCS, 2006–15[a]

[a] Adapted from PSA(N) (2016b). Injuries per million man-hours on permanently placed (fixed) and mobile facilities.

The result has been an impressive reduction in accidents on the NCS. Since the Alexander Kielland capsized (in 1980) and until 1990, 13 people were killed on the NCS. Seven of these were divers—a group that was heavily exposed to danger before the development and perfection of remotely operated underwater vehicles (ROVs). After 1990, there were many years without a single fatal accident, and the number of accidents (per hour worked) has dwindled to a fraction of what it was in the early years. For most of the last decade, the number of personal injuries has been below 15 incidents per million hours worked, as shown in Figure 9.5.

POLLUTION CONTROL ACT

From the very start of Norway's oil adventure, the authorities have been acutely aware of the environmental threats associated with working on the NCS. These threats are all the more serious, given Norway's heavy reliance on a vibrant offshore fishing industry. It was for this reason that petroleum activities were originally restricted to south of the 62nd parallel; the government wanted to ensure that the industry had enough competence before migrating father north into more dangerous (and environmentally sensitive) waters.

The fishing industry has been a strong and vocal critic of the industry's relentless march northward. As such, it has managed to protect some of the more fertile and fragile areas. Still, there has been a long-standing tug-of-war between those who want to protect the fish and those who want to access the petroleum that lies under them. In an attempt to find a compromise between these two positions, Norway adopted a management plan for the Barents Sea and the waters around Lofoten in 2006 (updated in 2015), which allows for exploration in parts of the Barents Sea, but bans drilling around Lofoten (*Meld. St. 20 (2014–2015)*).

Norway's broader environmental policy rests on the same set of basic principles as are found in the Petroleum Act (e.g., precaution, risk-reduction, continuous improvement, best available techniques, management systems, communication, and cooperation), but buttressed with a strong "polluter pays" principle. In particular, firms are required to submit management plans (where the operator is held responsible for all pollution), environmental impact assessments (EIAs), and discharge permits; and pay the requisite emission taxes.

The 1981 Pollution Control Act (PCA) provides a framework law to protect the environment—regardless of the source of pollution. The PCA regulates emissions both on land and at sea (industrial CO_2 emissions are considered to be pollution under the act, and are generally prohibited), and its baseline assumption is that all pollution is prohibited, unless it is expressly allowed by a permit or a regulation. Hence, companies that emit any form of pollution need to obtain an emissions permit pursuant to the PCA (and fulfill the formal terms of that permit).

In the petroleum sector, environmental regulation occurs at each and every stage of activity: from the collecting of seismic data through exploration drilling, operations, and decommissioning (*Meld. St. 28 (2010–2011)*, 115–16). For example, each exploratory well requires a drilling permit from the NPD, and these permits presume that the company has already received permission under the PCA, and consent from the NEA and PSA(N).

Before production commences, a development plan (a Plan for Development and Operations, or PDO) has to be filed for each field, in which the operator studies the environmental consequences of subsequent development.[27] This plan includes an EIA, which is then distributed for comment to the relevant authorities and industrial organizations (e.g., the Norwegian Oil and Gas Association (*Norsk Oje&Gas*), NHO, the Federation of Norwegian Industries (*Norsk Industri*), and so on). The PDO helps the government ensure that all relevant impact assessments have been conducted, and that sufficient measures have been taken to mitigate possible adverse effects. The PDO also summarizes previously undertaken impact assessments and studies, as well as the (new) assessments and studies specifically directed at the field to be developed.

When in operation, a field's emissions and discharges are regulated by way of permits issued under the PCA, as well as by the economic incentives provided by the CO_2 tax, a quota obligation for CO_2, and the NOx tax (or payments to the industry's NOx fund). Companies need to apply to the NEA and the Norwegian Radiation Protection Authority (NRPA) for permits under the PCA for both air and water emissions, and the MPE for flaring permits (*Meld. St. 28 (2010–11)*, 116). In 1997, a zero-discharge target was established for petroleum activities (*St. meld. nr. 58 (1996–97)*)—in 2009 it was expanded to include radioactive substances (*St. meld. nr. 26 (2006–2007)*). Preparing for an eventual oil spill is the responsibility of the NEA, while the Norwegian Coastal Administration (NCA) is responsible for the operational side of such clean-ups. Both the NCA and the Norwegian Clean Seas Association for Operating Companies (*Norsk Oljevernforening For Operatørselskap*, or NOFO) plan for eventual clean-up activities.

As always, the largest environmental threat is seen to be associated with the opening of new areas for exploration and production. This is even truer today than it was in the 1960s. The 1996 Petroleum Activities Act (§ 3–1) details the stringent requirements for opening new oil fields for exploration, while the past several decades have witnessed intense political battles over opening new (and sensitive) areas. As the influence of petroleum interests grows, along with their capacity to extract these resources in a responsible and safe fashion, these interests are pressing up against the needs of the world's largest cod stocks (in the Lofoten islands), and the challenges of oil spill preparedness (given the long distances and the extremely harsh conditions—including the edge of the polar ice flows—involved). There are no easy trade-offs here.

As with worker safety, Norway's environmental record remains strong. The two largest acute spills (Bravo in 1997; Statfjord in 2007) dissipated before reaching the coast, and the country's zero-discharge goal continues to push the country's level of air and sea emissions down. As the MPE recently reported to the Norwegian parliament: "Forty-five years of petroleum activity on the Norwegian Shelf has proven that production of oil and gas can be compatible with environmental considerations" (*Meld. St. 28 (2010–11)*, 114). The question is whether this impressive record can be maintained as the industry pushes into even more challenging locations.

Conclusion

Most states are ill-equipped to deal with the regulatory burden that accompanies the petroleum industry. As we have already seen with respect to the skills and experience needed to explore for and extract the resource, IOCs tend to have much more experience and competence (than the host government) in securing a safe workplace and environment. Unlike the host government, IOCs have achieved this before, in a number of other contexts, and they will have a good sense of what works and what doesn't.[28] They will also have direct experience with how much different regulatory solutions cost.

This latter point is essential, as awareness of the best regulatory regimes does not always translate into a willingness to employ them. Even if an IOC employs the most sophisticated safety regulations in one context (e.g., in Norway), there is no guarantee that this competence will be exported to other contexts. Indeed, in the absence of public shaming/spotlight effects, or of regulatory requirement, IOCs have an incentive to use the cheapest technology and management systems possible. To be brutally honest, it costs an IOC a great deal more to risk a Norwegian life than it does to risk an Angolan life.

States need to employ regulations to reduce these negative externalities—to ensure that the HSE costs of these activities are borne by the producers. As such, the government faces two (related) problems. The first is a rapidly changing technology front, which means that it is difficult to keep abreast of regulatory developments. If the government doesn't have people that are familiar with every phase of the production process, and how it is constantly changing, then it will be difficult to ensure that proper regulations are in place. It is one thing to ensure that there are enough life preservers (or oil-retention gear) on board an offshore platform. It is quite another matter to know how to protect workers and the environment in a context that is rapidly adjusting to new technological challenges.

Second, the government needs to find a way to monitor/execute these regulations. The current approach has been to encourage firms to develop their own risk-management regimes. In Norway, this form of self-regulation has worked because workers are provided with the information and political leverage to secure better HSE regulations. This secures a regulatory environment characterized by a high degree of trust, competence, and control. In many less developed political contexts, it could be a very risky strategy to delegate so much of the regulatory burden to the industry itself. As Gerard Zwetsloot (2013) noted in a recent review of developments, despite these massive changes underfoot, there is little consensus over whether these new OHS MS reforms actually make a difference:

> [T]here are debates as to whether OHS MS have a positive effect on health and safety or not. On the one hand, proponents almost take it for granted that an occupational safety and health management system will automatically lead to better health and safety performance, as is suggested in the OHSAS Standard. On the other hand, critics have used words such as "scam", "fraud," "bureaucracy", and "paper tigers" to describe management systems and pointed out that workers lose influence. (Zwetsloot 2013)

This delicate dance between firms and states continues into Chapter 10, where we consider the role that ethics and CSR play in petroleum management.

10

The Ethics of Petroleum

So far, this book has considered several facets of petroleum management from the perspective of national policymakers. We have learned how state officials must diligently manage the petroleum industry in order to ensure that their resource is extracted safely, that its bounty is distributed justly, and that the rest of the country does not suffer as a result. While we have been critical of the oil industry's power and influence, we have emphasized the importance of developing trust and the mutual exchange of information among stakeholders.

In this chapter, we change our tone and focus, as we consider the ethical limits to petroleum production. In doing so, we adopt a more normative approach, and one that draws our attention to the important role of firms. In particular, this chapter challenges the reader to consider whether it is ethical to make money from petroleum, when we know that petroleum production and consumption contributes to global warming, increases environmental risk, and worsens the conditions for some of the world's poorest people (via the Resource Curse). We then consider how the Norwegian state and its NOC, Statoil, approach these ethical challenges.

The Ethical Dilemmas

Ethics concerns the moral principles that govern an individual's or a community's behavior. As such, ethical questions are inherently relative—what is ethical for one person or community may not be ethical for another. For example, some individuals and communities think it is acceptable to make money from usury (the charging of interest); others prohibit the practice as being immoral. In every community, there are ethical limits to what is considered to be acceptable economic activity—for example, trade in illicit drugs, weapons, slaves—but the nature of these limits tends to vary over time and across countries. The question before us is: does petroleum production (and consumption) transgress this type of ethical limit?

The answer to this question depends upon our degree of certainty about the underlying (often implicit) assumptions and consequences of economic exchange. This is always the case; we needn't limit our focus to the petroleum market. For example, most people would not consider it unethical to buy a used television set from an acquaintance; both the seller and the buyer enjoy increased utility and satisfaction from the trade. But the ethical nature of the trade can shift abruptly if we become aware that the TV for sale is actually a stolen good. For some people, no matter how good the price, it is unethical to buy a stolen TV. Similar dilemmas can be found when we consider the effects or consequences of market exchanges. Some people consider it wrong to sell something that can knowingly cause harm to others (e.g., tobacco, weapons, inferior products, and so on). The dilemma is hardly new—a similar example is employed in Plato's *Republic* to consider the justice of returning a weapon to a mentally deranged friend.[1] When communities share these ethical concerns, they are often (but not always) incorporated into law, and laws vary from one time and place to another. This sort of ethical relativism becomes increasingly problematic in a world that is shrinking owing to globalization.

When questioning the ethics of petroleum, we have to undertake a similar examination of underlying assumptions and consequences. The answer to our inquiry depends upon our relationship to the petroleum market and our understanding about the nature of that market. The first issue is partly a question about levels of analysis; we can consider the ethical limits to petroleum from many different perspectives (e.g., the consumer, the producer, the host nation in which the petroleum lies, future generations, and so on). Given our vantage point, we are most interested in the perspective of states and firms.[2]

It is possible to question whether it is ethical for states to exploit their petroleum resources. The answer to this question can vary significantly with the wealth of the country in question. For example, should a wealthy and developed state, such as Norway, continue to make a great deal of money by a process that contributes to global warming? Should a poorer country, which has contributed little to today's difficulty, be equally constrained? These are fascinating and important ethical issues that should draw our attention and concern.

It is equally possible to consider whether firms have an ethical responsibility to behave in ways that can minimize the environmental and social costs of oil. Indeed, in Chapter 9 we learned that firms are being asked to behave in a more socially and environmentally responsible manner, as the capacity of states to regulate and manage the industry has waned. The remarkable rise of the CSR movement is one indicator of the growing role that ethics plays in influencing petroleum markets (even if ethics may not be the primary motivation driving CSR measures).

We will proceed on the assumption that it is legitimate to question the ethics of both states and firms as they relate to the petroleum trade. The next stop in our inquiry is to consider the ethical consequences. While the petroleum market can have an ethical impact along many disparate fronts,[3] we will focus our discussion on two of the most evident: the effect of petroleum on global climate change and the effect of petroleum on the surrounding communities (where the petroleum is produced).

Consider, first, the relationship between petroleum and global warming. Michael Booth (2014, 193) describes the conundrum: "It is universally acknowledged that fossil fuels, and oil in particular, are bad news for our planet—they are unsustainable, pollute the atmosphere, and it seems likely they are slowly making the planet warmer. Norway sources much of its own energy from clean, renewable hydroelectric power, thus absolving itself of direct consumer guilt. It is the wily drug pusher who refuses to touch its own product." Consequently, today's petroleum-producing states find themselves in a bit of a pickle—but that pickle is shared by firms as well. In a 2005 report, the oil and gas sector was linked to 41 percent of the direct carbon emissions among the 100 largest companies in the UK stock exchange (the FTSE 100), exceeding both the electricity (21 percent) and the mining (13 percent) sector (HGI 2005). That study included both direct (e.g., emissions from the actual process of refining the oil) and indirect (e.g., emissions that result from the petroleum that is sold and subsequently burned in vehicles) effects, providing potential investors with a clear link between the petroleum industry and global warming.

This linkage is becoming increasingly evident as a global fossil fuel divestment movement begins to take shape. Beginning with Swarthmore College (Pennsylvania, United States) in 2011, we can see a growing effort to move out of these investment objects—especially coal, but also petroleum markets. AXA (the French insurance group), several universities (including Stanford and Syracuse), the Church of England, and the Rockefeller family (whose money originally came from Standard Oil!) are all promising to sell off their fossil fuel investments. As we shall learn later, in the section entitled "The Ethical State?", the Norwegian parliament wants the country's pension fund, the GPFG, to move in this direction. These investors believe it is wrong to invest in an industry that can have such detrimental effects on the world's climate.

We should be blunt. It is unlikely that the world can maintain its present level of petroleum production while simultaneously pledging to limit global warming to 2°C. A recent report in the journal *Nature* argued that "a third of oil reserves, half of gas reserves and over 80 per cent of the current coal reserves should remain unused from 2010 to 2105 in order meet the target of 2°C" (McGlade and Ekins 2015, 187). As one of the authors of that study (Paul Ekins) noted in a subsequent interview, "[i]n 2013, fossil fuel companies spent some \$670bn (£443bn) on exploring for new oil and gas resources. One might ask why they are doing this when there is more in the ground than we can afford to burn" (Carrington 2015).

In the light of these developments, it is becoming increasingly clear that many people share legitimate ethical concerns about the petroleum trade:

> Oil companies can make their refineries and production sites more eco-efficient by using less energy or generating less waste. On the other hand, the harmful nature of oil products contravenes the notion of sustainable development. The use of oil products is simply unsustainable, if judged by the definition of sustainable development as "meeting the needs of the present generation without compromising the ability of future generations to meet their own needs" (World Commission on Environment and Development, 1987). Ultimately, the only way of making oil companies "sustainable" would be to shift away from the business of oil and gas altogether. (Frynas 2009b, 98)

As already hinted, these sort of ethical arguments find better traction in the developed world, which bears the lion's share of the responsibility for the world's current predicament. But is it right to expect developing countries to forego their potential for economic development in the name of solving a problem that was not of their doing? Surely, we cannot hold a wealthy country like Norway up to the same standard as a relatively poor country, like Tanzania?

An ethical argument for extracting petroleum in poor countries turns critically to whether these countries can develop their petroleum resources in a responsible manner. This brings us to the second ethical dilemma related to petroleum: its relationship to poverty, corruption, and exploitation. For a poor state to develop its petroleum resources in an ethically responsible manner, we need to be assured that the environmental and social costs of petroleum production (both short term and long term) result in greater economic and political opportunities for the people who live in that state. In short, an argument can be made for the ethical production of petroleum resources in poor countries *if* those resources are used to improve the conditions of their residents.[4]

But this is *exactly* the problem, as we learned from previous chapters. The empirical evidence points to a serious Paradox of Plenty, where greater resource wealth does not translate into better conditions for the residents of underdeveloped countries. As we hinted at in Chapters 1 and 4, there is much evidence that resource abundance hurts the prospect for democracy, as it enables autocrats to repress or buy off potential opponents.[5] Recent research is extending that argument to show that natural resource wealth can also help to consolidate authoritarian rule (Wright et al. 2015). Similar lines of research have linked natural resource wealth to higher levels of corruption,[6] and human rights abuses.[7] If the money from petroleum is being spent to keep corrupt, unjust, and inefficient governments in power, the ethical justification for petroleum evaporates.

Petroleum companies are clearly aware of this ethical linkage, as Shell learned the hard way in the late 1990s. Royal Dutch Shell has made a tremendous amount of money from decades of productive activity in Nigeria, during which time Shell paid royalties to the Nigerian federal government for the right to develop that oil. Unfortunately, the local people of the Niger Delta, the Ogoni—who live where most of the oil is located—enjoy few of the spoils (see Box 10.1). Instead, their local communities and environment have been destroyed, their people abused and abducted, and all of the money generated from this misery quickly evaporated. "Nigeria became the poster child for the concept of the oil curse" (Spense 2011, 71). Herein lies Shell's dilemma. At the end of the day, Shell made a lot of money, paid it to corrupt politicians, and worked under environmental and safety regulations that were clearly inadequate (but which allowed Shell to secure the oil cheaply). Shell knew that the money it paid was not reaching the local population; indeed, Shell probably preferred that the money had trickled down to the people. But Shell initially assumed this was not *their* problem; this was a matter of internal (Nigerian) affairs.

These internal Nigerian affairs came to cost Shell a great deal of anguish when an Ogoni protest movement began to make international headlines after its founding member was imprisoned (and eventually hanged). In response to rising unrest in the region, Shell first refused to intervene in the matter, for fear of offending its host. After much pressure was applied, the company then changed its tune and pleaded for mercy on behalf of the protestors. The action had no effect and the protestors were eventually hanged. But Shell's actions (and inactions) raised international attention to the fact that the Nigerian process was inherently flawed and that an IOC was implicated because of the business it had conducted in the country.

This example is far from unique; petroleum companies continue to find themselves in politically sensitive situations, whether it is exploring for oil in the arctic waters in northern Alaska, or working under trying political conditions in, for example, Angola, Equatorial Guinea, or Azerbaijan. Like Shell's, the reputation of other IOCs has been equally tarnished by foreign exploits, whether it is ExxonMobil in Aceh Indonesia, Hunt Oil in Peru, ChevronTexaco in Ecuador, or someone else, somewhere else (see Spense 2011, 71–6).

As Shell's experience in Nigeria suggests, the petroleum industry faces a severe ethical challenge. To make its money, the industry must work with rather unseemly political allies, producing a product that threatens the globe's future climate. It is in this highly politicized context that oil companies came to recognize the utility of a CSR campaign.

Box 10.1 SHELL, SARO-WIWA, AND NIGERIA[a]

Royal Dutch Shell had been working in Nigeria since before the country gained its independence; its first commercially viable discovery came in 1956, and it began production in 1958. The country, especially its Niger Delta (Ogoniland), is rich in oil—but the reward from that natural bounty never made it back to the Ogoni people who live there. While Ogoniland is not the only source for Nigerian oil, it represents a substantial share (90 percent in the mid-1990s); by 1994 it was estimated that Ogoniland had produced something like $30 billion worth of oil. Visitors to the region would be hard pressed to find evidence of this economic bounty (Cayford 1996, 184).

Although Shell paid royalties to the Nigerian state, the money didn't (and doesn't) make its way past corrupt officials. Worse, Nigerian corruption resulted in very weak and poorly enforced health, environment, and safety regulations—so that oil production developed in a socially and environmentally irresponsible manner. Before 1991, it was estimated that there had been almost 3,000 oil spills across the delta, amounting to 2.1 million barrels of oil (Cayford 1996, 184).

By the 1980s, Shell and other IOCs were increasingly linked to the corrupt Nigerian government. In the court of public opinion, if not always in the court of law, Shell was being judged responsible for those human and environmental costs. Shell's production and delivery systems were frequently defended by government troops whose hard-handed techniques resulted in massive casualties. For example, after the June 1993 elections, large-scale fighting broke out across Ogoniland: over one thousand people were killed, and tens of thousands were left homeless when villages were destroyed.

Three years earlier, in October of 1990, several Ogonis, including the author and journalist Ken Saro-Wiwa, founded a Movement for the Survival of the Ogoni People (MOSOP). In addition to demanding more autonomy from Nigeria's federal authorities, MOSOP began to challenge Shell's right to operate within Nigeria. Although the MOSOP's manifesto affirmed the group's non-violent approach, some of its affiliated organizations were very active in the subsequent uprising.

By the middle of 1993, Shell pulled out of Ogoniland, complaining about the breakdown of law and order. In the following May, four (pro-government) Ogoni leaders were attacked by a mob and beaten to death. In the wake of this violence, Saro-Wiwa and eight other Ogonis were arrested and sentenced to death in a trial that "blatantly violated international standards of due process" (HR Watch 1999). Several international organizations called on Shell to intervene and stop the trial. Shell originally refused, claiming that this was an internal (Nigerian) affair. But after widespread consumer and public pressure, Shell's president pleaded for mercy on humanitarian grounds. The company's belated actions had little effect; Saro-Wiwa and the eight others were hanged.

In June of 2008, the Nigerian government revoked Shell's license to operate the oil concessions in Ogoni areas.

For more information, see Cayford (1996), HR Watch (1999), and ICG (2008).

Corporate Social Responsibility

CSR has become the poster child of accountable capitalism. In the United States, most postgraduate degrees in business administration now require students to take at least once course in CSR, while the vast majority of corporate recruiters advise business school graduates to be knowledgeable about the subject (Alsop 2005). The topic flourishes at workshops, conferences, and forums, while a veritable army of "social auditors" has sprung up to evaluate the social and environmental responsibility of firms. As the homepage for the UN's Global Compact (which now includes over 8000 firms) trumpets, "sustainability reporting has gone mainstream" (UN, n.d.). There is even a Corporate Social Responsibility Newswire (www.csrwire.com).

The oil and gas industry has been at the forefront of this CSR movement. The reasons for this are not hard to understand, and they relate to the sector's growing exposure to environmental and social risks. On the one hand, the very nature of oil and gas operations poses an environmental threat along every step of the process: exploration, construction, production, transportation, refining, and even the consumption of the product. On the other hand, a never-ending search for new sources of oil and gas has forced the industry to conduct business in some rather unsavory parts of the globe. To make money, IOCs need to cooperate with corrupt, autocratic, and opaque governments.

In working under these conditions, IOCs expose themselves to critical attention from a number of important stakeholders, including consumers, home governments, non-governmental organizations, and even shareholders. For this reason, IOCs desperately need to protect their reputations; they are well aware of how their activities can result in environmental and social disasters, and they need to find a way to minimize the reputational costs associated with those potential disasters. "For the companies claiming to have a broader social responsibility, e.g., 'a force for good' (BP) and 'building a better world' (Shell), the paradox of plenty directly affects their legitimacy as corporate citizens" (Skjærseth et al. 2004, 18).

WHAT IS CSR?

CSR is the industry's response to this challenge. In the petroleum sector, CSR is a means for firms to insure against growing reputational costs, and to secure greater influence over the regulations that govern their industry. In the process, IOCs strengthen their relative bargaining position (vis-à-vis host states), in that they gain expertise over these shared rules of conduct, and they don't need to adapt to a plethora of (varying) ethical norms and laws on the ground.

In the general literature on the subject, there is remarkably little consensus about what constitutes CSR. As Frynas (2009b, 5) points out, in a table listing recent work on the subject, CSR can include

- business ethics and morality;
- corporate accountability;
- corporate citizenship;
- corporate giving and philanthropy;

- corporate greening and green marketing;
- diversity management;
- environmental responsibility;
- human rights;
- responsible buying and supply chain management; and
- socially responsible investment.[8]

One of the most influential depictions comes from Archie Carroll's (1979) pyramid of CSR. Carroll argues that firms need to meet four kinds of responsibilities before they can achieve complete CSR (see Figure 10.1). At the base of Carroll's pyramid lies a firm's economic responsibility—every firm has an obligation (to its shareholders) to bring home a profit. But firms also have a legal responsibility (they must obey the law)—and this responsibility occupies the second layer of the pyramid. These two responsibilities are widely recognized as sufficient to secure legitimacy (although there are many firms that do not even manage to fulfill these bottom-layer responsibilities). The upper layers of the pyramid are more disputed, in that it is doubtful that firms will pursue ethical and philanthropic responsibilities if they come into conflict with their legal and fiduciary obligations. While there continues to be disagreement about why firms might want to act in an ethical and philanthropic manner,[9] it is now widely accepted that corporations have an obligation to do so (or at least, appear to do so).

At this point it is important to note that much of the CSR literature is written by students of general business and management, so their focus

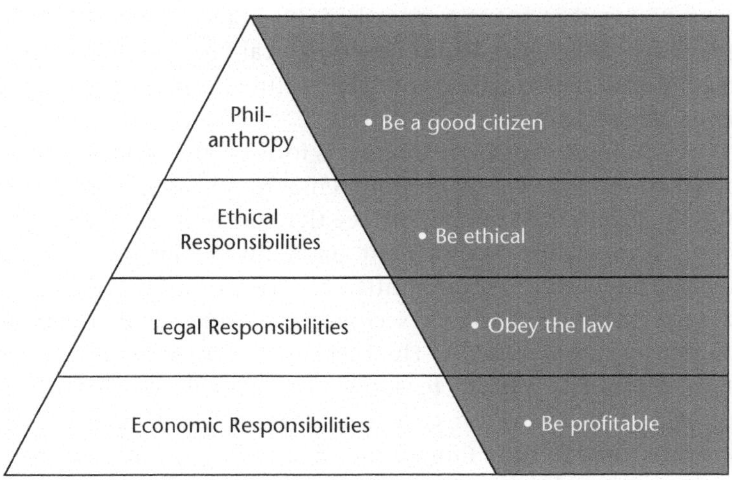

Figure 10.1 Carroll's pyramid of CSR[a]

[a] Adapted from Carroll (1991, 42).

tends to be on the profit-maximizing firm, usually an MNC. The oil industry is special in at least three ways. First of all, its relationship to the host state (and is residents) is different from other (more traditional) forms of FDI. As IOCs need to gain access to natural resources within a given country, and these are owned and controlled by the state, they have less need to placate domestic constituents and more need to befriend political authority.[10] This provides IOCs with fewer incentives to discourage corruption and/or advocate for the protection of human, social, and environmental rights (relative to MNCs in general).

Second, IOC stakeholders vary accordingly. Consumers of oil products find it more difficult to be brand savvy than other types of consumers: the link between the filling station and the exploration rig is long and tenuous. For this reason, CSR policies in the petroleum industry are more likely aimed at influencing shareholders, future (high-skilled) employees and governments (who will influence the next round of allocations), than at consumers, per se.

Finally, the petroleum industry is increasingly dominated by NOCs. While students of economics find it curious to see firms engaged in socially responsible behavior, students of politics are more likely to err in the opposite direction: we expect the state to act in a socially responsible manner. We are less surprised, then, when we see that NOCs show a stronger commitment (than private IOCs) to the principle of social responsibility. Large NOCs, such as PDVSA (Venezuela), Saudi Aramco (Saudi Arabia), and Gazprom (Russia) invest heavily in social programs as part of their government's broader social policy objectives. For example, in 2006, PDVSA reportedly spent US$1.3 *billion* on "social development." The same year, the four oil majors (Shell, Exxon, BP, and Chevron) together spent a mere US$500 million (Frynas 2009b, 105). In Chapter 8, we saw Statoil also supporting local firms and developing local competencies and infrastructure, when it still functioned as the long hand of the Norwegian government.

The NOCs' commitment to (and budgets for) social responsibility does not seem to extend to the realm of environmental responsibility, as evidenced in Table 10.1. In this limited sample, we see that all companies showed a commitment to community development programs, while the NOCs (at the bottom of the table) were less committed to other forms of CSR.

The rise of CSR in the petroleum sector can be explained by the interaction of two important developments. The first is the relative demise of the state. The state's decreased capacity to supply the benefits of citizenship plays a central role in the work on CSR (e.g., Matten and Crane 2005), where globalization is seen to have limited the capacity of governments to provide and defend traditional citizenship benefits (such as social, political, and civil rights). Corporations have flowed into the resulting vacuum. This trend is especially obvious in the sort of underdeveloped political contexts that are

Table 10.1 Summary of CSR policies and initiatives, by company[a]

	Reduction in CO_2 emissions	Community development programs	Government revenue transparency	UN Global Compact	Voluntary Principles on Security and Human Rights	Extractive Industries Transparency Initiative	World Business Council for Sustainable Development
Shell	YES	YES	YES	YES	YES	YES	YES
BP	YES	YES	YES	YES	YES	YES	YES
Exxon	YES	YES	YES	NO	YES	YES	NO
Chevron	YES	YES	YES	NO	YES	YES	YES
Petrobras	YES	YES	YES	YES	NO	NO	YES
Indian Oil	NO	YES	NO	YES	NO	NO	NO
PDVSA	NO	YES	NO	NO	NO	NO	NO
Kuwait Petroleum	NO	YES	NO	NO	NO	NO	NO

[a] Adapted from Frynas (2009b, 26).

now attracting the petroleum industry. When governments are unable or unwilling to provide social assistance, schools, infrastructure, education, and the protection of civil and even political rights, IOCs are increasingly called upon to fill the void.

The second development is the increased regulatory reach of IOCs. As we saw in Chapter 9, states have found it difficult to adapt to the petroleum sector's fast-moving and complicated regulatory front. As firms have risen to fill this need, CSR can be seen as an extended model of governance—one that includes the development of international standards and codes of conduct, providing global regulations and public goods.

Together, these developments generate a paradox: "government failure to deliver effective governance in the form of environmental protection and economic and social development often results in both government and local communities seeking to shift the burden of delivering such benefits to oil and gas companies" (Wagner and Armstrong 2010, 144).

In this new context, firms are embracing CSR as a cheap way to secure reputational insurance (Minor and Morgan 2011; Spense 2011), avoid the spotlight effect (Spar 1999) of poor publicity, and/or increase the likelihood of future petroleum licenses. In doing so, firms are able to extend their influence over the design and implementation of a new regulatory regime.

International Ethical Standards

As we saw in Chapter 9, these changes result in a dense web of international standards, codes of conduct, and reporting regimes. Table 10.2 provides a glimpse at some of the most significant of these, as they apply to the petroleum sector. All these standards claim to be universal, yet they compete with one another for the attention of firms and governments, and have remarkably little power or authority to punish deviants. In effect, the world has substituted a system with hard law standards that vary by national context for a soft law standard that is presumed to be universal (e.g., the protection of human rights), but in practice varies across monitoring or standards organizations.

The list of partnerships, standards, codes, and reporting measures is astonishingly varied. One way to simplify this activity is to think of it in terms of different forms of social responsibility. For example, Frynas (2009b) distinguishes between environmental, developmental, and governance challenges. We have already focused much attention on environmental policies, and the use of international standards and guidelines that apply to environmental protection (in Chapter 9), but we should point out that this area of corporate responsibility seems to be more developed in the petroleum industry (Frynas 2009b, 68).

Table 10.2 International ethics standards, codes, and guidelines[a]

Name	Impetus	Est.	Objective
OECD Guidelines for Multinational Enterprises (OECD GME)	OECD	1976	Recommendations providing (legally non-binding) principles and standards for responsible business conduct for MNCs operating in or from countries adhering to the OECD Declaration on International Investment and Multinational Enterprises.
Foreign Corrupt Practices Act (FCPA)	US	1977	US law that addresses accounting transparency requirements under the Securities Exchange Act of 1934 and bribery of foreign officials. Makes it unlawful for certain classes of persons and entities to make payments to foreign government officials to assist in obtaining or retaining business.
World Business Council for Sustainable Development (WBCSD)	World Industry Council for the Environment (WICE) and BCSD merge	1995	CEO-led global business association aimed at sustainable development. Provides a platform for companies to communicate about sustainable development and to share knowledge, experiences, and best practices.
Global Reporting Initiative (GRI)	Coalition for Environmentally Responsible Economies (CERES) and the Tellus Institute; United Nations Environment Program (UNEP) was also involved	1997	The most common framework employed for environmental and social reporting. An international independent organization that helps businesses, governments, and other organizations understand and communicate the impact of business on critical sustainability issues (such as climate change, human rights, and corruption).
UN Global Compact	UN Secretary-General Kofi Annan	1999	An initiative to encourage businesses worldwide to adopt sustainable and socially responsible policies, and to report on their implementation. A framework for businesses that employs ten principles in the areas of human rights, labor, environment, and anticorruption.
Convention on Combating Bribery of Foreign Public Officials in International Business Transactions	OECD	1999	An OECD convention aimed at reducing corruption in developing countries by encouraging sanctions against bribery in international business transactions carried out by companies based in the Convention's member countries.
Voluntary Principles on Security and Human Rights	Co-sponsored by US and UK, NGOs and IOCs	2000	A multistakeholder initiative involving governments, companies, and non-governmental organizations that promotes implementation of a set of principles that guide oil, gas, and mining companies on providing security for their operations in a manner that respects human rights. Specifically, participating companies agree to a set of voluntary principles in three areas: (1) risk assessment;

(*continued*)

Table 10.2 Continued

Name	Impetus	Est.	Objective
			(2) interactions between companies and public security; and (3) interactions between companies and private security.
Publish What You Pay (PWYP)	CAFOD, Global Witness, Open Society Foundation, OXFAM, Save the Children and Transparency International	2002	Shaming campaign aimed at increasing transparency in the extractive sector so that citizens can hold governments and companies accountable for the ways in which natural resources are managed.
Social Responsibility Working Group (SRWG)	IPIECA	2002	Oil industry group for sharing good practices on social responsibility issues, including human rights, social impact assessment in countries where they are active, and community outreach. Has three objectives: (1) enhance member understanding of social responsibility issues; (2) contribute to and monitor external initiatives; and (3) develop tools and guidance to encourage continuous improvement of the industry's performance.
Extractive Industries Transparency Initiative (EITI)	UK	2003	An international standard for openness around the management of revenues from natural resources. Governments disclose how much they receive from extractive companies operating in their country and these companies disclose how much they pay. Governments sign up to implement the EITI Standard and must meet seven requirements. See Box 10.2.
Oil and gas industry guidance on voluntary sustainability reporting	IPIECA, API and OGP	2005	The principal industry-specific framework for use by oil and gas companies reporting on environmental, health and safety, and social and economic performance. Updated in 2010.
Environmental, Social and Health Risk Impact Management Process (eSHRIMP)	OGP	2007	Oil industry process. Provides a flexible approach to environmental, social, and health stewardship associated with oil and gas projects throughout the upstream asset lifecycle. Updated in 2015, with Report 529, *Overview of IOGP's Environmental-Social-Health Risk and Impact Management Process.*
Extractive Industries Transparency Initiative Plus Plus (EITI++)	World Bank	2008	Extends the EITI by assisting resource-rich countries (being assisted by the World Bank) to design natural resource contracts, monitor operations, collect taxes, and spend natural resources more effectively.
UN Guiding Principles on Business and Human Rights (UNGP)	UN Human Rights Council	2011	Global standard for preventing and addressing the risk of adverse impacts on human rights linked to business activity. Based on three pillars outlining how states and businesses should implement the framework: (1) the state's duty to protect human

			rights; (2) the corporation's responsibility to respect human rights; and (3) access to remedies for victims of business-related abuses.
National Action Plan to implement UNGPs	UK	2013	Requires listed companies to disclose information to the public about environmental matters, social, community, and human rights issues, to the extent necessary for an understanding of the company's performance or position.
Directive 2013/34/EU	EU	2013	EU-Directive for disclosing non-financial and diversity information by certain large undertakings and groups. Requires large EU companies to publish a non-financial statement containing information on human rights, bribery, and anticorruption matters. Amended in 2014: Directive 2014/95/EU (October 22, 2014).
ISO	International Organization for Standardization	Varies	Using national standards institutions, technical committees of experts produce specific industry standards. For the purposes of this chapter, the most relevant are Social Responsibility (ISO 26000); Environmental Management Systems (ISO 14000); and Quality Management (ISO 9000). The ISO only produces the standards; it does not certify compliance.

[a] Listed by year established.

The industry has been less effective in dealing with the social challenges of corporate responsibility. Much public attention is aimed at measures for fighting corruption (although this attempt is in no way limited to the petroleum industry). The active use of national (home state) legislation for the prosecution of corruption cases that occur in secondary (host state) contexts is changing attitudes and behavior among corporate elites. The US's Foreign Corrupt Practices Act (1977) is the best known of these legislative efforts, but several other countries have followed suit. More recent developments, such as Section 1504 of the US Dodd-Frank Act and the EU Transparency Directive, offer hope and promise, but we will not know their effect until they kick into action.[11]

There has also been a growing focus on the need for firms to avoid abuses of human rights, which can be understood as "fundamental moral rights of the person that are necessary for a life with human dignity" (Forsythe 2012, 3). This focus can be traced back to the 1948 Universal Declaration of Human Rights (UDHR)—a non-binding declaration of the UN General Assembly, consisting of 30 articles, that have subsequently made their way into countless international treaties, national constitutions, and voluntary codes of conduct.[12] This tradition was extended in 2011, when the UN Human Rights Council issued the UN Guiding Principles on Business and Human Rights (UNGPs). Since that time, several states (e.g., the United Kingdom, Denmark, the Netherlands, and Italy) have instituted national legislation to implement the UNGPs (Ruggie 2013; Rivkin et al. 2015, 134). Similar types of "national contact points" can be found in several countries with regard to the OECD Guidelines for Multinational Enterprises (OECD GME). See Table 10.2 for examples (and Chapter 6 of *St. meld. nr. 10 (2008–9)* for elaboration with respect to Norway).

Finally, we would be remiss not to mention the concerted effort aimed at trying to increase transparency. At the end of the last millennium, inspired by the growing academic literature on the Resource Curse, a number of NGOs (such as Global Witness, Human Rights Watch, Oxfam, the Open Society Initiative, and so on) began to draw attention to the lack of transparency in the extractive industries. The result was the development of the "Publish What You Pay" (PWYP) slogan (which was institutionalized in 2002).

In February 2001, in response to this rising awareness for transparency, BP announced that it had paid a (US$111 million) signing bonus to the government of Angola, in exchange for an offshore license. None of the other oil companies followed BP's lead, and the company was hung out to dry by the Angolan government—promising there would be consequences for others who followed suit. Realizing that individual oil companies could be picked off, one by one, the UK government launched the Extractive Industries Transparency Initiative (EITI) in June of 2003, as a means to encourage more transparency in the industry (see Box 10.2). By 2015, 31 countries were implementing the seven EITI criteria.

Progress on the transparency front has been slow and labored. The details of IOC–host country contracts still remain veiled in secrecy,[13] and the capacity of the international community (both public and private, economic and

Box 10.2 EXTRACTIVE INDUSTRIES TRANSPARENCY INITIATIVE[a]

The Extractive Industries Transparency Initiative (EITI) was first suggested by the UK government, and launched in June 2003. It describes itself as "a coalition of governments, companies, civil society groups, investors and international organizations" that aims to "improve governance in resource-rich countries through the full publication and verification of company payments and government revenues from oil, gas and mining."

Each EITI-implementing country commits itself to eight requirements (EITI 2016a, 13–38):

1. Effective oversight by the multistakeholder group.
2. Transparent legal and institutional frameworks.
3. Transparent exploration and production activities.
4. Comprehensive reconciliation of company payments and government revenues from the extractive industries.
5. Disclosure of information related to revenue allocation.
6. Disclosure of information related to social expenditures and the impact of the extractive sector on the economy.
7. Ensure that stakeholders are engaged in dialogue about natural resource revenue management.
8. Compliance with the deadlines and the requirements for EITI implementation and reporting.

EITI consists of both Supporting Companies and Implementing Countries. Supporting Companies are not expected to disclose what they earn—but *all* companies (EITI Supporting Companies or not) that work in an Implementing Country are required to disclose how much they pay to the government. Implementing countries disclose their revenues from natural resource extraction. Companies report what they have paid to the government, and the government reports what it has received; then an independent evaluator compares the figures and publishes an EITI Report. We used one of these reports in Chapter 5 (Deloitte 2014).

In July 2015, there were 27 EITI-implementing countries: Albania, Burkina Faso, Cameroon, Chad, Democratic Republic of the Congo, Ghana, Guinea, Iraq, Côte d'Ivoire, Kazakhstan, Kyrgyz Republic, Liberia, Mali, Mauritania, Mongolia, Mozambique, Niger, Nigeria, Norway, Peru, Republic of the Congo, Sierra Leone, Tanzania, Timor-Leste, Togo, Trinidad and Tobago, and Zambia. Four additional countries (Central African Republic, Guatemala, Indonesia, and Yemen) are still formally compliant, but currently suspended (for either political or technical reasons).

At the same time, there were then over 80 supporting companies—up from just 16 in 2008 (EITI 2016b; Frynas 2009b, 142)—and 21 international partner organizations registered, both private and public in nature (EITI 2016c), such as the World Bank, API, IMF, and so on.

[a] Adapted from https://beta.eiti.org/

political) to invoke change has proven very limited. For the author of one of the most detailed studies on the subject matter, Jędrzej Frynas, the lesson is clear: "The evidence suggests that CSR has the greatest potential for addressing environmental challenges... In contrast, the evidence suggests that CSR has less potential for addressing problems related to community development and governance" (Frynas 2009a, 194).

Unfortunately, we concur with Frynas's conclusion. Perhaps the best anecdotal evidence for this argument is provided by an example we first introduced in Chapter 4—on the Chad–Cameron pipeline. At that time, we considered how the government of Chad used the OBM to its advantage—to change the terms of an agreement with the IOCs (once they had sunk their investments in the project). We can now turn our attention to the nature of the World Bank's assistance to this project, as it was heralded as a groundbreaking policy initiative to avoid the Resource Curse and to secure just returns for the people of Chad (see Box 10.3). In particular, the World Bank acted as a "moral guarantor," providing the IOCs with CSR cover and securing broad political support for the project. Despite its strong focus on good governance and transparency, the project proved an embarrassing and dismal failure.

There was one sliver of a silver lining to this dark cloud. The World Bank's Revenue Management Plan in Chad did seem to have a limited effect on deterring the Dutch Disease; when its oil export growth was highest (2004–05), the country's real exchange rate only appreciated by two percent (Frynas 2009b, 144). But the people of Chad, and its "priority sectors," benefited very little from the massive revenues generated by the project. Observers are left to ponder: what good has the oil done for the people of Chad? What hope is there for people in other poor countries, when a World Bank-sponsored program failed so miserably?

As Jaén (2010) argues, the failures of this World Bank project demonstrate (among other things) the limits to focusing on good governance and transparency. Both components were in ample supply in the case of Chad: information regarding revenue inflows, the allocation of resources from these inflows, information about the level of extractive activity (and its socio-economic impacts)—these were all relatively plentiful and easy to access. What we should learn from this example is that good governance is extremely difficult to export and apply successfully, and that the temptations offered by large petroleum revenues don't make it any easier.

> The point of all this is that although informative transparency is desirable in itself, it does not automatically pre-empt the problems highlighted in the literature concerning the "resource curse". Admirable initiatives such as Publish What You Pay and the Extractive Industries Transparency Initiative (EITI) are necessary, but not sufficient. (Jaén 2010, 4)

Box 10.3 THE REVENUE MANAGEMENT PLAN IN CHAD[a]

In 1998, the government of Chad signed its first contract with a consortium of oil companies, including Esso, Chevron, and Malaysia's NOC (Petronas). The intent was to tap into the impressive Doba oil fields in southern Chad, but political instability and concerns about government corruption nailed the project to the drawing board.

In June of 2000, the World Bank broke the impasse by providing political and moral cover for the IOCs. In particular, the World Bank offered a Revenue Management Plan, in which the oil companies were not directly involved. In describing its support for the project, the World Bank announced:

> The Board of the World Bank Group today overwhelmingly agreed to support the Chad-Cameroon Petroleum Development and Pipeline Project—an unprecedented framework to transform oil wealth into direct benefits for the poor, the vulnerable and the environment. In addition to the financing, the package of support includes a first-of-its-kind program to direct new revenues to support economic and social development programs in Chad, which is one of the world's poorest countries.
> (World Bank 2010a)

The World Bank's financial contribution was minimal. The total project costs were estimated at $4.1 billion, and the private sponsors financed about 85 percent of these costs (World Bank 2010b)—it was the World Bank's moral support and political backing that helped legitimate and secure the deal.

To secure that objective, it was agreed that 10 percent of Chad's oil revenues would be placed in a London-Based Future Generations Fund. Of the remaining revenues, between 80 and 85 percent would be channeled into another London-based escrow account for the so-called "priority sectors," such as education, health and social services, rural development, and infrastructure.

As part of the agreement, the World Bank encouraged the creation of the *Collège de Contrôle et Surveillance des Ressources Pétrolieres*. This body was meant to supervise the revenues that were being channeled into the priority sector. The World Bank also helped Chad improve its public resource management by training staff at the Ministry of Finance.

By 2004, the program began to fall apart. A parallel Chad–IMF agreement collapsed because of the government's macroeconomic non-compliances. In response, the IMF froze access to parts of the oil revenues, and this triggered fiscal tensions which led (eventually) to an attempted coup. Chad unilaterally reneged on earlier agreements on priority spending and in December 2005 abolished the Future Generations Fund. The government then began to define security as part of the priority sector (allowing the government to increase military spending). The World Bank suspended all payments and froze the oil revenue account.

As the situation deteriorated, the government in Chad was rescued with the help of French military intervention. The fear of eminent collapse led the World Bank to accept Chad's proposed reforms to allow the Chad government to commit 70 percent of the 2007 budget to priority sectors. Over time, this figure was watered down, so that by August of 2008, the World Bank was forced to express its concern regarding compliance with the Memorandum of Understanding. On September 5, 2008, Chad paid back its debt to the World Bank, and this ended the Bank's involvement.

[a] Adapted from Jaén (2010); Amnesty International (2005); Frynas (2009b, 144); and World Bank (2010c).

This is also the main lesson derived from Frynas's (2009a and 2009b) extensive work on the subject—that the focus on transparency is good, and easy, but it is not really where the problem lies.[14] We need more transparency on how the money is *spent*, not just on how it is being collected. Frynas (2009b, 133) quotes a World Bank official on this very point: "CSR is missing the boat, the real issues are the fiscal issues... oil companies have a role to play in the improved management of revenues." A Transparency International report (2008, 24) concurs.

The problem is that IOCs don't seem to be aware that they play an important role in this tragedy. A 2004 study of four IOC majors (Exxon, Shell, BP, and Total) demonstrated that company executives are hesitant to acknowledge the Resource Curse phenomenon; they continue to emphasize the benefits that their investments will bring to the host country. "None of the companies accept responsibility for 'the paradox of plenty' problem, but they all support the Universal Declaration of Human Rights, they work with NGOs and IOs when necessary and they all claim they are fighting corruption and bribery" (Skjærseth et al. 2004, 18).

While revenue transparency is an important first step, what we really need is good governance—citizens need better oversight and control over how the money is spent. The problem is that CSR is undermining that effort, as it is part of a larger movement to move away from government regulation and capacity. It is for this reason that the former US Secretary of Labor (under President Bill Clinton) Robert Reich (2008) argues that CSR can do significant damage as it removes the incentives for governments to enact laws and undertake actions that maximize social welfare.

Norwegian Ethics

By most accounts, Norwegians and Norwegian policymakers maintain high ethical standards. There are few overt signs of corruption in Norway—the country consistently scores high on Transparency International's Corruption Perception indicator—and the state plays a very active role ensuring that domestic companies behave in an environmentally and socially responsible manner. As demonstrated in Chapters 1 and 3, Norwegian policies deliver strong welfare systems that protect society's weakest members and secure an egalitarian society and a full employment economy. Norway's strong, effective, and judicious state makes CSR rather superfluous.

Norway plays a leading moral role in global politics, but its largesse is funded with money steeped in oil and gas. In light of the ethical dilemmas sketched in the first part of the chapter, we want to examine the extent to which Norwegian authorities are explicitly concerned about their Faustian bargain. To do this, the next part of the chapter looks first at the ethical content of Norwegian state policy (in general), and the particular instruments used by the state to try to encourage ethical outcomes. The second part considers Statoil as a socially responsible corporation. The Norwegian state and its NOC appear to be more ethical than most—they convey an honest and sincere attempt to do the right thing. But neither state nor firm should be absolved from their ethical responsibilities, and each has already succumbed to hubris and the lure of liquid gold (and sometimes corruption).

The Ethical State?

The Norwegian state prides itself in being an active and ethical member of the global community. In countless fora around the world, Norway brands itself as a force for peace, development, and environmental responsibility. After all, Norway is the host of the Nobel Peace Prize, and the country uses its good offices to encourage dialogue between partners in conflict (e.g., the Oslo Accords, as a mediator in the Sri Lankan civil war, and so on). Among wealthy countries, Norway is one of a very few that is committed to delivering one percent of its GDP to Official Development Assistance (ODA).[15] Finally, Norway plays a very active and important role in the campaign to stop global warming. It does all of these things out of an authentic desire to make the world a better place.

In some areas, Norway's ethical conscience is especially evident. For example, the government's Oil for Development (OfD) program was explicitly designed to help new petroleum countries strengthen their environmental, governance, and financial aspects of petroleum management. Through that program, the Norwegian government wants to help developing countries combat the Resource Curse, generate sustainable economic growth, promote the welfare of the whole population, and minimize the environmental impact of the petroleum industry. These are laudable goals, even if they can also serve as a springboard for Norwegian firms in a growing (foreign) market.

Norway's ethical credentials are most evident in the stringent criteria that steer investments in the GPFG. As noted in Chapter 7, the fund has observed strict ethical guidelines since 2004; these guidelines provide the fund's managers with two tools for securing more ethical outcomes: the exercise of ownership rights and the exclusion of companies. The ownership rights are directed more at securing good rates of return in the long term, by promoting sustainable development. Exclusion is a tool of last resort and is used to prevent the fund from being complicit in serious violations of ethical norms.

Even though exclusion is used sparingly, it has proven an effective tool. This is because of the public attention that it draws to firms that are shunned. The guidelines provide a number of grounds for exclusion; for example, if a firm produces weapons "that through their normal use may violate fundamental humanitarian principles"; is associated with unacceptable health risks (e.g., tobacco companies); or acts in ways that otherwise constitute a serious violation of fundamental ethical norms.

For example, the GPFG sold all of its holdings in Wal-Mart stock, for its "serious or systematic human rights violations." Kristin Halvorsen, who was then Finance Minister, explained that Wal-Mart (and others) were excluded "because, in view of their practices, investing in them entails an unacceptable risk that the fund may be complicit in serious, systematic or gross violations of norms." The Finance Ministry then reported that the GPFG's Council of Ethics had found "an extensive body of material" that indicated Wal-Mart had broken norms, including employing minors against international rules, allowing hazardous working conditions at many of its suppliers, and blocking workers' efforts to form unions. It also listed other alleged Wal-Mart abuses, including pressuring workers to work overtime without compensation, discriminating against women in pay, and blocking "all attempts to unionize," and that Wal-Mart employees were "in a number of cases unreasonably punished and locked in" (Acher 2006).

In June of 2015, the Norwegian parliament ordered the GPFG to sell off its investments related to coal. In particular, the fund is now required to shift its holdings out of billions of dollars of stock in companies whose business relies by at least 30 percent on coal (Schwartz 2015). This is a rather remarkable sentiment for a fund that is based on the extraction of petroleum! Nevertheless, it signals an awareness on the part of the Norwegian authorities of their ethical predicament in a context of growing public awareness about global warming.

As already hinted, the GPFG's investment guidelines are administered by a Council of Ethics, which is made up of a small group of economists and lawyers. What is most important, we think, is the degree to which the decisions are transparent, at all levels of activity. For example, it is possible to research how the fund has voted on its shares, which companies are under observation or excluded from the fund, who sits on the Council of Ethics, and their recommendations (as well as, of course, the content of the guidelines).[16]

These ethical guidelines are supplemented with basic investment principles, to protect the long-term financial interests of the Norwegian people (NBIM 2011). These principles highlight the importance of good corporate governance and maintaining dialogue with companies, investors, and market authorities. In particular, NBIM has six strategic focus areas for their ownership activities,[17] and it is expected that their investment targets will follow internationally recognized corporate governance standards. As in the area of ethics, the NBIM employs an external advisory board, the Corporate Governance Advisory Board, to provide advice on ownership issues and how to fulfill NBIM's ownership mandate.[18]

A commitment to ethical behavior is also evident in the government's efforts to influence the rising number of Norwegian firms in global markets. In the wake of Norway's oil adventure, a flotilla of Norwegian firms—with expertise and experience from Norway's challenging production environment—is now setting sail for foreign markets. Although Statoil leads the armada (and will be considered in more detail in the section "Statoil: The Ethical Firm?"), there are many other smaller firms that are increasingly exposed to foreign business environments. Most Norwegians pride themselves in believing that Norwegian firms act ethically in the global economy; the Norwegian government is committed to ensuring that these firms act in an ethically responsible manner when operating abroad.

To facilitate this, the government introduced a report to parliament on "Corporate Social Responsibility in a Global Economy" (*St. meld. nr. 10 (2008–2009)*). This report shows Norway to be an enthusiastic contributor to the sort of international framework of soft law standards we described in the first part of the chapter. In it, the government commits state-owned firms to follow the recommendations of the OECD GME, and recommends that all Norwegian firms follow suit. The report describes its intent to establish a "National Contact Point," and promises to provide financial support for several international CSR efforts, such as the UN Global Compact, the Global Reporting Initiative (GRI), the ILO, and (of course) the OECD GME.

The existence of this framework, and Norway's strong support for it, does not seem to be enough to constrain the global ambitions of Norway's business elite. There is widespread recognition among these elite that corruption is endemic in many parts of the world, and to be a real competitor in these contexts requires that they turn a blind eye to corruption. These corporate executives, we are told, don't want to engage in corruption—but they must if they intend to compete. In short, despite the impressive network of voluntary agreements, guidelines and standards, and domestic legal restrictions, Norwegian-based MNCs still seem to be caught in a collective action dilemma.

Evidence of this was especially strong in 2014, when three serious corruption cases (each of which concerns a firm with significant state ownership), reminded the Norwegian public (and their representatives) that Norwegian firms act very differently when working abroad. As Guro Slettemark, head of Transparency International in Norway, said: "Business too often comes before ethics in Norway's business culture. Politicians need to wake up" (Crouch 2015). The corruption cases come from three very different sectors, none of which are directly related to petroleum:

- **Telenor (telecoms: 54 percent owned by government).** In December 2014, the chief executive of Telenor, Jon Fredrik Baksaas, resigned from the board of Russian telecom VimpelCom after media allegations that Telenor had implicitly accepted VimpelCom's bribery payments in Uzbekistan.
- **Yara (fertilizer: 36 percent owned by government).** Four former senior Yara managers, including a former chief executive officer (CEO), were charged with serious corruption, and received the largest fine in Norwegian corporate history for paying bribes in Libya and India. In July of 2015, a Norwegian court sentenced the managers to prison; the former CEO, Thorlieif Enger, received a three-year jail sentence. Enger appealed the decision and avoided his jail sentence.
- **Kongsberg (weapons/defense: 50 percent owned by government).** In 2014, the offices of Kongsberg were raided and the company was charged with suspicion of gross corruption relating to deliveries (worth around 1.4 billion NOK) of communication equipment to Romania during the period 1999–2008.

One reason for this increase in corruption charges may be the strength of recently revamped Norwegian legislation relating to both corruption and transparency. Like its better-known US counterpart, the Foreign Corrupt Practices Act (FCPA), Norway's current Penal Code prohibits all forms of corruption—for Norwegian nationals and persons living in Norway who are involved in activities abroad.[19] It is this Penal Code that was used to convict Yara's former CEO. Norway is also the first country to require firms in the extractive industry to report on their payments to governments—although we can hope that other countries in Europe will follow suit in 2017, when the EU Transparency Directive comes into effect (and that the US SEC will not water down the relevant sections of the Dodd-Frank Act). This Norwegian requirement is the reason we have detailed information on Statoil's activities (reported in the next section).[20]

In all these ways, the Norwegian state signals its concern about Norwegian firms that are increasingly exposed to the temptation of corruption. But there is remarkably little reflection about the special ethical challenges related to petroleum for either Norway or Norwegian firms. For example, in the 106-page CSR report introduced earlier in this section (*St. meld. nr. 10 (2008–2009)*), oil is mentioned only nine times (petroleum 18 times), often only vaguely. When the report does mention natural resource extraction activities (in section 4.3), there is a strong focus on the need for transparency (and explicit reference to the efforts of EITI), but nothing on whether it is ethical to make money at the expense of future generations. Although much of

Norway's economy relies on petroleum, and there are unique ethical dilemmas associated with that industry, the government is not keen to dwell on the fact.

The same sort of posturing can be detected in Norway's policies regarding climate change. Each and every Norwegian government, regardless of political color,[21] has been a strong advocate for global agreements to reduce industrial greenhouse gasses, along with binding commitments for states to secure those reductions. But Norway itself is nowhere near to meeting its own commitments (see Box 10.4). The reason for this, obviously, is that much of Norway's greenhouse gas emissions come from petroleum activities,[22] and Norwegians have no intent on killing the goose that lays their golden egg.

Norway's industrial greenhouse gases emissions in 2012 were approximately 4.5 percent *above* their 1990 levels. The Climate Action Tracker (2015) website's analysis of Norway's record reads as follows: "Currently-implemented policies are projected to result in domestic emissions in 2030 above 1990 levels by 4[%] and hence far from being on track to reduce domestic emissions to levels implied by the 40% reduction goal, even after accounting for LULUCF [Land Use, Land-Use Change, and Forestry] credits and possible use of international emissions units." As a rough indicator for Norway's commitment to

Box 10.4 NORWAY'S CLIMATE COMMITMENTS[a]

Kyoto Protocol (KP):

- Member of KP Commitment Period 1 (2008–12)
  - +1 percent target (below base year)
- Member of KP Commitment Period 2 (2013–20)
  - −16 percent target (below base year)

Copenhagen Pledge:

- 30–40 percent reduction from 1990 emissions.
- Agreed to a higher pledge level if there is a global and comprehensive agreement after 2012, with major emitting parties agreeing on reductions in line with achieving the 2° Celsius target.

Intended Nationally Determined Contribution:*

- Domestic reduction of at least 40 percent by 2030 (from 1990).
- Economy-wide greenhouse gas coverage.

Long-Term Goals:

- Carbon neutrality by 2050 (2030 in the case of a global agreement).

[a] Adapted from Climate Action Tracker (2015). *In preparation for the adoption of the Paris Agreement (in December 2015), governments "in a position to do so" were asked to submit an "intended nationally determined contribution."

climate action, that organization characterizes Norway's commitment as merely "medium."

When the Norwegian parliament ratified a second commitment period for the Kyoto Agreement in 2013, the only parliamentary member from the country's tiny Green Party (*Miljøpartiet De Grønne*, or MDG, Rasmus Hansson, pointed to this sort of hypocrisy when noting that "Norway has failed completely to fulfill our modest requirements with regard to the first Kyoto Agreement. If the Labor Party and the Conservatives have thought to continue with today's climate polices, they might as well vote against the new agreement that will be ratified in the Parliament today" (MDG 2013).

To be fair, successive Norwegian governments have been committed to policies aimed at reducing greenhouse gas emissions (for example, using taxes to encourage electric car use and discourage the use of combustion engines; rainforest conservation efforts; and loud advocacy for international climate agreements). But its ability to play this role depends entirely upon its dirty little secret: the money comes from oil.

This may be changing, as is evident by recent debates at the country's universities.[23] Students at the University of Bergen started a national debate about the ethics of petroleum after their university signed an "akademia-agreement" with Statoil. Other Norwegian universities have similar agreements; our university, NTNU, has an "akademia-agreement" with Statoil worth 60 million NOK over five years (Mikkelsen 2014). The Dean of the University of Bergen asked the Norwegian national research ethics committee for natural science and technology (NENT [*Den nasjonale forskningsetiske komité for naturvitenskap og teknologi*]) to consider the ethical consequences of petroleum research.

The question asked of NENT was whether the research supported by this type of "akademia-agreement" will prolong the oil age, and whether this might stand in contrast to the university's ethical guidelines. NENT's response was diplomatic, yet still revealing:

> It is NENT's opinion that the universities, in light of their comprehensive collaboration with the petroleum industry, should be more aware of their responsibility as knowledge-producers in an environment characterized by conflicting interests. In this context, the committee would like to see more thorough reflection from the universities about their possible conservation role in their collaboration with the petroleum industry. (NENT 2014, 4)

In conclusion, they noted:

> It is NENT's opinion that it is indefensible, from a research-ethical perspective, for petroleum research's framework conditions and research activities to obstruct the process of adjustment in ways that makes it impossible for Norway to meet its UN climate goals, to which the country is committed. (NENT 2014, 11)

It is too early to know what this opinion will mean, if anything, for future petroleum research in Norway. It is highly unlikely that it will have an effect on the country's willingness to produce more petroleum. On the other hand, Norwegian youth are clearly concerned about the ethical dimensions of Norway's reliance on petroleum, and the increased public awareness on the issue is probably part of the reason why the number of students applying to NTNU's prestigious petroleum engineering department dropped by 73 percent in 2015 (Sjøberg 2015)—another reason was a radical drop in the price of oil (prompting significant lay-offs in the petroleum sector).

To conclude, we see a Norwegian state that tries very hard to do the right thing. Indeed, in many areas, Norway plays a leading role in demonstrating strong ethical principles of investment, development, and climate management. But all of these efforts are expensive, and Norway's ability to play this role depends critically on its sale of petroleum. Our point is not to criticize the Norwegian state. Norway labors more than most states in trying to reconcile its reliance on a dirty, dangerous industry with its image as a global advocate for peace, development, and environmental responsibility. As Norway is often held up as a model, however, there is a widespread reluctance to criticize its behavior. Our point is that every state, including Norway, needs to be exposed to critical reflection if it hopes to better itself.

Statoil: The Ethical Firm?

As we saw in Chapters 4 and 8, Statoil benefited greatly from a number of local content protections early in its history. As it began to grow, Statoil's political masters became increasingly concerned about its inordinate power and influence. When a Conservative (Willoch) government pressed to split up Statoil in 1985, creating the SDFI, one of their main objectives was to provide the firm with freer reins to expand in foreign markets. As the energy analyst, Jarand Rystad, noted many years later (in 2011): "When seen from a 100-year perspective, once the oil fairy tale dies out, what should remain is a large international oil and gas company stationed in Norway, such that we can pursue oil and gas activities for many years into the future" (Nøra 2011). Statoil, it seems, is well on the way to fulfilling these aspirations.

From its humble beginnings in 1972, Statoil has become a global petroleum actor, with business operations spread across 37 different countries (see Table 10.3). There is a remarkable breadth of activity revealed in this table, even if the focus of Statoil's development and production activities lies in just eight countries (far right-hand column). Indeed, in 2014 Statoil generated 39 percent of its total equity production outside of Norway (Statoil 2015a, 3). But much of this activity comes from some very risky projects. In addition to this geographic variation, Statoil has invested in a number of challenging

Table 10.3 Statoil's international presence[a]

	Business operations	Production	Main development & production projects
1	Algeria	✓	
2	Angola	✓	✓
3	Australia	✓	✓
4	Azerbaijan		
5	Bahamas		
6	Belgium		
7	Brazil	✓	✓
8	Canada	✓	✓
9	China		
10	Colombia		
11	Cuba		
12	Denmark		
13	Faroe Islands		
14	Germany		
15	Greenland		
16	India		
17	Indonesia		
18	Ireland		✓
19	Kazakhstan		
20	Libya	✓	
21	Mexico		
22	Mozambique		
23	Myanmar		
24	Netherlands		
25	New Zealand		
26	Nicaragua		
27	Nigeria	✓	
28	Norway	✓	✓
29	Russia	✓	
30	Singapore		
31	Suriname		
32	Tanzania		
33	Turkey		
34	United Arab Emirates		
35	United Kingdom	✓	✓
36	United States	✓	✓
37	Venezuela	✓	

[a] Adapted from Statoil (2015a, 3) for production and main development and production projects; the full list of business operations comes from Statoil (2014a).

project types, including deep-water drilling in Brazil, shale gas in North America, oil sands in Canada, and renewable (wind) energy in the United Kingdom.

Statoil's international success can be explained, in part, by three significant advantages it has relative to the IOC majors: (1) with the Norwegian state as a dominant actor, Statoil is less pressed to focus on short-term returns, and can emphasize more strategic, long-term investments in more treacherous markets (e.g., Iran, Venezuela); (2) Statoil benefits from the technological competence it developed in building out the challenging North Sea (e.g., Statoil is a leading expert in capture and storage of CO_2);[24] and (3) Statoil benefits from being Norwegian—as Norway was never a global power in the same way that the United States and United Kingdom were; and many postcolonial states are now looking for investment partners. In short, many of the markets that Statoil is courting would rather choose a company that grew out of a small social democratic country with a strong background in resource nationalism than a formal colonial power (Noreng 2008, 31–3).

As a result of the Norwegian transparency requirements, as noted in the section entitled "The Ethical State?", we can now get a solid look at Statoil's global activities, as the company is required to publish an annual report on the payments it makes to governments. As such, Statoil was one of the first major oil companies to start disclosing revenues and payments to governments on a country-by-country (and to a certain extent, a project-by-project) basis.

Table 10.4 provides the summary figures from Statoil's first report (Statoil 2015a). Here we can clearly see that Statoil's tax-based contribution to the Norwegian state approaches 90,000 million NOK! Beyond Norway, however, the three largest recipient states have been governed by some remarkably dicey regimes: Angola (22,952 million NOK), Azerbaijan (10,066 million NOK), and Nigeria (5,181 million NOK).[25] This activity should raise a number of red flags. For example, each of these countries finds itself near the bottom of Transparency International's Corruption Perception ranking: in 2014, Angola is ranked 161st; Azerbaijan is ranked 126th; and Nigeria is ranked 136th, out of 175 countries in total (Transparency International, 2014b).

Despite working in such risky markets, Statoil has maintained a reputation for corporate social and environmental responsibility. Indeed, the company prides itself on its social responsibility and the fact that it was ranked first among the 2014 Global 100 Index of the world's most sustainable energy companies, and came fourth among all corporations (Corporate Knights 2014). When introducing itself to the world, Statoil highlights its sustainable and ethical responsibility:

> We focus on contributing to sustainable development via our core activities in the countries in which we operate. We are committed to openness and anti-corruption work, and to respect for human rights and employee rights. That applies both to our own activities and to those parts of the value chain over which we have significant influence. (Statoil 2015c)

The company's massive online presence includes subsections on both ethics and sustainability, and the two are interlinked. In *The Statoil Book*, for example, we are told Statoil believes "that ethical conduct is a necessary condition for sustainable business" (Statoil 2016, 70). In order to promote

Table 10.4 Statoil's payments to governments[a]

	Taxes	Royalties	Fees	Bonuses	Host government entitlements (Value)	Host government entitlements (Mmboe)	Total (Value)
Algeria	*1,568.1*		*0.4*		*3,524.2*	*9.2*	*5,092.8*
Angola	*4,870.9*				*18,080.6*	*29.1*	*22,951.5*
Australia			2.6				2.6
Azerbaijan	*1,288.6*				*8,777.6*	*14.3*	*10,066.2*
Brazil	*49.9*	*865.2*	*318.0*				*1,233.1*
Canada	*1.9*	*768.2*	*16.5*				*786.7*
Faroe Islands			19.3				19.3
Greenland			0.1				0.1
Indonesia	0.2		1.9				2.1
Iran	0.5						0.5
Libya	*277.6*				*314.2*	*0.5*	*591.8*
Nigeria	*2,948.0*		*360.0*		*1,873.4*	*3.0*	*5,181.4*
Norway	*89,679.7*		*648.8*				*90,328.5*
Russia					*728.9*	*1.2*	*728.9*
Sweden	16.1						16.1
United Kingdom	*87.1*		*11.3*				*98.4*
United States	*245.6*	*585.5*	*50.4*	*75.7*			*957.2*
Other	0.2		1.2				1.4
Total	**101,034.5**	**2,218.9**	**1,430.5**	**75.7**	**33,298.9**	**57.3**	**138,058.5**

[a] Adapted from Statoil (2015a, 8). Millions NOK, 2014. Countries in italics are those where Statoil is currently engaged in production.

ethical behavior, the company has developed an impressive array of instruments, including an extensive ethics code of conduct (Statoil 2014b); an anticorruption compliance program (Statoil 2012); a mandatory e-learning program for employers and suppliers (Statoil 2014c); a global ethics hotline (Statoil 2014d); and an extensive system of ethics committees. Among the latter is the "Safety, Sustainability and Ethics Committee," whose task is to assist the company's Board of Directors in these matters (Statoil 2015d).

The same sort of image is projected in a section committed to sustainability (Statoil 2015b), where the company tries to downplay its contribution to global warming. Here readers can find the company's responses to hypothetical questions from concerned investors/consumers. Along both fronts (ethics and sustainability), it is impossible for us to gauge the sincerity of the company, as it employs modern marketing techniques to obscure its ethical obligations and to promote its brand.

Despite its very public commitment to corporate responsibility, and its reputation for being a clean, transparent, and socially responsible oil company, Statoil is not without its blemishes. The most serious of these was its conviction on corruption charges (in both Norway and the United States) for

Box 10.5 THE RAFSANJANI–STATOIL CORRUPTION CASE[a]

Statoil opened an office in Tehran (Iran) in 2001. In June 2002 Statoil signed a ten-year, US$15 million consulting agreement with Horton Investments to exploit Iran's oil and natural gas. Horton Investments is a company owned by exiled Iranian Abbas Yazdi, and was alleged to be a financial intermediary for Mehdi Hashemi Rafsanjani, son of Iran's former president, director of the National Iranian Oil Company (NIOC), and an influential figure in the country's energy sector.

In June 2002 and January 2003, Statoil paid bribes to an Iranian government official in order for him to use his influence to assist Statoil in obtaining a contract to develop three phases of the South Pars oil and gas field in Iran, and to open doors to additional projects in the Iranian oil and gas exploration industry.

In the autumn of 2002 Statoil was awarded operational control of South Pars, the world's largest offshore gas field, in partnership with an NIOC subsidiary. Statoil committed to spending US$300 million to construct three production platforms and a pipeline. After completion, control of South Pars was to revert to NIOC and Statoil was to recover its project costs and a share of NIOC sales.

In September 2003 Statoil's internal auditors reportedly questioned a US$5.2 million payment to Horton Investments' account in the Turks and Caicos Islands, leading to a nine-month probe by the Norwegian National Authority for Investigation and Prosecution of Economic and Environmental Crime (*Økokrim*). This investigation led to an indictment, for which Statoil was eventually found guilty of bribery and fined 20 million NOK (US$3 million) by the Norwegian authorities in June 2004. The former director of Statoil's international department, Richard John Hubbard, was also fined 200,000 NOK, or 20 days' imprisonment in default of payment, for his role in the affair. In September, Hubbard resigned, along with Statoil's Chairman (Leif Terje Løddesøl) and the company's CEO (Olav Fjell).

Subsequently, in 2006, Statoil agreed to a US$21 million fine in violation of the US Foreign Corrupt Practices Act for the same act of bribery. In paying the fine, Statoil took "responsibility for bribery" and agreed it had "improperly" accounted for the payments in its books. Statoil's fine is one of the largest ever imposed by the United States for violations of the Foreign Corrupt Practices Act.

[a] Adapted from Nooryani (2006) and STAR (2015).

bribes it made to Iranian officials (see Box 10.5). But there have been many and frequent allegations and suspicions, given Norway's activities in such opaque and corrupt markets. Most recently, for example, a Norwegian newspaper (*Dagens Næringsliv*) article questioned the legitimacy of Statoil's "mysterious partners" in Angola (Sæter 2014), where Statoil has invested more than NOK 60 billion (roughly US$10 billion). Angola is Statoil's most important source of production outside of Norway. Despite such a significant commitment, Statoil is not willing (or able) to find out who actually controls the companies, with which it has partnered. On the surface, every indication is that the companies are merely fronts for corrupt politicians (Sæter 2014; Burgis 2015), causing Ricardo Soares de Oliveira, an Angolan expert who teaches at Oxford University, to warn: "Since Statoil has a better reputation than many other oil companies, or at least it tries to give the impression of a company that has higher standards, this type of questionable connection will hurt Statoil's reputation" (Sæter 2014).

Indeed, Statoil is often caught in situations where it is less than transparent. For example, in July 2014, a PSA between Statoil and the Tanzanian government was leaked to the public, revealing that the split of "profit gas" between the Tanzanian government and Statoil was about 20–30 percent lower than what was used in the country's model contracts.[26] The public discussion in Tanzania turned around the size of this additional windfall (above the published model contract forms), and how the increased revenues to the Norwegian government from this deal could be more than twice the total of Norwegian aid given to Tanzania *since independence* (Taylor 2014).

Finally, as Statoil searches for new profit opportunities in different corners of the world, it exposes itself (and its host country) to complex and deep-running global conflicts. As already hinted, Statoil often benefits from being associated with the Norwegian government. But the flip side of this relationship is that Norway is also associated with the activities of Statoil abroad. This makes it more difficult for Norway to play the role of an honest broker, and the relationship exposes Norwegian foreign policy to new challenges. The example in Tanzania shows how Norway's generous aid program can be dwarfed by the amount of revenue that Statoil is taking out of these developing countries. In this light, Norwegian ODA might appear as the cost of admission to access lucrative gas/oil fields. If so, this cannot end well for Norway's reputation.

A more tragic example of this link between Norway's and Statoil's activities abroad occurred when 40 people were killed, including five Statoil employees, in a terrorist attack on the Tigantourine gas facility in In Amenas, Libya on January 16, 2013. Statoil shared operatorship in this project, and its employees were exposed to an attack on their facilities, when an al-Qaeda-linked group affiliated with a brigade led by Mokhtar Belmokhtar demanded an end to French military operations in northern Mali (Statoil 2013b). In response to the crisis, the Norwegian government was pressed into high alert in an effort to secure the release of the Norwegian hostages, and provide support to their friends and family back in Norway.

This development raises yet another ethical dilemma for the Norwegian people, as their state is increasingly asked to support Statoil's activities in dangerous, corrupt, and unstable parts of the globe. Statoil works hand in hand with regimes that Norway's foreign ministry would rather isolate. While Statoil reaps the financial gains from such risky investments, the Norwegian taxpayer ends up paying for the costs, as Norway is drawn deeper into messy conflicts far from home.

It is probably not possible to be a global oil company and avoid critical attention. Statoil is trying to walk a very fine line, between maintaining its reputation as a clean and responsible actor (from a neutral and unassuming country), while working with regimes that lack legitimacy (and often, it would seem, morals), to develop a product which is increasingly seen as tainted. In short, both Norway and Statoil are immersed in a pool of ethical dilemmas—whether they wish to acknowledge it, or not.

Conclusion

This chapter is unavoidably different from the others, as we hope to push our readers to consider seriously the ethical challenges associated with petroleum development. In so doing, we have introduced the concept of CSR in a critical light—as we are concerned that it is undermining the state's capacity to develop the regulatory expertise and experience that states need to manage their resources.

CSR is clearly good for business. It may even be good for consumers. In certain sectors, and in particular ways, CSR can prove helpful. For example, in some sectors, consumer pressure can be more effective than lobbying public officials in facilitating change. In the oil industry, CSR seems to be most effective at limiting environmental abuses among both firms and states, as solutions to environmental problems tend to require the same sort of (technical) skills and ways of thinking that are common in the petroleum industry (Frynas 2009b). But CSR has not proven very effective in terms of developing better forms of governance or strengthening respect for human, social, and political rights. In these arenas, the problems (and solutions) are much more complex. Indeed, there is a growing concern that CSR may be undermining the sort of skills and institutions that states need to secure greater human and social development. If this is the case, then the ethical argument for developing petroleum in these countries evaporates quickly.

Oil companies and policymakers need to wrestle with the sort of ethical dilemmas we have introduced in this chapter, and be prepared to defend their conclusions in an open and honest fashion. We cannot continue to pretend that these dilemmas don't exist, or that the petroleum industry is not somehow related to them. This need is especially acute with respect to the Resource Curse; helping countries to develop their resources can introduce more problems and challenges that it solves, and the industry (in partnership with host states) needs to do a better job of making sure that the wealth produced by petroleum trickles down to the broader population.

As we saw in the Norwegian case, increased critical reflection is changing attitudes and policy with regard to the prominence of oil in Norway's economic future. Eventually, Norway will have to decide which legacy it wishes to bequeath its future generations: Norway the oil power, or Norway the global champion for peace, development, and environmental sustainability. In the end, Norway cannot keep its cake and eat it too.

In this light it is fitting that Norway is host to the Nobel Peace Prize—the political legacy of a Swedish inventor who made a great deal of money from ethically questionable activities. Alfred Nobel is today remembered most for his Peace Prize, not for his invention of dynamite or his contribution to the modern armaments' industry (that funds the prize that honors his name). We wonder if Norway will be as successful in caressing its legacy for future generations.

11

Conclusion

Since its inception, the fate of the petroleum industry has been tied to its mastery of technology. At every step, the industry has developed better and more efficient means for reaching, improving, and delivering its products to market. One cannot look at the network of subsea templates producing oil and gas in the North Sea (or the army of ROVs that service it) and not marvel at the technology. Although (perhaps) less romantic, hydraulic fracturing technologies offer another example of the important role that technology plays in this industry—technology allows us to grasp opportunities that seemed unimaginable just a few decades earlier.

Given its history, and its access to enormous pools of capital, it is not surprising that the industry focuses so heavily on technological solutions. It is relatively easy to invest in technology—as these sorts of investments have a proven record of accomplishment (and return). But for people accustomed to working with public sector budgets, our heads spin at the thought of how much money, energy, and time is made available to solve these enormous technological challenges.

While we stand in astonishment of the technology frontier, we worry that our technological awe (and the industry's focus) may be misplaced. If we think critically about petroleum's future, in light of the ethical challenges introduced in Chapter 10, the most important inhibitors to petroleum development are probably political, not technical, in nature. If framed as a question, we might ask it in another way: can petroleum companies secure the (political and financial) capital they need to pursue such marvelous technological fixes if their industry is covered by a dark ethical cloud? Investments in technology are relatively uncontroversial—they offer a safe (and proven) bet; investments in governance and social policy are anything but.

There are at least two reasons for this deficit of interest in governance issues. First of all, the industry does not want to be held responsible for the social and economic impacts of its activities. While it has been opportune to invest in CSR measures that improve its public image, the petroleum industry has not tried to understand (let alone solve) the political and ethical challenges that are largely a byproduct of their substantial investments. Quite simply, and until very recently, IOCs have rejected the notion that they should actively address macro-level governance issues (Frynas 2009a, 191).

Second, we know remarkably little about what states can do to improve governance and avoid the Paradox of Plenty. This ignorance is partly because of the complexity of the problem at hand. But complexity lies at the heart of the technical challenges as well—so this cannot, and should not, be used as an excuse to ignore the problem. If this complexity is easier to handle in the realm of engineering and natural science—it is partly because we have invested much more time, energy, and money in unraveling and understanding that complexity. The social science research frontiers lie decades behind those of the natural sciences.

Research and education will play a central role in reducing this ignorance; we need to study, systematically, what works (and what doesn't work) to encourage sustainable investments, maximize government rents, encourage local content, deter Dutch Disease, and spread wealth in ways that benefit the entire population. The social scientific literature in this area is remarkably underdeveloped, given the dramatic effects it has on people's lives. Much of the work is carried out by parties with vested interests, using case studies that are seldom representative, indicators that do not travel well across countries and time, and in tests that lack systematic controls.

This book was written to try and fill two voids: to show that states can avoid the Paradox of Plenty by building and deploying capable institutional infrastructures and employing careful management techniques to achieve national developmental goals; and that it is in the interest of oil companies to help states secure that objective. Indeed, we have tried to demonstrate that states and IOCs share a great deal of common ground.

Middle Ground

This section considers the degree to which the state's interests overlap with those of the IOCs. We begin by emphasizing the need for IOCs to recognize their complicity in the problem. IOCs have always played, and will continue to play, an important political role in the development process. We then point to the lack of interest in, and support for, governance issues. This resistance is partly ideological, but mostly the result of inadequate information and study (hence this book). Finally, although we recognize the differences in interest and temporal horizons that separate state and industry, we conclude that there is more than enough room for overlap to secure constructive compromises.

The Industry

As we learned in the previous chapter, industry leaders are either unaware of, or afraid to, address the challenges associated with the Recourse Curse or the Paradox of Plenty (Skjærseth et al. 2004). The petroleum industry chooses to see the social, economic, and political turmoil generated by its investments as externalities—something outside its realm of responsibility. MNCs, including IOCs, like to think they hover above domestic politics; they see themselves as primarily economic actors, skeptical of claims that they are somehow responsible for macro-level challenges concerning the society-wide impact of their industry. From the traditional perspective of IOCs, it is solely the host government's job to address these challenges.

This is a myopic, perhaps even dishonest, view of their role. IOCs are consummate political actors. As we have seen, they intervene actively and expertly in the political process to secure their objectives, whether it is lobbying *for* new legislation, lower tax rates, laxer regulations, stabilization clauses, or lobbying *against* the unionization efforts of their workers, environmental organizations, and so on. As a result of these sorts of activity, IOCs can be complicit in governance failures. Worse, IOCs can even benefit (commercially) from governance failure. Shell's experience in Nigeria (described in Box 10.1) is a clear example of this—but Shell's experiences are far from unique. All too often, the industry is seen to benefit from the ineffective enforcement of health and safety regulations, or the use of corrupt channels to secure profitable agreements.

In the face of growing environmental pressures, and increased demand for alternative energy sources, the industry will find it difficult to continue conducting business as usual. In addition to worrying about how to best access the resource, refine it, and get it to market, IOCs now face a growing host of political and social challenges. To make things even more difficult, IOCs have precious little insight or experience in dealing with these challenges—their area of expertise has always lain elsewhere.

To survive, the petroleum industry will need to demonstrate that it can develop these resources in an environmentally, as well as a *socially*, responsible manner. The first challenge can be resolved with the sort of technological solutions and innovations to which it is accustomed; the latter requires radically different tools (and attitudes). Political challenges are far more difficult for the industry to solve—if only because social problems are inherently more intractable and complex (and we understand them less completely).

An ethical, politically sustainable market for petroleum exists if we can ensure that the economic wealth that is generated from petroleum production *improves* lives, rather than impoverishes them. While nearly everyone will agree on this need to develop the resource in a socially responsible manner, the actual practice of petroleum extraction has proved the task to be challenging.

The State

Although IOCs need to play a strong supporting part, the state must play the main role in resolving the Paradox of Plenty. This is not an easy argument to make, for at least two reasons. First, for many of today's elites, embracing the state is akin to blasphemy. Policymakers and IOC executives have come to depend upon liberal approaches to economic development, and they are leery of increased state intervention. The problem is that laissez-faire attitudes about resource development only worsen the problems associated with the Resource Curse and Dutch Disease. To solve these sorts of problem, and to ensure that a nation's economic bounty is equally (and justly) distributed, we need to rethink the role of the state in economic development. This is an (ideologically) awkward thought for many of those involved.

Second, the Paradox of Plenty is partly a problem of too much state. Hence, it is important to emphasize that it is not the state per se that is the problem—but the character and behavior of that state. In the preceding chapters we have demonstrated the necessity of a state that is both developmental and democratic to provide the basic infrastructure that the industry requires (education and research facilities, domestic suppliers, qualified labor force, competent bureaucracy). But a Democratic Developmental State is also required to secure fair and adequate contract systems and levels of government take, as well as a competent and sufficient regulatory framework—both of which are necessary to secure a level playing field for fair competition. When the international community tries to shrink the role (and capacity) of the state, it ends up throwing the baby out with the bathwater.

In particular, we need national policymakers to become more responsive to their constituents, even when their revenues arrive in the form of resource rents, extracted by third parties. The solution to this challenge is anything but easy, but pressure from significant interest groups, both internally and externally, can help bring it about. In the Norwegian case, a strong and autonomous labor movement played an important watchdog role, ensuring that policymakers and firms did not overstep, or abuse, their authority.

Unfortunately, most of today's preferred solutions are focused on voluntary agreements and codes of conduct that are limited to (reporting) transparency and encouraging anticorruption drives. These issue areas are important, but hardly sufficient. In some cases, the expansion of parallel "soft law" responses to common challenges may even be undermining the power, authority, and competence of national authorities (as Robert Reich (2008) has argued with respect to CSR). In other cases, the focus has been diluted by other concerns, as IOCs and states demonstrate an unwillingness to adopt the challenge as their own.

Other venues have begun to recognize the larger macroeconomic and political challenges associated with petroleum's development. Indeed, Norway's Oil for Development (OfD) program was specifically created to assist with this sort of challenge. Established in 2005, the OfD program was designed to redirect Norwegian foreign aid to the oil sector so that it can help developing countries deal with the resource, financing, and environmental dimensions of natural resource extraction, along with the need to develop competence and capacity in the public sector. Rather than focusing attention on the narrow issue of extraction (for which there is more than enough private consultants), this program was touted to draw the attention of policymakers toward broader issues of management—including revenue and environmental issues, and the need to include civil society and secure a gender balance, and to strengthen anti-corruption efforts. The explicit intent of the OfD is to "transfer Norwegian experience with petroleum governance/management in a way that contributes to lasting reduction of poverty in developing countries, and that the extraction of resources is done in an environmentally friendly way" (*UD Prop. 1S (2010–2011)*, 52).

Unfortunately, recognizing a need is not the same thing as satisfying it. While the OfD program offers a step in the right direction, it hasn't come very far. In its 2012 evaluation of the program, Norad noted "that OfD has generally not given the governance problems sufficient attention" (Norad 2013, xix). Worse, the allocation of OfD funds does not indicate any real change of direction: the lion's share of the funds is still going to help resolve technical challenges related to mineral resource/mining, with too little money and interest being invested in the challenge of governance.

FINDING COMPROMISE

The trick to successful management rests in the capacity to elicit information from companies and local communities, and to encourage compromise and cooperation. IOCs and government officials often live in two very different worlds, with apparently conflicting interests. These differences are most evident when we compare their core interests and time frames. IOCs need to focus on the here and now; they don't expect to be around when any long-term problems come home to roost. Even though the time horizons for oil companies tend to be longer than those in other sectors (after all, it may take decades before an oil basin is fully drained), shareholders compare the quarterly returns from their IOC holdings against returns from firms in other sectors. To placate these investors, IOCs need to deliver quick and substantial returns on their investments. Consequently, IOCs prioritize high levels of development early on, and focus myopically on project profitability.

The host state, by contrast, needs to ensure that the resource is fully utilized. Doing so requires the adoption of a longer time horizon, as quick solutions tend to offer inefficient outcomes (not just in terms of well efficiencies, but also in terms of economic and social efficiencies). Like IOCs, the state aims to optimize the revenues it can generate from petroleum resources, but revenues are just one of several items in the state's optimization basket. Engineers already recognize the importance of proper reservoir management to extend the life and value of a resource. The same sort of consideration is needed to maximize the social and political bounty from that resource. In effect, policymakers need to balance short-term fiscal rewards against the potential long-term costs to employment, social welfare, and the environment. Ultimately, the state should prioritize the long-term view and the broader social consequences, while the IOC will prioritize the short term, and focus on project profitability.

In the Norwegian case, this balance was achieved in a way that minimized the Paradox of Plenty. By encouraging a moderate pace of development, the Norwegian authorities created a breathing space, in which they could better learn the tools of the trade, and expand their bargaining power relative to the IOCs (by exploiting the OBM). By controlling the speed and content of their licensing rounds, political authorities provided domestic suppliers with enough time to adapt and retool. The same subtle advantages were secured by the state's active use of local content policies and the development of an NOC which shared the state's longer-term perspective. As a result, the Norwegian economy was able to absorb the new economic activity, and use it to advantage.

This willingness to go slow takes enormous courage and foresight. Political authorities are under significant pressure to develop the resource quickly. We have already noted that IOCs have a strong interest in speeding up the returns on their investments; their stockholders expect early and strong profits. But local residents will also pressurize the authorities to move fast, in hopes of securing their slice of the pie, as quickly as possible.

Embracing a moderate pace of development can generate very substantial long-term rewards, among them the successful management of the resource and the wealth it generates—but doing so entails significant short-term costs (both political and economic). Consequently, the state needs to provide full information about its intent, and manage expectations accordingly. It needs to act carefully, so as not to alienate the important interests that desire a speedy development of the resource. Policymakers need to inform the public about the potential for both curse and reward that accompanies a new-found resource. They should provide reasonable and honest expectations about the appropriate timeline, channels of influence, and expected outcomes. This benefits all actors: IOCs, local firms, and constituents.

Lessons from Norway

We have endeavored to show how Norway avoided the Paradox of Plenty, and used its petroleum revenues to the advantage of the nation and its citizens. The Norwegian experience teaches us that petroleum need not always be a curse, and that responsible actors (both public and private, national and international) can work together to produce desirable (and collectively optimal) outcomes.

The Norwegian case is meant to be an illustration, not a model. By focusing attention on Norway, we do not mean to suggest that the Norwegian approach can (or should) be mimicked by others. Nations aren't like the molecules in oil that react in a consistent and predictable fashion. Still, Norway offers useful lessons to other oil-producing countries that wish to maximize the benefits of oil and gas for national development.

We began the book by distinguishing between those features of Norwegian experience that are not transferable to other countries, and those that are. This is because Norway's political and economic points of departure were quite different from those found in many of today's developing countries. These national characteristics, and the context in which Norway's original policies were made, cannot be replicated in today's developing world. Each country faces a unique set of opportunities and constraints, in both domestic and international contexts, within which it must find its own solutions.

Still, there are lessons to be gleaned from Chapters 2 and 3. We have seen that the international environment is constantly changing, requiring frequent updates to Norway's policy mix. This sort of institutional dynamism provides both challenges and opportunities for policymakers. While we can hope that policymakers will endeavor to change the international context—to one where there is more scope for policy autonomy—states can still find room for sovereign maneuver. States can avoid international agreements that limit their scope for policy autonomy, or only accept these constraints grudgingly (when better alternatives are unavailable, and in return for improved terms elsewhere). Too often, states back their way into binding arrangements that limit their policy freedom (e.g., an agreement is signed by one department of government that constrains the scope for policy autonomy in another department of government). States need to be made aware of the nature of these constraints, and then work actively to avoid them, or at least minimize their impact.

On the domestic front, Norway's example demonstrates the importance of state institutions and infrastructure that can encourage broad-based political and economic development. Even if other states do not enjoy these benefits at the outset of their petroleum quest, they can use the revenues and influence generated by petroleum resources to build up the required institutional capacity. This is only possible, of course, if states recognize the long-term nature of the project, the utility of a moderate pace of development, and are willing to prioritize accordingly. As we described in Chapter 8, there is much that the state can do to encourage local suppliers, but doing so takes time and resources. In the absence of these endowments (good education, infrastructure, capital markets, and so on), the state needs to place the horse before the cart; it should develop the basic infrastructure before the oil resources are unleashed on the domestic economy.

The heart of our book, Chapters 4–9, focuses on the institutional frameworks and policies that were developed in Norway, and which might be adopted fruitfully by other countries. Each chapter addressed a particular concern, such as developing an institutional framework for administering the resource, deciding on an appropriate operating agreement and fiscal regime, creating an effective NOC, dealing with the macroeconomic consequences of increased wealth, building an investment fund, maximizing local content, and regulating the industry in ways that can protect workers and the environment. For each issue area, we began with a general introduction to the options available, and then directed attention to the particular Norwegian solution to the problem at hand.

In presenting the information in this way, we hope that readers will better understand the nature of the tasks facing resource-rich countries, and the logic of their diverse policy responses. Policymakers need not, and should not, mimic the Norwegian model (or any other model) blindly. Policy choices and institutional solutions may not work in different (domestic and international) contexts—but understanding *how* they work will allow policymakers to develop tailor-made policies for the particular needs of individual countries. In this way, readers can determine the appropriate constellation of policies and institutions when considering their own country's response to the challenges of resource abundance and wealth.

From the Norwegian case, it is possible to derive three central lessons. The first, and most important, lesson is that politics matters. In the real world, experience demonstrates that unfettered oil markets lead to Dutch Disease and the Resource Curse. It should be clear to the reader that these two ailments benefit none of the relevant stakeholders. To cure these ailments, the state must play an active role. This means—first and foremost—that the state must determine an appropriate pace of resource development: finding a workable compromise between the short-term interests of the industry and the long-term interests of the community.

A larger role for politics means recognizing that the state plays a central role in securing the institutions and infrastructure that the industry requires for subsequent development. The state must be able to stand above the fray of competition. This means developing the competencies needed to play an active (and impartial) role as regulator, as well as a stern (but just) manager of the wealth that is generated. When contrasted against the Norwegian case, it is easy to see why so many countries are handicapped by their resource abundance; they suffer from poor or ineffective government, not too much government.

Policymakers need to be aware of the role that power plays in the industry, and address it head-on. This was clearly the view of Arve Johnsen, CEO of Statoil from 1972 to 1988, who recognized, "This [the oil industry] is the most powerful industry in the world... Power is not something you get, but something you got to take" (Johnsen, quoted in Moen et al. 2009, 2191). When reflecting on the lessons learnt in response to the 1970s energy crisis, one US congressional committee (1975, 18) drew a similar conclusion:

> [I]n a democratic society, important questions of policy with respect to a vital commodity, like oil, the lifeblood of an industrial society, cannot be left to private companies acting in accord with private interests and a closed circle of government officials. (quoted in Rabe 1982, 181)

The second transferable lesson from the Norwegian example concerns the issue of deliberative democracy and the role of trust. Trust lies at the heart of the Norwegian model; and it is nurtured by familiarity, born of shared institutions and experiences. IOCs were at first surprised by this open, pragmatic, and constructive approach—they expected a socialist government to punish international capitalists and abscond with the profits. What they found was a state that wanted to work together with the industry to maximize their (collective) opportunities over the long haul. In the process, IOCs and the Norwegian state found that they both benefited from working together to create research programs that increased the size of the pie (e.g., IOR), reduced costs (e.g., NORSOK), and improved HSE, among other things. In the end, both Norway and its IOCs made a great deal of money, and they learned to trust one another.

The Norwegian system of trust is a product of a long history of democratic participation, fused with corporatist relations and partnerships. This history was born of conflicting interests, but time and a long line of successes have nudged those conflicting interests into a collaborative context; the Norwegian approach encourages trust between workers and employers, between state and firms, and between private companies and its NOC. Over time, this trust allows the parties to secure synergistic outcomes that exceed what could be achieved by any individual component working in isolation. In the absence of this environment of trust—and the collaborative institutions that secure it—neither HSE, wage restraint, nor local content policies would have been so successful. This does not imply that countries cannot hope to succeed in the absence of such corporatist partnerships; our point is simply that policies need to be tailored to their context, based on competence, and designed to promote the people's best interests.

The third and final lesson is to recognize the dynamism of petroleum management policies. Norway's oil adventure began in the late 1960s, but it has experienced a remarkable number of policy shifts in the intervening years. Norway's policy mix, institutions, and objectives have constantly adapted to changing environments (both internally, but most importantly, externally). As we have seen, many of the tools that Norway employed to build local competence and protect the domestic economy are no longer available to states today—either because these states lack an experienced administrative apparatus, an organized system of industrial (corporatist) arrangements, or because international agreements have made these tools obsolete.

The most radical change is evident in Norway's own position in the international oil hierarchy. This change has affected both the Norwegian state (and more noticeably, its NOC—Statoil). On starting out, Norwegians were acutely aware of their lack of experience and expertise. To bring their resource to market, Norwegians depended on foreign competence; to access that competence they had to invite some very large and powerful firms into their kingdom. As the sovereign authority in that kingdom, the Norwegian government (and subsequently Statoil) worked hard to protect domestic interests, and to leverage expertise from the more experienced international firms to develop their own domestic competence. Until Norway and Statoil were strong enough to compete with these international firms, they tilted the playing field to their (domestic) advantage. In short, they employed a classic infant industry protection strategy. But when Statoil became too powerful, and began to threaten political power, the state moved quickly to decrease its power and influence.

It can appear as though history is sometimes forgotten (or ignored), as both Norway and Statoil work to pry open foreign markets in order to access their natural resources. The future of the Norwegian petroleum industry depends on its ability to secure access to other people's resources. Just as wealthy states have changed their position on international trade (and protection) after they developed, Norway and Statoil seem to be kicking away the ladder (Chang, 2005), after they have obtained a certain level of international competitiveness.

In conclusion, the Norwegian example teaches us to think about political and economic outcomes in a more dynamic fashion. Usually we think of the petroleum sector in zero-sum terms—where the size of the revenue pie is fixed, and where different actors fight over preferable distributions. The Norwegian approach tries to encourage cooperation in a way that increases the size of the revenue pie, so that each slice grows in size (but not at the expense of the other slices).

We hope that foreign readers will be inspired by the Norwegian case, and find useful examples of what can be done in order to ensure that broad swaths of the domestic population can enjoy the wealth generated from petroleum revenues. For our Norwegian readers, we hope to remind them of their past, and the need to recognize that developing states must secure their own expertise and capacities, in order that their own people can benefit—as Norwegians did—from oil. In the end, petroleum resources belong to the people who live atop them. So, too, does the wealth that these resources generate.

Glossary

Appreciation An increase in the value of an asset over time. An appreciated currency is more valuable, and therefore it can buy more foreign produced goods that are denominated in foreign currency. When applied to exchange rates, appreciations can be measured in real or nominal terms. The nominal exchange rate is defined as the number of units of the domestic currency that can purchase a unit of a given foreign currency. This is the price quoted in the daily newspapers. The real exchange rate is defined as the ratio of the price level abroad and the domestic price level, where the foreign price level is converted into domestic currency units via the nominal exchange rate. Hence, a real appreciation of the exchange rate refers to a level of domestic inflation that exceeds the level of inflation abroad. A real appreciation is when domestic goods become more expensive than foreign goods.

Achnacarry A small hamlet and castle in the Lochaber region of the highlands of Scotland. Home of the "Achnacarry Agreement" or the "As Is" Agreement of 1928. At Achnacarry Castle, representatives for the three major oil companies at the time (Standard, Shell, and Anglo-Persian) met to agree on how international oil markets should be governed to avoid price competition when Middle Eastern oil flooded the market. The other majors quickly signed on to the agreement, and the settlement held until the 1950s.

Aukrust Model An example of a Scandinavian Inflation Model, from Odd Aukrust. A two-sector (exposed/sheltered) model that recognizes that the developments of prices and incomes in small and medium-sized economies are strongly affected by events in the outside world. The model is used to demonstrate how prices in the exposed sector are set by the world market, so that wages in the sheltered sector cannot push these higher without undermining international competitiveness. From Aukrust (1977).

Back-In The point at which a carried partner is allowed back into the regular obligations of a working interest partner. See Carried Interest. Usually set after the point of commercial viability, an NOC is expected to begin contributing its share of the development and operation costs.

Booking Barrels The accounting practice of adding barrels of oil to a company's list of assets.

Bretton Woods System Narrowly conceived, this refers to the postwar monetary order negotiated at Bretton Woods, New Hampshire in 1944. More generally, it refers to the network of international institutions that were a result of those negotiations (and which covered more than monetary policy): the IMF, The International Bank for Reconstruction and Development (part of the World Bank), and the International Trade Organization (eventually GATT, then the WTO).

Budgetary Rule An informal agreement among Norway's political parties to not transfer more than 4 percent of the expected real rate of return from the GPFG to the central government's budget. *Handlingsregel* in Norwegian.

Carried Interest When a working interest partner (most usually a NOC) in the exploration or development phase of a contract pays a share of costs and expenses that is disproportionally lower than its working interest share. Typically, NOCs are carried through the exploration phase, at which point the government takes up a working interest (or "backs in").

Cartel An association of countries, manufacturers, producers, or suppliers whose purpose is to maintain higher prices by restricting competition and controlling supply.

Collective Action Dilemma A situation in which multiple individuals would all benefit from a certain action, but that action has an associated cost that makes it implausible that any one individual can or will undertake and solve it alone. In short, even though everyone would benefit from a certain action, each individual finds it suboptimal to act—so the desired collective action is thwarted.

Concession The granting of rights, land, or property by a government, local authority, corporation, individual, or other legal entity. With a concession, the government grants a private interest the exclusive right to explore, operate, or exploit a public resource (e.g., an oil field) for a given period of time.

Concession Law of 1906 Norwegian Law of 1906 to protect the country's waterpower from foreign ownership. Also known as the Panic Law. The law included a provisional concession law (subsequently extended and tightened) that applied to waterpower as well as to forests and minerals. After 50–60 years, ownership reverted to the state free of charge (right of escheat), regardless of whether the owner was a foreign or a Norwegian private company.

Corporate Social Responsibility (CSR) Sometimes called corporate conscience or corporate citizenship. CSR is a form of corporate self-regulation integrated into a business model, where a given business is trusted to monitor and ensure its active compliance with the spirit of the law, ethical standards, and national or international norms.

Corporatism A means of organizing society by major interest groups, or corporate groups, such as agricultural, business, ethnic, labor, military, patronage, or scientific affiliations, on the basis of common interests. In our context, corporatism refers to the tripartite organization of economic policy, involving labor organizations, employer organizations, and the state. This is sometimes referred to as neocorporatism. Corporatist interest representation exists alongside other forms of representation inherent to representative democracy (e.g., political parties).

Cost Oil In production sharing contracts (PSCs), Cost Oil is the oil used to reimburse a contractor for its exploration, development, and operating costs. Different from Profit Oil.

Decommodification There are many things for sale in the market that are not real commodities. For example, labor is more than the time paid to a worker for her efforts. Decommodification refers to social entitlements and programs that allow people a degree of freedom from the market, that is, that their existence does not rely solely on market performance. The concept originates with Polanyi (2001 [1944]).

Devaluation Used in reference to fixed exchange rate regimes. A deliberate downward adjustment to the value of a country's currency, relative to another currency, group of currencies, or standard.

Developmental State Reference to the type of state-led macroeconomic planning found in East Asia's Newly Industrialized Countries, but subsequently used to described state-based development policies in other countries. Developmental States are states that play an active role in achieving economic development; these states develop the necessary regulatory, planning, and policy measures/institutions necessary to accomplish that objective.

Downstream One of three main sectors in the petroleum industry (upstream, midstream, and downstream). Downstream refers to the refining of petroleum crude oil and the processing and purifying of raw natural gas, as well as the marketing and distribution of products derived from crude oil and natural gas.

Dutch Disease Negative economic consequences, mostly a real appreciation, that arise from large increases in a country's income. Named after Dutch experiences from discovering natural gas in the 1950s, Dutch Disease is primarily associated with natural resource discoveries, but can result from any large increase in foreign currency, including foreign direct investment or foreign aid.

Economic Rent Any payment to a factor of production in excess of the cost needed to bring that factor into production. In classical economics, economic rent is any payment made (including imputed value) or benefit received for non-produced inputs such as location (land) and for assets formed by creating official privilege over natural opportunities (e.g., patents).

Embedded Autonomy This term originates in the work of Peter Evans (1995) and refers to the complicated relationship between government and non-governmental actors in facilitating economic development. Evans argues that government agencies cannot effectively involve themselves in economic matters without obtaining detailed information from non-governmental actors (they need to be embedded), but these same agencies need to be independent of private interests (autonomous).

Escheat A common law doctrine that transfers the property of a person who dies without heirs to the crown or state. Escheat is used to ensure that property is not left in "limbo" without recognized ownership.

Ethics A set of principles of right conduct and/or the study of the general nature of morals and of the specific moral choices to be made by a person; moral philosophy.

Exchange Rate The exchange rate allows us to denominate the cost or price of a good or service in a common currency. A currency's nominal exchange rate is the price you pay in one currency for another currency (e.g., 1 US dollar buys 7 Norwegian kroner). The real exchange rate is an accounting notion used to compare the level of domestic prices in a country relative to the level of global prices. For example, the real exchange rate (RER) is equal to the nominal exchange rate (e), multiplied by the world price level, divided by the domestic price level, or:

$$(\text{RER}) = \frac{(\text{e})^{*}\text{World price level}}{\text{Domestic price level}}$$

Exposed Sector That segment of the national economy that is vulnerable to international competition: exporters, importers, and those that produce goods/services that compete directly with imports—for example, television manufacturers. Sometimes referred to as the "Tradables Sector."

Externality An unaccounted for consequence (either positive or negative) of an economic activity that is experienced by unrelated third parties.

Factor Central component of economic production. In economic analyses, there are three primary factors of production: land, labor, and capital.

Geneva Convention of 1958 Colloquially called the Convention of the High Seas. Asserts that national governments have exclusive rights to exploit resources found on their onshore and offshore territories.

Gini Coefficient Sometimes referred to as the Gini index, or Gini ratio. The most commonly used measure of inequality. The Gini coefficient is a measure of statistical dispersion that is used to capture the distribution of inequality. Although it most frequently compares income distributions, it can be used with other variables as well (e.g., wealth). In the case of an income Gini coefficient, 0 corresponds to perfect income equality (everyone has the same income) and 1 corresponds to perfect income inequality (one person has all the income, while everyone else has zero income).

Goodwill Agreements Non-binding agreements in Norway where IOCs declare their intent to conduct as much of their petroleum-related research and development in Norway as possible. See also Technology Agreements.

Government Take The government's share of economic rents, or the total government share of production or cash flows from royalties, taxes, bonuses, profit oil, and so on. The most common means to operationalize government take is in terms of the government's cash flow, divided by the gross project cash flow.

Handlingsregel See Budgetary Rule.

Hjemfallsrett Norwegian legal concept that ensures that all exploitation rights return to the state, without compensation, after a given period of time.

Hotelling's Rule Accounting rule that holds that the most socially and economically profitable extraction path of a non-renewable resource is one along which the price of the resource, determined by the marginal net revenue from the sale of the resource, increases at the rate of interest. From Hotelling (1931).

Import Substitution Industrialization A strategy for economic development where industrialization is encouraged by promoting domestic production by restricting imports. This strategy is seen as a temporary measure, until a country is able to compete fairly with international producers. See Infant Industry Protection and Terms of Trade.

Incomes Policy Policies aimed to affect wage and price developments, often instituted as a response to inflation. Under corporatist arrangements, organizations representing workers, employers, and government will meet regularly to establish the basic framework for future wage/income (and hence price) developments.

Infant Industry Protection New industries find it difficult to compete with established competitors abroad. To help new industries get established, governments often use subsidies and tariffs to protect them from international competitors, until they are able to compete on their own. The origins of the policy can be traced to the first US Secretary of the Treasury, Alexander Hamilton.

Internal Control Principle Norwegian control principle introduced in the latter half of the 1970s to the HSE sector to ensure that license holders are held responsible for compliance with the rules and regulations in their activities.

Investor State Dispute Settlement (ISDS) An ISDS is an instrument of international law that grants an investor the right to use dispute settlement proceedings against a foreign government. These instruments are included in a growing number of treaties, including bilateral investment treaties, international trade treaties, and international investment agreements. If an investor from one country invests in another country, and both of which have agreed to ISDS, and the host state violates the rights granted to the investor under public international law, then that investor may bring the matter before an arbitration tribunal.

Joint Venture A form of partnership between two or more parties (usually between an IOC and an NOC) that allows for the sharing of risk and reward.

Laissez Faire An economic doctrine associated with the French physiocrats that encourages commercial activity with minimal government interference. Often used as short hand for "free trade" or "liberal economy." The entire term, usually ascribed to Quesnay, is "*Laissez faire et laissez passer*" (Let do and let pass).

Lifting Share of production that various parties are allowed to take in physical and legal possession (that is, what IOCs can "book").

Local Content The additional value to the national (local) economy from the purchase of goods and services, for example wages and benefits, materials, equipment and plant, subcontracts, and taxes.

Midstream One of three main sectors in the petroleum industry (upstream, midstream, and downstream). The midstream sector involves the transportation (by pipeline, rail, barge, oil tanker, or truck), storage, and wholesale marketing of crude or refined petroleum products.

Non-Tradables Sector Also called the Sheltered Sector. The sector of the economy that is not directly exposed to foreign competition (imports and exports), usually with reference to locally rendered services, including health, education, retail, and construction. See also Tradables Sector.

Norm Price The norm price corresponds to the price that petroleum could have been traded for between independent parties in a free market. It is a price posted by the Norwegian Petroleum Price Board for tax assessment purposes and to ensure that oil companies don't use Transfer Pricing to avoid their fiscal responsibilities.

Obsolescing Bargaining Mechanism (OBM) A mechanism that affects the balance of power between MNCs and host governments when negotiating over fixed investments. Initially, the partners reach a bargain that favors the MNC; with time (as the MNC's fixed assets in the country increase), bargaining power shifts to the government. The term originates with Vernon (1971).

Offsets Offset transactions are contractual conditions that require the seller (usually a foreign supplier) to transfer additional economic benefits to the buyer (usually a host government or domestic company) as a condition for the sale of a base good or service.

Organization of the Petroleum Exporting Countries (OPEC) OPEC is an intergovernmental organization of 14 nations, founded in 1960 and headquartered in Vienna. OPEC members account for a little less than half of global oil production, and roughly three-quarters of the world's proven oil reserves. When acting as a cartel, this could give them significant influence on global oil prices. See Table 2.4.

Panic Law See Concession Law of 1906.

Policy Maker A country, usually a large country, that can influence the outcome of international agreements or norms. These states can act strategically in international contexts. See also Policy Taker.

Policy Taker A country, usually a small country, that is unable to influence the outcomes of international agreements or norms, so is forced to respond to those changes. See also Policy Maker.

Primary Sector Sector of the economy making direct use of natural resources, for example agriculture, forestry, fishing, mining, and petroleum. Usually contrasted against the Secondary Sector and Tertiary Sector.

Product Sharing Contract (PSC) Also called a Product Sharing Agreement (PSA). An agreement between one or more investors and the government in which rights to the extraction of natural resources from a specific area over a specified period of time are determined.

Profit Oil In a PSC, the share of production left over, after the "royalty oil" and "cost oil" have been removed.

Public Tender A bidding process that is open to all qualified bidders. This is usually carried out in the form of sealed bids that are opened in public, and where the eventual decision is decided on the basis of price and quality. Sometimes referred to as Competitive Tender or Open Tender. See also Tender.

R Factor A factor used in establishing royalty rates, to ensure that the company doesn't walk away with an unwarranted share of the profit. R stands for the ratio of cumulative cash inflows (net of royalties) that the contractor has received at any given time, divided by the total investment to date (including operating costs). The ratio is initially zero. If R is less than 1, the contractor has yet to recover its capital investment and operational costs (i.e., total expenditure is more than total receipts). The more R exceeds 1, the more profitable for the contractor. The authorities can curb excessive profitability by stipulating that the royalty level, or government share of profit oil, shall increase along with the R factor.

Rechtsstaat A German term that translates roughly as "legal state," "state of law," or "state of rights." Although originally from German jurisprudence, it has become a doctrine in continental European legal thinking. A *rechtsstaat* is a constitutional state, where government power is constrained by the rule of law, but also includes a conception of justice.

Regional Trade Agreements (RTAs) RTAs are preferential trading arrangements between countries or groups of countries. They are designed to encourage open and competitive markets for businesses in the participating countries, but stand in contrast with the multilateral agreements preferred by the WTO. Examples include the EU, Mercosur, and NAFTA.

Rent See Box 3.1.

Rentier State A state that derives all or a substantial portion of its national revenues from the rent of indigenous resources to external clients. First postulated by Hossein Mahdavy (1970). See also Rent and Economic Rent.

Reputational Risk A risk of loss resulting from damages to a firm's reputation, consequent to an adverse or potentially criminal event (even if the company is not found guilty). Examples include lost revenue; increased operating, capital or regulatory costs; and destruction of shareholder value.

Resource Curse An observation that countries with an abundance of natural resources, specifically point-source non-renewable resources (such as petroleum), tend to have lower rates of economic growth.

Royalty A payment to an owner for the use of property (here natural resources); royalties are designed to compensate the owner for use of the asset. In effect, royalties are levies where the owner is entitled to a given percentage of the wealth produced.

State's Direct Financial Interest (SDFI) Norwegian SDFI. A portfolio of the Norwegian government's directly owned exploration and production licenses for petroleum and natural gas on the NCS.

Secondary Sector That economic sector of the economy that produces a finished, usable product, for example production and construction. Usually contrasted against Primary and Tertiary Sectors.

Service Contract A contract between IOCs and the host government that does not transfer the rights of ownership to the resource. In this contract form, the IOC never actually gains ownership, or "title," to the resource—it is simply paid a fee for its services in extracting the government's resource.

Seven Sisters Moniker used to describe the most important and dominant oil companies in the first half of the twentieth century. These were Anglo-Persian, Gulf,

Royal Dutch, Standard Oil of California, Standard Oil of New Jersey, Standard Oil of New York, and Texaco. See Table 2.2.

Sheltered Sector That segment of the national economy that is not vulnerable to international competition, for example local service providers (e.g., barbers). Sometimes referred to as the Non-Tradables Sector.

Sliding Scale A formula used by the Norwegian government to ensure that the government's participation share increased with the size of a field's productive capacity.

Social Democracy A political ideology whose aim is the establishment of democratic socialism through reformist and gradualist (non-revolutionary) means. The underlying belief is an understanding that effective political democracy requires a level of economic and social democracy. Social democratic regimes employ universal welfare states and corporatist bargaining arrangements to secure more egalitarian economic and more democratic political outcomes.

Soft Law Refers to quasi-legal instruments which do not have any legally binding force, or whose binding force is somewhat "weaker" than the binding force of traditional law. Often contrasted with "hard law."

Sovereign Wealth Fund A state-owned investment fund that invests in global markets. The content and objective of the fund can vary significantly, but they usually invest in real and financial assets such as stocks, bonds, and real estate.

Stabilization Clause A means for foreign investors to mitigate or manage the political risks associated with their project, where the host country promises that it won't change the underlying regulatory or fiscal framework governing the investment.

Stakeholder A person, group, or organization that has a vested interest or concern in an organization, and can be affected by that organization's actions or polices. Sometimes stakeholders are allowed a formal say in the organization's decision-making, but more often their influence is less formal. Stakeholders in the petroleum industry might include a firm's shareholders and workers (and their unions), creditors, the government (local and national), social organizations, and community groups in the vicinity of the firm's operations, and so on.

Sustainability "Meeting the needs of the present generation without compromising the ability of future generations to meet their own needs," as defined by the World Commission on Environment and Development (1987).

Technology Agreement Norwegian legal agreements for securing local content in research activities. Two basic types: 50% Agreements mandate that half of all research must be conducted in Norway; Offer Agreements require firms to employ Norwegians in defined areas and periods of time to secure a concession. See also Goodwill Agreements.

Tempoutvalget Also known as the Pace Committee. Norwegian government committee from 1982, assigned with the task of assessing the future development of petroleum activities in the country.

Ten Oil Commandments Effort to ensure political control over the resource by Norwegian politicians. Provides an explicit account of fundamentally Norwegian values as they apply to the petroleum sector. Introduced in a government white paper in 1971. See Box 4.2.

Tender An offer of money to someone or something that holds a claim or demand. Usually, a tender is seen as an unconditional offer to perform and/or produce the subject matter under tender, that is, a promise to pay money for the services/products offered. See also Public Tender.

Terms of Trade Generally refers to the relative price of exports in terms of imports and is defined as the ratio of export prices to import prices. Among development scholars, Terms of Trade is used to refer to the unequal effects of trade for those countries who trade in manufactured goods as opposed to commodities. Countries that export manufactured goods enjoy prices that rise faster than those in countries that export commodities.

Tertiary Sector That economic sector of the economy that produces services. Usually contrasted against Primary and Secondary Sectors.

Texas Railroad Commission (TRC) The TRC was established in 1891 to regulate the rail industry of the 1800s. Since that time the Commission is responsible for overseeing the activities of many different industries. It became the main instrument for regulating the petroleum industry, in response to the 1930s oil boom, when prices plummeted. The TRC controlled the output of oil, and with it the price and subsequently became a model for the development of OPEC.

Tradables Sector Also called the Exposed Sector. Usually refers to the sectors whose output in terms of goods or services are traded internationally, or are in competition with international producers of goods and services. See also Non-Tradables Sector.

Transfer Pricing The internal (non-market) pricing of goods and services when transferred within an organization, but across affiliates in different countries, with different tax regimes. Transfer pricing allows an organization to shift its value creation to low-tax jurisdictions, minimizing its tax burden in high-tax jurisdictions.

Triangular Diplomacy Realization that diplomacy is no longer restricted to interstate relations in a globalized economy. Diplomacy is now conducted between states and states, states and firms, and firms and firms. From Stopford and Strange (1991).

Unconventionals In contrast to conventional fields, where the petroleum lies in underground caverns and can be accessed by way of a well, unconventional fields refer to a myriad of alternative sources (and techniques) for obtaining petroleum, including oil sands, shale oil, tight oil, thermal depolymerization, coal and gas conversion, and so on.

Unitization Contract Unitization is the joint, coordinated operation of a petroleum reservoir by all the owners of rights in the separate tracts overlying the reservoir. A Unitization Contract defines the terms of that join operation.

Upstream One of three main sectors in the petroleum industry (upstream, midstream, and downstream). Upstream includes the searching for potential underground or underwater crude oil and natural gas fields, drilling of exploratory wells, and subsequently drilling and operating the wells that recover and bring the crude oil and/or raw natural gas to the surface. The upstream sector is also known as the exploration and production (E&P) sector.

Wildcat A person or well. A wildcat well is an exploratory well drilled where little or no geological information is available. It is a very high-risk (but potentially high reward) endeavor. Hence, wildcatters have a reputation for being enormous risk-takers.

Notes

Chapter 1

1. We use the term international oil companies (IOCs) in a very loose fashion, when referring to a group of large and influential (international) oil and gas companies (e.g., BP, Chevron, ConocoPhillips, ExxonMobil, Shell, and Total). These companies are sometimes referred to as Integrated Oil Companies, leading International Operators, Oil Majors, or Super Majors; they are integrated enterprises, with a global reach, capable of operating in every part of the supply chain.
2. A giant field is here defined to be a recoverable reserve of 500 million barrels of oil equivalent (MMboe) or greater. Conventional fields refer to fields that can exploit traditional (oil well) methods of extracting the petroleum resources. Unconventional fields refer to alternative means/technologies for producing or extracting petroleum (e.g., oil sands, shale oil, and so on). It is worth noting that a recent and rapid rise in unconventional field production in democratic countries (United States and Canada) has had a massive influence on the market.
3. Billion barrels of oil equivalent.
4. In particular, using the POLITY IV indicator, where democratic states are those that score between 6 and 10 on the index. See Polity (2014) for a description of the index.
5. As early as 1859, John Elliot Cairns described something akin to Dutch Disease when accounting for the effects of the Australian gold rush in the 1850s (see Ross 1999, 306, fn.29). Even Adam Smith (1776, IV.7.29), in his *Wealth of Nations*, noted the role that American bullion played in providing mercantilism with a "degree of splendour and glory which it could never otherwise have attained." Perry Anderson (1979) is the seminal authority on the rise of absolutist states. North and Thomas (1973) were some of the first modern political economists to tie together these political and economic strings of thought. For a mapping of this modern literature, see Ross (1999). For a more up-to-date survey of the Resource Curse literature, see Frankel (2012).
6. A definition of "appreciation," along with many other important concepts, is located in the book's glossary.
7. The figures in Sachs and Warner's original (1995) work did not include country labels. This figure is from a subsequent article (2001).
8. Recall that American revolutionaries declared independence from Britain on the grounds that there would be no taxation without representation. The opposite is also true: without taxation there is no representation.

9. It should be noted that Norway is not as reliant on petroleum revenues as other states (e.g., Nigeria). Skeptics might suggest that this implies Norway is not as exposed to the Resource Curse, because it is simply not as dependent upon the resource. However, it is also possible to argue from the other end of the causal chain: Norway has managed to maintain vibrant economic alternatives, so that it need not rely so heavily on petroleum. In other words, Norway's lack of reliance can be seen as proof that it has avoided the Resource Curse.
10. Since 2014, Norway's unemployment level has risen, as international oil prices have fallen. As of March 2016, the Norwegian unemployment rate was estimated to be 4.7 percent of the labor force—uncomfortably high for most Norwegians, but still very low, compared to other developed economies.

Chapter 2

1. Belgium, France, Italy, Luxembourg, the Netherlands, and West Germany formed the European Coal and Steel Community (ECSC) in 1951. With the signing of the Treaty of Rome in 1958, the community became the European Economic Community (EEC). In 1993, the Maastricht Treaty transformed the EEC into the European Union (EU). Hence, we will refer to the EEC for developments prior to 1993, and the EU for more recent developments.
2. In June 1970, when voting on whether Norway should join the EEC, only 17 members of parliament were opposed (132 supported the motion). Only two small parties formally opposed membership: the Farmers' Party (*Senterpartiet*) and the Socialist People's Party. Of the two largest parties, the Conservative Party was united in support of EEC membership, while the Labor Party was split between party elites (in support) and a smaller grassroots group that opposed EEC membership. Still, the elite position ruled in the Labor Party.
3. For a good overview of their work, see chapter 5 in Toye and Toye (2004).
4. This hypothesis is often explained with reference to two different mechanisms. The first is Engel's Law, which holds that as income rises, the demand for manufactured goods rises faster than the demand for primary goods (Kindleberger 1943; 1989). The second explanation recognizes the institutional differences inherent to the two markets (e.g. cost-plus pricing and unionized labor). For example, the benefits of increased productivity are distributed differently in (manufactured) product markets in the developed world and primary goods markets operating in the developing world. In product markets, organized labor will bargain over the gains from any increase in productivity. Hence any increase in productivity (for whatever reason) will generate higher wages, and not necessarily a cheaper price for the product manufactured. In primary sector markets, however, labor is less able to secure its share of the productivity increase, so that an increase in productivity usually leads to falling prices (and wages).
5. The early work can be found in Folke Hilgerdt (1945), along with the work of Prebisch (1950) and Singer (1950). In a footnote, Michael Ross (1999, 303) provides a good review of the literature before the turn of the millennium. For more recent

work, see Bloch and Sapsford (2000), Harvey et al. (2010), Arezki et al. (2014), and Yamada and Yoon (2014).

6. See also Yergin (2009, 246–7), Taverne (2008, 54–6), and Al Jazeera (2013).
7. The moniker, "Seven Sisters" was coined by the head of the Italian state oil company, Enrico Mattei, but was made famous by Anthony Sampson's (1975) book, *The Seven Sisters*.
8. Pérez Alfonzo came to Washington in early 1959 to propose the creation of a western hemisphere oil system, run by governments (not the IOCs). Like an existing sugar cartel, the oil cartel would provide producing countries with a quota (in effect a guaranteed share of the US market). In doing so, IOCs would no longer be able to decide which producing countries would deliver the petroleum. See Yergin (2009, 495) and Rabe (1982, 159).
9. Article 102 of the Charter of the UN, the agreement creating OPEC, was registered with the UN Secretariat on November 6, 1962.
10. On April 29, 1958, as recorded in the Final Act (A/CONF.13/L.58, 1958, UNCLOS, Off. Rec. vol. 2, 146), the United Nations Conference on the Law of the Sea introduced four conventions and an optional protocol for signature: the Convention on the Territorial Sea and the Contiguous Zone; the Convention on the High Seas; the Convention on Fishing and Conservation of the Living Resources of the High Seas; the Convention on the Continental Shelf; and the Optional Protocol of Signature concerning the Compulsory Settlement of Disputes. See UN (1958).
11. A plurality agreement in the WTO gives member countries an option as to whether to agree to new rules (on a voluntary basis). It is not binding for all members.
12. In addition to the 28 member states of the EU, the signatories include Armenia, Canada, Chinese Taipei, Hong Kong, Iceland, Israel, Japan, Liechtenstein, the Netherlands, Norway, Singapore, South Korea, Switzerland, and the United States. See WTO (2015b).
13. The EFTA is currently composed of just four countries: Norway, Iceland, Liechtenstein, and Switzerland. The EEA does not include Switzerland (whose economic relationship to the EU is defined by another treaty).
14. As the number of directives being passed by Brussels (and quietly adopted by Norway) increased, and as the EU's regulatory hand began to reach deeper into Norway's political norms, the rank and file in Norway's Labor Party revolted against its leadership and demanded that the EU's third Postal Directive be vetoed. In 2011, for the very first time, Norway reclaimed the sovereignty it was said to have protected in opposing EU membership.
15. Article 18 TFEU (ex Article 12 TEC); Article 49 TFEU (ex Article 43 TEC); and Article 56 TFEU (ex Article 49 TEC), respectively.
16. Directive 2004/18/EC of the European Parliament and the Council of 31 March 2004 on the coordination of procedures for the award of public works contracts, public supply contracts and public service contracts (OJ L134, 30.4.2004), as amended, and Directive 2004/17/EC of the European Parliament and of the Council of 31 March 2004 coordinating the procurement procedures of entities operating in the water, energy, transport, and postal services sectors (OJ L134, 30.4.2004), as amended.

17. Article 2 of the Procurement Directive and Article 10 of the Utilities Procurement Directive.
18. Judgment in Case E-2/06 *EFTA Surveillance Authority* v. *The Kingdom of Norway:* Conditions for acquiring rights to waterfalls for energy production incompatible with the EEA Agreement.
19. In particular, public ownership was maintained by establishing that: (1) private owners cannot get concession to build or buy hydroelectric power stations, except very small hydroelectric stations; (2) concessions on waterfalls that are returned to the state cannot be sold back to the old, or new owners; and (3) private persons can still own up to one-third of a power station.

Chapter 3

1. Developmental nationalism is a concept used in the developmental state literature to describe the ideological disposition of elites in those states. For further elaboration on the role and nature of developmental nationalism in the developmental state literature, see Edigheji (2005 and 2013).
2. For an introduction to the most influential works on the Developmental State, see Evans (1995) and Senghaas (1985). For its application in Norway, see Gran (1994).
3. Today's developed economies grew with substantial state support, behind protectionist barriers. On reaching positions of power, these states came to embrace free trade, and they now press all other states (rich and poor) to do likewise. In essence, the developed world tries to "kick away the ladder" by which it climbed to its position of power; in so doing, it prevents developing countries from using similar policies and institutions to climb out of poverty (List 1885, 295–6; Chang 2005; see also Leftwich 2000).
4. See, for example, Amsden (1989); Deyo (1987); Johnson (1982); and Wade (1990).
5. As is often the case, this politician was stretching the point, to leverage popular support. Other states at the time had similar "systems."
6. In 1814, Denmark handed Norway over to Sweden, as a sort of war booty in the Treaty of Kiel. From 1814 to 1905, Norway was under Swedish rule (although it enjoyed a great deal of political latitude). Norway is a remarkably young country; it has only been independent since 1905.
7. Although this is partly a result of the fact that many other European constitutions were in liberal retreat in the wake of Napoleon's defeat.
8. For example, the use of directorates in Norway goes back to 1820, as a reaction against the dominance of lawyers in the country's bureaucracy: "It was stated then that where the state has rights to strategic resources—waterfalls, forests, minerals, ocean, agricultural land—it is also obligated to ensure that it possesses the expertise to exploit these resources" (Al-Kasim 2006a, ix–x).
9. For good English introductions, see the relevant sections on Norway in Esping-Andersen (1985) and Katzenstein (1985).
10. As Mjøset and Cappelen (2011, 173) point out, this is not unique to Norway. The Norwegian policy was influenced by the British Radcliffe Committee (1959).

11. In the postwar period, and prior to the Great Recession, Norway has suffered only three recessions of significance (1958, mid-1970s, and 1993) and the state's ability to use active countercyclical stimuli was restricted by external developments in each case. See Moses (2000).

Chapter 4

1. For example, Mehlum et al. (2006) show how the Resource Curse is mostly associated with oil states that have "bad institutions," while Andersen and Aslaksen (2008) find the Resource Curse in democratic presidential countries, but not in democratic parliamentary countries. See also Atkinson and Hamilton (2003); Brunnschweiler and Bulte (2008); and Jones Luong and Weinthal (2010).
2. For example, Eifert et al. (2003); Lahn et al. (2007); and NRGI (2014).
3. The underlying logic of this type of argument is one of checks and balances, which draws implicitly on the liberal constitutional tradition, inspired by Montesquieu and the Federalist Papers. In this tradition, government power is divided into branches (executive, legislative, judicial); when these powers are institutionally segregated (that is, each has specific duties/responsibilities, and each draws from an independent source of power), it minimizes the chances of a tyrant capturing power. More to the point, the powers invested in each of these branches can be set off against one another by a system of checks and balances—where one branch of the government is designed to limit, or check, the others. In this approach, states limit the potential for corruption and abuse of power by creating a constellation of institutions that check and balance one another.
4. Precept 5 of the 2009 *Natural Resource Charter* reads "National Resource companies should be competitive and commercial operations. They should avoid conducting regulatory functions or other activities" (NRC 2009, 3). The current version waters this down; its Precept 6 reads: "Nationally owned companies should be accountable, with well-defined mandates and an objective of commercial efficiency." See Thurber et al. (2011, 1–2).
5. A prominent example is the edited volume produced by the Stanford-based Program on Energy and Sustainable Development, which focuses on the nature of government–NOC interactions—albeit with a less clear delineation between regulatory and policymaking functions (see Victor et al. 2012).
6. Governments can employ different kinds of instruments to determine how easy (or difficult) it is to change the rules. Petroleum *laws* are difficult to change, since the legislature will need to become involved with every proposed change. Government *regulations* and *ministerial decrees* are more easily repealed and replaced by new regulations or decrees (although government regulations sometime require parliamentary approval). Hence, governments enjoy more freedom of movement when they use regulations or decrees (as opposed to laws) to secure their objectives.
7. At the most general level, governments decide whether to create NOCs for exploration, development, and production (e.g., Saudi Arabia, Mexico, Venezuela, and Iran); invite private firms to develop the resource (e.g., the United Sates, the United

Kingdom, Russia, and Canada); or decide to employ a combination of the two (e.g., Indonesia, Nigeria, Azerbaijan, and Kazakhstan).

8. This relationship between the NOC and its state is also tension-filled, as we shall see in the case of Statoil. As the NOC grows in power and influence, it can be seen as a "state within a state"—a potential threat to the authority and power of policymakers.
9. Most of the literature does not make a clear distinction between the policy and regulatory responsibilities, and many states combine these features in a single institution. The analytical focus of this literature, then, is on the degree of autonomy that NOCs enjoy from their political authorities (combined in one policy + regulatory body).
10. This relationship is similar to the notion of embedded autonomy, described in Chapter 3 (by Peter Evans (1995) and the developmental state literature): states need to develop the competence to make important decisions, but remain independent from the providers of that competence.
11. This is the classical view of the relationship between IOCs and states—one that emphasizes the initial bargaining power of the IOCs in a global economy. This classical view still dominates, but the nature of the IOC's relative bargaining power is changing rapidly. Large IOCs are increasingly forced into a role of providing procurement services for the government (rather than actually raising the capital, skills/technology, and market access that they traditionally provided). Although most IOCs still employ impressive geological and geophysical talents, and maintain the capacity and expertise to run large E&P sites—they increasingly act as middlemen, procuring the technology and skills from third parties on behalf of the government. After all, most IOCs no longer own their own drilling rigs—as this technology has become highly evolved and specialized, and this work is usually subcontracted out to specialty firms. At the same time, states have found it easier to finance their development objectives independently, while developing their own domestic expertise in NOCs. For a description of the strategic consequences of these developments, see Leis et al. (2012) and Bagheri and Di Minin (2015).
12. For a recent argument about why political leaders employ NOCs, see Mahdavi (2014).
13. There are also important legal benefits associated with NOCs. As states enter into agreements with IOCs (see Chapter 5), the state can be held liable for breeches of contract (and its assets can then be seized to cover compensation costs). For this reason, the government itself is usually not included as a direct commercial partner in commercial agreements, and the NOC becomes the government's intermediary. This is yet another attraction of the Tripartite Model.
14. In February 1958, the Norwegian Geological Survey had written a letter to the Ministry of Foreign Affairs, noting that: "The chances of finding coal, oil or sulphur on the continental shelf off the Norwegian coast can be discounted" (Gundersen 2007).
15. Quoted in *Aftenposten* (August 19, 1965), cited in Ryggvik (1992, 241).
16. In a Norwegian documentary, *Olje!* (NRK 2009 (I), 39:59), we learn that Nils Gulnes, Jens Evensen (both at the Norwegian Petroleum Council, or NPC), and

the Esso chief in Norway, Dick Loeffler, were informally discussing the consequences of the recent Cod Field discovery. Gulnes wondered if it wasn't reasonable for Norway to request better terms, now that it was known it had offshore oil. Rather surprisingly, Loeffler agreed.

17. By 1974–75, the government had introduced a new Petroleum Act that raised the tax rate on oil companies significantly (see Chapter 5). Many of the IOCs were irate over the change, as demonstrated by the fact that Phillips Petroleum filed a formal complaint with the Norwegian parliament:

    > As the oil companies were awarded production licenses in 1965, the Norwegian authorities were aware that the exploration, development and exploitation costs in the Norwegian part of the harsh weather exposed North Sea would be higher than in other places. Acceptable tax legislation was adopted to balance the considerable costs and risk which the operations in the North Sea entailed. Now, when we are ready to commence production on a large scale, and even before we have made any profit, we are faced with the proposal of significant changes in the tax rules. We had not expected such changes because we used Act No. 3 of 11 June 1965 on Taxation of Subsea Petroleum Resources as a basis, which led us to believe that we, in respect of taxation, would be treated in the same manner as other activities in Norway.
    >
    > (cited in, and translated by, Jansen and Bjerke n.d., 14)

18. Norway also introduced a sliding scale (in 1974), where state participation increases with the size of production. See Chapter 8.
19. This competitive context also helped to limit Statoil's domestic power: "the Norwegian licensing process, Statoil's original position in all projects and the presence of domestic (Hydro and Saga) and international competitors might have given Norway's authorities an informal edge in assessing its NOC's relatively narrow mission" (Boscheck 2007, 375).
20. See Hanisch and Nerheim (1992, 156), along with NOU (1972/15) and NOU (1974/40).
21. "If most finds in the Norwegian sector had been in blocks which were allocated in the first round of concessions, Norwegian oil policy would hardly appear as a model for other countries. Luckily for Norway, the majority of oil deposits in the Norwegian sector (Frigg, Statfjord, Gullfaks, Snorre, Troll, etc.) were to be found north of the area, which was advertised in the first round. The most important general lesson for countries in the same situation must be 'Don't give too much away in the first round. Take the time needed to set up a legal framework. And make sure that this framework is sufficiently flexible that the state can tighten the rules when the conditions change'" (Ryggvik 2010, 21).
22. The original recommendation referred to different institutions (e.g., the Ministry of Industry, the "Shelf Directorate," and a Norwegian holding company (Norwegian Petroleum, AS), as Norway's contemporary institutions did not yet exist.
23. Note that Al-Kasim describes the third leg of the triangle as "commercial" rather than "operational." We have chosen to focus on operational (rather than commercial)

competencies in recognition of the fact that a strong state may want to develop this expertise in an organization that has no commercial ambitions. Although this decision is unlikely, it is possible in theory.

24. Note that we have amended Figure 4.3 to include the labor market partners. The original organization chart comes from the NPD, which does not explicitly recognize the labor component to the Norwegian model.
25. For the previous (left) government organization, see MPE (2004).
26. See also Box 9.3.
27. This was the result of deliberation concerning *St. meld. nr. 17 (2002–2003)*.
28. There were several reasons, in addition to Labor Party opposition, why Norsk Hydro would have made an awkward NOC. At the time (autumn 1970), 52 percent of Hydro's shares were in private hands, and many of the shares were owned by foreigners. Worse, Hydro had contractual obligations as part of the Petronord Group, which made it difficult for it to act independently—as a state actor. Finally, Hydro had a tradition of reacting negatively to state influence.
29. This distinguishes it from many other state oil companies in OPEC, where there was no clear division between the state administration and the state's operative oil company (that is, no clear division of administrative responsibilities, a là Tripartite Model).
30. Actually, the Labor Party's position changed on this point. In the earliest years, the Labor Party, the Conservative Party, and the government's nascent administrative cadre in charge of oil all agreed on the desirability of an NOC that was to be a non-operative holding company. When the Labor Party took over from the Borten government in 1971, however, it changed its position. See Skule and Grytli (1997, 29).
31. In the end, Norway ended up with what Dag Harald Claes (2002, 12) refers to as "the fundamental credo that Norway should have three oil companies[:] one state owned (Statoil), a semi-state owned (Norsk Hydro) and one private (Saga Petroleum)." Saga was created in 1972, by assimilating a handful of smaller (private) Norwegian companies in the oil industry.
32. In 1972, without any knowledge of the industry at all, Statoil was even awarded a 50 percent partnership in the transmission company, Norpipe, which was established to transport oil from Ekofisk to Teesside in the United Kingdom and gas from Ekofisk to Emden in Germany (Austvik 2012, 325).
33. For a detailed description of these reforms, see Claes (2002, chapter 4).
34. There are many who are generally skeptical of adopting governance strategies that work in the developed world into countries that lack certain institutional endowments. "To the extent that this is true, reformers would be well-advised to carefully consider the attributes of a specific oil sector context before encouraging the use of a particular 'best practice' like the Norwegian Model" (Thurber et al. 2011, 2).

Chapter 5

1. At the most general level, there are three possibilities. First, the blocks might be opened up to *competitive bidding*, where companies compete with one another to offer the most attractive bid for the right to develop a given block. Second, the

government might accept *ad hoc negotiations*, where a company (e.g., an IOC) approaches the government, unsolicited, about securing a contract on a particular parcel of land. This might happen in places where the state had no prior expectation about petroleum reserves, and has yet to develop a plan for development. Finally, states might adopt *a first-come, first-served* approach, where the government lays out the necessary terms and requirements, and the first company to meet those terms wins the contract.

2. See Johnston (2009) for a critical review of the role played by the changing practice of reserve disclosure (and booking barrels).
3. In other words, the NOC is made a partner to the development of a block, but it does not pay any of the exploration costs associated with that block (it is just along for the ride). There is no guarantee that the exploration phase will produce commercially viable fields, hence there is much risk associated with this part of a field's development. Once a well is commercially viable, the government (usually the NOC) enjoys an option to "back in" to the partnership. From that point forward, then, the NOC will pay its fair share of the development and operation costs (like any other working interest partner) and enjoy its fair share of the partnership's rewards.
4. There is growing international pressure to force IOCs to publish the terms of their contracts, but progress has been remarkably slow. In 2015, Statoil disclosed the nature of economic transactions it had made with 18 different governments, under a Norwegian law that requires all large oil, gas, mining and logging companies registered or listed in Norway to publish the payments they make to all governments (NRGI 2015). See Chapter 10 for more details.
5. Iran employs a unique "buyback" scheme with shorter time horizons, greater government control over the resource, and more restricted profit rates. But this type of contract is not widespread. Unitization contracts are more common, but are somewhat different in nature, so we will not describe them in any detail here. These contracts are used to redistribute shares and work obligations among license holders in blocks that are placed over a common field.
6. These agreements are sometimes referred to as royalty or tax systems.
7. Traditionally, the royalty (e.g., 12.5 percent) payment was based on the tonnage of crude oil produced over a long period of time (e.g., 50–60 years). In their modern form, concessions tend to be for shorter periods (e.g., 3–5 years for exploration, 30–40 years for exploitation), and the government is paid with both royalties (e.g., 11–15 percent) and taxes (e.g., net income).
8. Al-Emadi (2010) distinguishes between contractual joint ventures (CJVs); joint venture corporations (JVCs); and joint venture partnerships (JVPs). It is more common to distinguish between incorporated joint ventures (e.g., JVCs) and unincorporated joint ventures (e.g., CJVs and JVPs).
9. Sometimes referred to as production sharing agreements (PSAs).
10. Total Oil = Cost Oil + Profit Oil. Negotiations between the state and the IOC determine what costs can be reimbursed, and when. Once the costs of E&P are covered, the remaining (profit) oil is split between the IOC and the government/NOC. This split (e.g., 65 percent goes to the state; 35 percent goes to the IOC) is also the result of negotiations.

11. Johnston (2007, 58) argues that service agreements are most common in Latin America and the Middle East, while PSCs are most common in Africa and the former Soviet Union. See Johnston (2007), Al-Emadi (2010) and Bagheri and Di Minin (2015) for useful overviews.
12. In 1943, Venezuela amended its concessionary agreements by introducing a 50/50 "profit sharing formula," which lasted until 1948. The government share was subsequently increased to 65/35 and 70/30 (in 1958). Although there was much resistance from the IOCs, other oil producing countries (e.g., in the Middle East) took notice and followed suit (see Rabe 1982).
13. See Leis et al. (2012), who divide oilfield service companies (OSCs) into two types: surface OSCs provide facilities-related services; while subsurface OSCs provide reservoir-related services (among these are the "Big Four": Schlumberger, Halliburton, Baker Hughes, and Weatherford).
14. See Johnston (2007) for a description of the different means for estimating the government take, and Johnston (2008, 34) for an innovative graphic depiction of variations in government take, across countries. For an alternative metric, see API (2012).
15. R stands for the ratio of cumulative revenues/cumulative costs. See the Glossary for additional clarification.
16. See Baunsgaard (2004, 85) for a comparative snapshot of the different fiscal petroleum regimes used in several developing countries.
17. The legal complexities are evident in the fact that it took some 40 years for Norway and Russia to agree on the maritime boundary that separates these two countries in the Barents Sea and Arctic Ocean. The foreign ministers of these two countries signed a treaty establishing that border on September 15, 2010.
18. This check was introduced with the twentieth concession round.
19. In the early concession rounds, the criteria were somewhat different. Then, the Ministry of Industry was advised by the NPC, and considered: (1) the applicant's financial strength and operational experience in oil exploration; (2) whether the applicant had contributed (or would contribute) to strengthening the Norwegian economy; and (3) the extent to which Norwegian interests were participating in the application group (Al-Kasim 2006a, 19).
20. These contracts are not disclosed; only the standard licensing agreement is published (see Norwegian Government 2014a). But information on the license groups and their operators is made public at NPD (2016a).
21. Like the concession agreements for waterfalls, the duration of the production licenses is limited (in principle to 46 years), and the principle of escheat is evident in various rules. Extraction permits are initially given for a period of just six years. After the first three years, firms agree to divest themselves of a quarter of the allocated block. If they want to retain a block beyond his initial six-year period, they have to give up another quarter of their original allocation after the second three-year period. Those parts of a block that are retained after a concession extension, however, can be held for another 40 years (Ryggvik 2010, 17). On the other hand, after nine years, the state is able to recover half of the area originally granted (Al-Kasim 2006a, 16).

22. Esso, Shell, Gulf, and Syracuse were granted licenses on their own; the remaining five licenses were allocated to joint ventures, consisting of two to nine different (including Norwegian) companies. Norwegian interests were represented by the Norwegian Oil Consortium (NOCO) and Norsk Hydro, with Hydro being the biggest beneficiary. It was participating in the Petronord group (which included Elf, Hydro, Cofranord, Eurafrep, Coparex, Total, and Aquitaine), and was allocated ownership shares in nine blocks. Apparently, the IOCs were surprised by how little Norwegian companies seemed to benefit from the deal (Johannesen et al. 2005, 268–9).
23. Both the carrying clause and the sliding scale were dropped in subsequent concession rounds (eleventh and twelfth, respectively) and later by parliamentary decree (in 1991 and 1994, respectively).
24. In Norwegian, TFO, for "*Tildeling i Forhåndsdefinerte Områder.*" This area includes large parts of the North Sea, Holtenbanken (Norwegian Sea), and areas around the Snow White Field (Barents Sea).
25. See Norsk Olje&gas (2010) for a clear and brief English introduction to the Norwegian licensing system.
26. For example, if capital expenditure is $100 million, the company can recover $130 million. In 2013, in a much debated and controversial reform, the government reduced the total uplift allowance from 30 to 22 percent. See Ministry of Finance (2013).
27. For current lists of norm prices, and an overview of the system, see Norwegian Government (2014b).
28. A current list of licenses (and the SDFI's share) can be found at Petoro (2016).
29. In 2013, discrepancies were identified in the reports from 42 percent of the Norwegian entities! See Deloitte (2014, 6).
30. See EITI (n.d.). Similar efforts have been launched by George Soros's Open Society Foundation (2016); the Publish What You Pay Initiative (PWYP 2016); the IMF's Good Practices in Fiscal Transparency (IMF 2007); and Open Oil (2015).

Chapter 6

1. For example: "Since Dutch Disease involves an inter-temporally sub-optimal over-expansion of the resource extraction sector with a corresponding sub-optimal contraction of the manufacturing sector, any management targeted to avoid the disease must limit the transmission mechanisms" (Røed Larsen 2005, 81).
2. The impetus for the inflation determines whether this is a case of Dutch Disease or what is called a Balassa-Samuelson Effect (BSE). In a BSE approach, rapid economic growth results from higher productivity growth in the tradables sector, relative to the non-tradables sector. The larger the productivity growth rate difference between these two sectors, the faster the real exchange rate appreciation of the economy. With Dutch Disease, strong foreign demand generates higher export prices, which moves the domestic prices of tradables upward. Subsequent rapid economic growth, which produces a consumption boom, extends price hikes to non-tradables.

3. As the value of a resource left in the ground increases along with its market price, we can expect owners to extract less of that resource as the price goes up (that is, they would prefer to let it grow in value as a stored resource while investing their capital elsewhere). A higher price will lessen demand, and—when demand falls—the price of the resource will eventually drop. Once the price falls below the prevailing rate of interest, owners will want to invest in increased production (as they will only lose money by storing it for the future).
4. It is interesting to note that Hotelling was writing at a time of growing conservationism and environmentalism in the United States, and he was responding to the growing political pressure to minimize overexploitation of non-renewable natural resources. The opening paragraph of his article addresses this point directly. As pressure for regulation declined, however, Hotelling's Rule fell out of fashion until Robert Solow (1974) reintroduced it again, 40 years later (in the wake of OPEC's formation). Both Solow's and Hotelling's efforts can be seen as highly ideological attempts to prioritize market mechanisms for regulating petroleum and other natural resource industries, in the face of increasing government intervention. See Mitchell (2011, 195–6).
5. Here the oil is sold, the money enters the government coffers, and then it goes out again in the form of foreign investment—this is different from an investment fund, where the monies are kept outside the host country, for fear of their inflationary consequences.
6. Although the Resource Curse literature always attributes this quote to Jean Bodin, it would seem that few people have actually read it. In the original text (as provided below), Bodin is clearly paraphrasing Livy (Titus Linius):

   > And therefore Titus Linius said, That men of a fat and fertile soil are most commonly effeminate and cowards; whereas contrariwise a barren country makes men temperate by necessity and by consequence careful, vigilant and industrious. (Bodin 1962 [1576], V, I, 565)

7. Several earlier studies found slightly more support. Bye et al. (1994) found Norway experienced some symptoms in the 1970s and 1980s, while restructuring its economy, but that its economy had regained a well-functioning tradables sector. Hutchinson (1994), on the other hand, found indicators of adverse effects from the oil boom on the manufacturing sector. A couple of years later, Brunstad and Dyrstad (1997) found weak manufacturing performance and a rise in the cost of living in oil-near sectors, as a result of Norway's petroleum sector. Bjørnland (1998), by contrast, found that manufacturing output in Norway seemed to have benefited from higher oil revenues. Cappelen et al. (2000) found the traded goods sector in Norway contracted owing to oil, but the manufacturing sector did not.
8. Originally, the pace was established by setting a ceiling on production, but it turned out that production was proceeding at a much slower pace than expected, so the ceiling eventually turned into a goal. Then, in the 1980s, the authorities changed tack and aimed at influencing the rate of investment, rather than the rate of production, in the effort to affect the pace. It is now recognized that the most effective way to control the pace of development is to control the pace at which licenses/blocks are released.

9. In addition to "*Tempoutvalget*", this committee is sometimes referred to as "Committee for the future of the petroleum industry" (*Utvalget for petroleumsvirksomhetens framtid*).
10. Named after the then Finance Minister, Per Kleppe, the Kleppe-packages refer to the government's intervention to bring about successful income policies in the 1970s (in order to avoid conflict, and to stave off inflation).

Chapter 7

1. There is another, common, type of fund employed by developing nations with natural resources. In these funds, money is stored away for prioritized spending areas, such as social and physical infrastructure. This type of fund works like a parallel budget—money is diverted from the national budget and sent to the fund, and the money is earmarked for, for example, infrastructural development. In effect, this fund is little more than a fancy budget post.
2. The earliest forewarnings are found in the so-called "Oljemelding" (*St. meld. nr. 25 (1973–74)*). Then Finance Minister, Per Kleppe argued that the money needed to be invested abroad, and not used to increase wages, or purchase material goods. In particular, he hoped the money could be used to build a more egalitarian, greener, more rural-friendly, and gender-neutral country (Hanisch 1999, 24).
3. One of the committee members, Tora Haug, dissented further by advising against the construction of a larger foreign investment fund, rather than spending the money in Norway, where it belonged (NOU 1983/27, 100).
4. Among these are the former Prime Minister, Jens Stoltenberg, who wrote his master's thesis on the subject, and one of his more popular finance ministers, Sigbjørn Johnsen. See, for example, Reinertsen (2009).
5. For some time, the populist Progress Party was very critical of the rule, advocating the need to spend more money in Norway, on Norwegians. But when that party entered its first coalition government (2013), and was rewarded with the Ministry of Finance post, it quickly changed its tune and delivered a budget that respected the rule. Indeed, the opening sentence of Chapter 3 (On "Economic Policy") in its first independent budget reads: "The government places the budgetary rule as the foundation for its fiscal policy" (*Regjeringen legger handlingsregelen til grunn for budsjettpolitikken*) (*Meld. St. 2 (2014–2015)*, 33).
6. In March of 2016, the GPFG has 59.8 percent of its investments in equities, 37 percent in fixed income investments, and 3.1 percent invested in real estate.

Chapter 8

1. This is paraphrasing the Nobel laureate Joseph Stiglitz (2003), who advises developing countries: "Don't do as the US tells you, do as the US did." See also Reinert (2007).
2. Paragraphs 1(a) and 1(b) on the Illustrative List. See WTO (2016b).
3. In the Global Procurement Agreement (GPA), the type of institutions covered is determined by each member country, and can include state-owned entities and

public utilities in the energy sector. For example, South Korea includes the Korea National Oil Corporation and Taiwan includes the Chinese Petroleum Corporation of Taiwan. This can help to explain why Brazil and other petroleum countries have been reluctant to join the GPA, as there is no guarantee that a country will be able to negotiate an exception to the rule, and this country could find itself prohibited from requiring minimum levels of domestic content in its award of contracts (Warner 2011, 34–5).

4. See Chapter 2 of Tordo et al. (2013), for a more detailed listing of LCP pros and cons.
5. To see how individual states rank in terms of their level of business-friendliness, see the World Bank's "Doing Business" database (World Bank 2016).
6. The first (semi-submersible platform) rig to drill on the Norwegian Continental Shelf (NCS), the Ocean Traveler, was designed and built to the latest technical standards (in the Gulf of Mexico) and then towed across the Atlantic. Once in place, a supply ship soon knocked into one of its supporting columns while being tossed around at sea. The crew had to be evacuated, the rig was repaired, and it became immediately evident that something more substantial was needed for North Sea conditions. In 1965, Odeco farmed out a second (twin) platform, Ocean Viking, this time to Norwegian contractors—and these plans were soon delayed, so that necessary changes could be made. Still, Odeco was not willing to trust the locals; it delivered the drawings, and monitored the process with its own (American) experts.
7. In the government white paper announcing the first concession round, the Ministry of Industry had mentioned only two criteria for reward: the applicant's financial strength and practical experience with oil exploration. But when the Minister of Industry, Karl Trasti, announced to the press the results of the first round, he said that they "had emphasized the applicant's financial strength and practical experience with oil exploration. But we have also considered the degree to which the applicant has considered marketing in Norway, building refineries in Norway, using Norwegian ships or other ways in which the applicant has or will contribute to strengthening Norway's economy in general" (Kvendseth 1988, 16).
8. While it is embarrassing to note, the first Norwegian platform workers were referred to as "white niggers" because of their low pay and status, and horrible working conditions (Smith-Solbakken 1997, 86 and 111).
9. "The Ministry considers it a significant advantage that Norwegian interests can increase their participation in the further exploration and extraction of petroleum on the Norwegian continental shelf... It is the Ministry of Industry's view that Norwegian participation should continue to be taken into account when granting licenses, where such participation is seen as a deliberate policy of building up a Norwegian integrated oil environment. This type of [domestic] oil environment must be given the widest possible targets, both nationally and internationally... That is why the Ministry believes that the state should contribute to the coordination and concentration of the Norwegian interests within the Norwegian petroleum industry" (*St. meld. nr. 76 (1970–71)*, 20).
10. Translation from Al-Kasim (2006a, 60). See also Søbye (2001, 71).

11. Many of these licensing agreements encouraged IOCs to train Norwegian personnel, whether they worked in Statoil, the Ministry of Industry, or the NPD. Locals were invited to in-house workshops and courses, alongside IOC employees, or special courses were designed specifically for government officials.
12. In this way, numerous Statoil employees were trained on the Frigg field. The best example is perhaps the agreement between Statoil and Mobil, who were paired as partners in the Statfjord field. In exchange for a 15 percent ownership share in the licensing group, Mobil promised to let Statoil take over the operator role after ten years. In the meantime, the two companies produced duplicate (Statoil/Mobil) leadership teams. In allocating the exploration license for the Varg Field in 1974, Statoil was chosen as operator, but Esso became the "technical operator"—allowing Statoil to build up its exploration competence.
13. But the government's influence was also used to secure other objectives, including increasing oil recovery objectives. A good example is provided by the Troll-Oseberg Gas Injection (TOGI) project. Because the MPE must approve field development plans, the government was able to instruct Statoil to vote against its own position, for the benefit of the country. The Troll field consisted mostly of gas, and was controlled by Statoil. The neighboring Oseberg field was mostly oil, and controlled by Norsk Hydro. The government wanted to pump gas from the Troll field, into the Oseberg field, to flush out more oil. This would obviously benefit Hydro over Statoil, as Statoil had to suboptimize their field development plan, but Statoil was forced to play ball.
14. For example, in 1975, Statoil initiated an effort to bring together the Norwegian engineering community, into a single, large Norwegian engineer company: Norwegian Petroleum Consultants (NPC). As in the rest of the oil economy—where IOCs were encouraged to help develop Norwegian operator and exploratory competences—NPC got a strong helping hand. Statoil pressed Mobil (as Statoil controlled the majority of votes in their joint venture) to use NPC in developing Statfjord B. Mobil agreed, but wanted to ensure sufficient rigor, so they teamed up NPC with the American company, Brown and Root, where the latter took a leading role, and trained up the Norwegian junior partner (Nerheim 1994).
15. See Kemp (1987, 39ff) and NOU (1983/16, 42–4).
16. In particular, this report emphasized five specific R&D areas: social impact and management problems; deep-water technology; petroleum technology; petroleum economics; and data on sea and oceanic conditions (*St. meld. nr. 53 (1979–80)*, 67).
17. The results were impressive. It is estimated that the goodwill agreements on their own generated 5.39 billion kroner before 1990, and 1500 diverse projects were initiated. The Institutt for Kontinentalsokkelsundersøkelser; SINTEF; Rogalandsforskning; and Christian Michelsens Institute received over a 100 million kroner in the 1980s (Skule and Grytli 1997, 32).
18. Where production, drilling, and living quarters are all located on the same structure.
19. Nerheim (1996), referenced in Skule and Grytli (1997, 43).
20. There were other tax-related changes too. On January 1, 1987, several regulations concerning depreciation were changed so that firms could write off their costs from

the first year of investment (rather than from the start of production). Then, in 1988, the authorities dropped the production tax on new fields (Skule and Grytli 1997, 56).

Chapter 9

1. This, despite the fact that the number of spills, on average, has declined since the 1970s. Between 1970 and 1979, there were 24.5 spills per year on average. In 2000–2009, that average had dropped down to just 3.4 spills per year (ITOPF 2014).
2. In addition, some gas is combusted through flaring (for safety reasons), and there can be evaporation of light oil components in connection with storage and loading of crude oil.
3. Porter and van der Linde (1995, 128) provide six reasons why environmental regulations are necessary: (1) to create pressures that push companies to innovate; (2) to introduce environmental improvements; (3) to educate companies about better opportunities for research/technology use; (4) to improve the environmentally friendliness of products and processes; (5) to stimulate demand for environmental improvements; and (6) to level the playing field.
4. There are many acronyms used to describe this subject matter. The most common is probably OHS (Occupational Health and Safety). Although environmental issues often fall under the rubric of OHS regulations, we wanted to focus attention on the commonalities of OHS and environmental regulatory regimes, hence HSE.
5. As we have limited space to develop these arguments, we are describing this change as a simple transition from an old regime to a new regime. The evolution of modern regulatory regimes is more complex than that. For example, Hale and Hovden (1998) suggest that HSE regimes have developed and evolved through three "ages of safety": (1) a technical age; (2) a human factors age; and (3) a management systems age. Hudson (2007) depicted a similar development, but focuses on three waves: (1) a technical wave; (2) a systems wave; and (3) a culture wave. See also Saksvik and Quinlan (2003).
6. A certified OSH MS has become an essential business-to-business requirement. According to the British Standards Institute (BSI), more than 56,200 companies worldwide employ certified OSH management systems (Zwetsloot 2013).
7. In 1995, the ILO launched a campaign to achieve universal ratification of these eight conventions. There are currently over 1,200 ratifications, representing 86 percent of the possible number of ratifications. For a country-by-county overview, see ILO (n.d.).
8. There are two main types of stabilization clauses (and hybrid forms). First, "freezing clauses" freeze the applicable host country law at the time of the original contract. Under these clauses, legislation adopted after the date of the contract does not apply to the foreign investors or the project unless the investors agree. Likewise, "consistency clauses" protect the investment contract, by requiring that host country legislation applies to the project, only if it is consistent with the investment contract (that is, the contract is prioritized over the law). Second, "economic equilibrium clauses" tie changes in the terms of a contract to a renegotiation of the

contract to restore its original economic equilibrium (or to pay compensation). In short, the latter clause stabilizes the economic equilibrium; the former stabilizes the regulatory framework itself.

9. Royal Decree of 1967 on safety in exploring and drilling for subsea petroleum deposits.
10. The fourth commandment is most explicit as it refers to the need for the industry to develop in a way that doesn't challenge existing livelihoods and the natural environment. But environmental concerns can be seen in the fifth and ninth commandments as well, as they concern flaring issues and the more challenging working environment north of the 62nd parallel.
11. The IOCs continued to argue for as little regulation as possible, while Norwegian shipowners advocated for a system similar to the ones they used onboard their ships (and which the British authorities adopted): a certification system, where a private company, Det Norske Veritas (DNV), would conduct inspections.
12. For those who can read Norwegian, wonderfully detailed accounts are found in Smith-Solbakken (1997) and Ryggvik and Smith-Solbakken (1997).
13. These data are problematic, as they include strikes in both the mining and petroleum sector, and it is clear from several accounts that much of the early strike activity is not being captured in the statistics. Consider Ryggvik and Smith-Solbakken's (1997, 473, note 429) caveat: "We have not found any systematic approach to what kind of strikes are included in the SSB's numbers. For some strikes, the employer may have wanted to keep quiet about events. Most conflicts were discussed in detail in several of the country's newspapers. We therefore assume that fuzzy categorization and poor reporting practices are the main cause of the deficient numbers."
14. In the summer of 1974, the Petroleum Directorate's Head of Inspection [*kontrollsjefen*], Arne Flikke, resigned in protest over the lack of resources at his disposal. During the hectic "Ekofisk" period, the authorities had only three to four active technical inspectors (Hanisch and Nerheim 1992, 320).
15. Recall from Chapter 6 that the government had imposed a wage freeze in 1978, in an attempt to limit the spread of inflation. One of the main targets of this wage freeze was the booming oil sector. Because of the strong bonds of cooperation between the Norwegian Labor Party and the LO, the LO was perceived to be more interested in carrying the government's policy than representing its workers' interests. OFS, as a non-affiliated labor federation, became increasingly attractive, and for some time there was an active battle over membership between the OFS and the LO's NOPEF.
16. NOAF was the predecessor to today's Oil Industry's National Association (*Oljeindustriens Landsforening*, or OLF).
17. This movement for increased worker participation was a product of the Industrial Democracy Program, inspired by ideas from the Tavistock Institute in London, which was launched in the late 1960s. NAF, the LO, and the government supported the idea, on the expectation that industrial democracy and worker participation would lead to increased productivity and an improved work environment—benefiting both workers and employers (Rosness and Forseth 2014, 313).

18. The Ministry of Industry and the IOCs lobbied to keep the WEA from being applied offshore, but the strikes and unrest noted above—in the wake of many serious accidents—created the political leverage needed to apply the WEA to fixed installations on the NCS. Mobile installations (and diving) were initially exempted, but fell under the law in 1993.
19. For more detailed descriptions, see Vogel (1998), Karlsen and Lindøe (2006), and Lindøe, Karlsen and Lie (2001).
20. The Framework Regulation of August 23, 2001.
21. This responsibility was transferred to the PSA(N) from the NPD on January 1, 2004.
22. In particular, the PSA(N), Klif (now NEA), the NDH and the NFSA.
23. Nor is "internal control" a Norwegian invention. It was first employed in the US car and aviation industry. But Norwegian regulations go a step further by making it a formal part of the state safety regulatory regime (not just a matter of industry's internal organizing of activities). Even in Norway, the concept of internal control was in play before it was used in the petroleum sector; it was applied in Norway's nuclear power sector as early as the 1950s (Bang and Thuestad 2014, 245).
24. In the white paper, responsibility for this culture lay with the firm's leadership—who needed to set an example or tone for the dominant HSE-culture in an organization. The labor movement was upset by this, and demanded more influence (Andersen et al. 2009, 17). The result was an expansion of tripartite bodies, as described in Box 9.4.
25. In particular, firms are required to enter a work permit system that includes a safe job analysis. For details, see IOM^3 (2015).
26. "The licensee must have an organisation able to lead operations on an independent basis from Norway. To achieve this, the ministry can set specific requirements for organization and company capitalisation" (§48).
27. See the NPD's PDO guidelines (NPD 2010), for more details.
28. Still, it is worth commenting that the success of the Norwegian regulatory regime has influenced other regimes, in both the United Kingdom (Piper Alpha) and the United States (Horizon).

Chapter 10

1. "Suppose that a friend when in his right mind has deposited arms with me and he asks for them when he is not in his right mind, ought I to give them back to him? No one would say that I ought or that I should be right in doing so, any more than they would say that I ought always to speak the truth to one who is in his condition" (*The Republic*, Book I).
2. From the perspective of the individual we can assess the ethical behavior of consumers who rely on petroleum products. An ethical consumer who is concerned about global warming and development might abstain entirely from petroleum products, as it is very difficult to know with certainty the source and political lineage of the fuel in the tank. The ethical consumer does not need to interfere with the larger market for petroleum, but if enough consumers withdraw, the market will collapse on its own. While individual consumers may be dedicated

and informed enough to signal their preferences, we doubt their capacity to function as informed regulators.

3. The list is long, but alternatives include the environmental risks associated with petroleum extraction and delivery; the size of the profit margins enjoyed by the industry (and how they are secured); and the relationship between the trade in petroleum and the trade in violence.
4. If we are to be ethically consistent, we then need to find a way to transfer the increased carbon emissions from these poorer countries in exchange for a reduction in carbon emissions in wealthier countries.
5. Most studies find that oil (and gas) deters democratization. Spearheaded by Ross (2001), a host of studies finds oil wealth generating lower levels of democracy (e.g., Jensen and Wantchekon 2004; Smith 2004; Ulfelder 2007; Basedau and Lay 2009; Tsui 2011; and Ross 2012). Still, there are those that continue to challenge these findings, for example Herb (2005); Dunning (2008); and Haber and Menaldo (2011).
6. For examples of the literature linking petroleum to corruption, see Leite and Wiedmann (2002); Sala-i-Martin and Subramanian (2003); Isham et al. (2005); Hayman (2009); Kolstad and Søreide (2009); Bhattacharyya and Hodler (2010); Arezki and Brückner (2011); and Yates (2012).
7. On the link between petroleum and human rights abuses, see Letnes (2002; 2008); Watts (2005); IPIECA (2008); and DeMeritt and Young (2013).
8. See the original source for a list of references associated with each "interpretation."
9. At the eye of the storm lies an assumption about economic rationality. From the mainstream economic tradition, that stretches back to Adam Smith, there should be no need for ethical intervention. The market will allocate justly and efficiently, as if it were guided by an invisible hand. In this tradition, the relationship between shareholders and management in a firm can be modeled as a principal-agent problem; any decision that does not directly serve the interests of the shareholders is seen to be a shirking of the manager's fiduciary responsibilities, that is, it is a sort of theft to divert money from shareholders' profits to stakeholder projects. The archetypical example is Friedman (1970).
10. Investors in the secondary and tertiary sectors are more concerned with the wellbeing of the host-country residents (in general) or human rights abuses (in particular) than are investors in the primary sector. Petroleum investors, by contrast, are most concerned with maintaining good relationships with the host government (which controls their concession/license). See Spar (1999); and Letnes (2002; 2008).
11. Many states are tightening their legislation on corruption, partly as a result of the UN Convention against Corruption, the Council of Europe's Criminal Law Convention on Corruption, and the OECD's Anti-Bribery Convention. In the United States, the American Petroleum Institute (API) has lobbied the Securities and Exchange Commission to change relevant portions of the Dodd-Frank Act, and the eventual outcome remains uncertain at the time of this writing. The EU Directive requires that member states and EEA countries implement appropriate legislation no later than for the finance year beginning January 1, 2015, with reports being submitted in 2017.

12. It is unfortunate that we don't have more time to study the rise of the modern human rights campaign, as it started immediately after WWII and was heavily influenced by the Cold War (and its subsequent victor). There remain significant differences in interpretation, as evident in dueling UN Conventions—one on Civil and Political Rights; the other on Economic, Social and Cultural Rights (the former has not been ratified by China, the latter has not been ratified by the United States). There has always been (and continues to be) some discussion about whether it is appropriate to establish universal ethical standards that are imposed on unwilling states, as it clashes with the Westphalian principle of respect for the internal sovereignty of fellow states. This movement away from national notions of (ethical) sovereignty to universal (and foreign) principles is parallel to the movement we see with regard to social and environmental regulation of the petroleum sector, where domestic conceptions of ethical behavior (whether just or not) are trumped by international standards that are written and imported from abroad.
13. This is not because IOCs prefer corrupt business environments. The industry has some legitimate concerns about disclosure requirements at the project (if not the country) level, and the requirements of operators, as opposed to secondary investors. For example, when joint ventures prevail, observers are likely to think that payments reported by one party were made by other partners on the project as well.
14. "Previous research measured the transparency of individual countries according to quantitative indicators such as macro-economic forecasts, the publication of International Monetary Fund reports on the macro-economic performance of countries and the quality of government budget documentation. All of these studies imply that the quality of decision-making on spending is crucial, in terms of complying with international norms and accounting standards, publication and independent verification of government budgets and the actual outcomes of decision-making. Not a single study quoted in footnotes focused specifically on the transparency of revenues; indeed, there appears to be an assumption among researchers that transparency of revenues is a secondary concern. There is no scientific basis for the assertion that revenue transparency leads to better social or economic outcomes" (Frynas 2009a, 194).
15. Compared to other OECD countries, the Norwegian record is impressive, but it pales in comparison to the generosity of another oil state: Kuwait. The Kuwait Fund for economic development provided development assistance that represented 9.5 percent of the country's Gross National Income in 2013 (or 18,555/188,000 million USD)! These figures come from World Bank (2015) and Kuwait Fund (n.d.).
16. See Box 7.1 for a list of relevant links, to gain further information.
17. Equal treatment of shareholders; shareholder influence and board accountability; well-functioning, legitimate, and efficient markets; children's rights; climate change risk management; and water management.
18. Founded in 2013, the board consists of three international experts in the fields of corporate governance and board leadership: Professor John Kay, Tony Watson, and Peter Montagnon.
19. The Penal Code includes three sections that are particularly important in the fight against corruption. These are section 276a, on corruption; section 276b, on gross

corruption; and section 276c, on trading in influence. Those convicted of corruption face up to three years' imprisonment, while the penalty for gross corruption is imprisonment for up to ten years. Aiding and abetting carries the same penalty.

20. "Report on Payments to Governments" (*lov om rapportering om betalinger til myndigheter m.v.*), warranted under the Norwegian Accounting Act (*Regnskapsloven*) and the Norwegian Securities Act (*Verdipapirhandelloven*).
21. Although the country's current Finance Minister, Siv Jensen, publicly doubts whether global warming is manmade, her government continues to support global efforts for emission reductions (Fox News 2015). The dominant parties in Norwegian politics, Labor and the Conservatives, agree about the desirability of extracting Norway's petroleum resources as quickly as possible (in as safe a means as possible).
22. In 2011, the petroleum sector emitted 13.6 million tons CO_2 equivalents or about 29 percent of Norway's total emissions (Ministry of Climate and Environment 2014, 6).
23. Another example is the increased political pressure to power offshore oil installations with "clean" electricity generated by Norwegian hydroelectric installations on shore (and very long extension cords!), rather than generating the energy offshore by burning petroleum.
24. Statoil has stored 25 percent of the 80 million tons of CO_2 captured and stored worldwide (Statoil 2015b).
25. Using a rough average (2014) exchange rate (6.7 NOK/USD), this corresponds to US $3,425 million, $1,502 million, and $773 million, respectively.
26. The NRGI's analysis of the leaked contract holds that it was "not out of line with international standards for a country that had no proven offshore reserves of natural gas at the time when the original contract was signed" (Manley and Lassourd 2014, 1).

References

Acher, John. 2006. "Norway Dumps Wal-Mart from $240 Billion Investment Fund." June 6. Reuters/*USA Today*.

Al-Emadi, Tatal. 2010. "Joint Venture Contracts (JVCs) among Current Negotiated Petroleum Contracts: A Literature Review of JVCs Development, Concept and Elements." *Georgetown Journal of International Law: The Summit* 1: 645–67.

Al Jazeera. 2013. "The Secret of the Seven Sisters." Four-part documentary. Posted April 23. Accessed February 20, 2015. http://www.aljazeera.com/programmes/specialseries/2013/04/201344105231487582.html.

Al-Kasim, Farouk. 2006a. *Managing Petroleum Resources. The "Norwegian Model" in a Broad Perspective*. OIES 30. Oxford: Oxford Institute for Energy Studies.

Al-Kasim, Farouk. 2006b. "The Relevance of the Norwegian Model to Developing Countries." NORAD seminar: Olje for Utvikling. Petroteam AS presentation. Uploaded by Trond Kubberud. Last modified January 26, 2006. Accessed March 31, 2015. http://siteresources.worldbank.org/EXTNTFPSI/Resources/606764-1150299531473/FaroukAl-Kasim.pdf.

Al-Kasim, Farouk. 2015. *A Guide to Managing Petroleum Resources*. Stavanger: Petrad.

Al-Kasim, Farouk, Tina Søreide, and Aled Williams. 2008. "Grand Corruption in the Regulation of Oil." *U4* Issue 2:2008. U4 Anti-Corruption Resource Center working paper series. Bergen: Chr. Michelsens Institute. Accessed December 20, 2016. https://www.cmi.no/publications/file/3034-grand-corruption-in-the-regulation-of-oil.pdf.

Alsop, Ronald. 2005. "Recruiters Seek MBAs Trained in Responsibility." *Wall Street Journal*. December 13.

Amnesty International. 2005. "Contracting Out of Human Rights: The Chad–Cameroon pipeline project." London: Amnesty International, UK. Accessed December 11, 2016. http://www.amnesty.org.uk/sites/default/files/pol340122005en.pdf.

Amnesty International. 2006. *Human Rights, Trade and Investment Matters*. Amnesty International. London: Amnesty International, UK. Accessed June 22, 2015. https://www.amnestyusa.org/sites/default/files/pdfs/hrtradeinvestmentmatters.pdf.

Amsden, Alice H. 1989. *Asia's Next Giant: South Korea and Late Industrialization*. New York: Oxford University Press.

Andersen, Jørgen Juel, and Silje Aslaksen. 2008. "Constitutions and the resource curse." *Journal of Development Economics* 87(2): 227–46.

Andersen, Rolf K., Mona Bråten, Brita Gjerstad, and Jorunn Tharaldsen. 2009. "Systematisk HMS-arbeid i norske virksomheter. Status og utfordringer 2009." FAFO- Report 51. Oslo: FAFO.

Andersen, Svein S. 1993. *The Struggle over North Sea Oil and Gas: Government Strategies in Denmark, Britain and Norway*. Oslo: Scandinavian University Press.

Anderson, Perry. 1979. *Lineages of the Absolutist State*. London: Verso.

Anderson, Richard G. 2009. "Resolving a Banking Crisis, the Nordic Way." *Economic Synopsis*. No. 10. St. Louis: Federal Reserve Bank of St. Louis. Accessed December 11, 2016. https://research.stlouisfed.org/publications/es/09/ES0910.pdf.

API [American Petroleum Institute]. 2012. "Composite Index: Global Rating and Ranking Fiscal Systems for Oil and Natural Gas Upstream Investment." Press release. American Petroleum Institute. Accessed June 21, 2016. http://www.api.org/~/media/files/news/2012/12-march/composite-index-onepager.pdf.

Arezki, Rabah, and Markus Brückner. 2011. "Oil Rents, Corruption, and State Stability: Evidence from Panel Data Regressions." *European Economic Review* 55(7): 955–63.

Arezki, Rabah, Kaddour Hadri, Prakash Loungani, and Yao Rao. 2014. "Testing the Prebisch–Singer Hypothesis since 1650: Evidence from Panel Techniques that Allow for Multiple Breaks." *Journal of International Money and Finance* 42(April): 208–23.

Arezki, Rabah, Kirk Hamilton, and Kazim Kazimov. 2011. "Resource Windfalls, Macroeconomic Stability and Growth: The Role of Political Institutions." IMF Working Paper 11/142. May. Washington DC: International Monetary Fund.

Arnesen, Finn. 1995. *Statlig styring og EØS-rettslige skranker*. Oslo: Universitetsforlaget.

Atkinson, Giles, and Kirk Hamilton. 2003. "Savings, Growth and the Resource Curse Hypothesis." *World Development* 31(11): 1793–1807.

Aukrust, Odd. 1977. "Inflation in the Open Economy." In *Worldwide Inflation. Theory and Recent Experience*, edited by L. B. Krause and W. S. Salant, 107–53. Washington DC: Brookings Institution.

Aukrust, Odd, and Petter Jakob Bjerve. 1945. *Hva krigen kostet Norge*. Oslo: Dreyer.

Austvik, Ole Gunnar. 2010. "EU Regulation and National Innovation: The Case of Norwegian Petroleum Policy." In *Innovative Regulatory Approaches Coping with Scandinavian and European Union Policies*, edited by Noralv Veggeland, 103–30. New York: Nova Science Publisher.

Austvik, Ole Gunnar. 2012. "Landlord and Entrepreneur: The Shifting Roles of the State in Norwegian Oil and Gas Policy." *Governance: An International Journal of Policy, Administration, and Institutions* 25(2): 315–34.

Auty, Richard M. 1993. *Sustaining Development in Mineral Economies: The Resource Curse Thesis*. London: Routledge.

Bagheri, Seyed Kamran, and Alberto Di Minin. 2015. "The Changing Competitive Landscape of the Global Upstream Petroleum Industry." *World Energy Law & Business* 8(1): 1–19.

Bai, Guoing P., and Yan Xu. 2014. "Giant Fields Retain Dominance in Reserves Growth." *Oil & Gas Journal* 112(2). Accessed December 11, 2016. http://www.ogj.com/articles/print/volume-112/issue-2/exploration-development/giant-fields-retain-dominance-in-reserves-growth.html.

Baker III Institute. 2007. "The Changing Role of National Oil Companies in International Energy Markets." James A. Baker III Institute for Public Policy, Policy Report No. 35. Accessed May 21, 2015. http://bakerinstitute.org/files/420/.

Bang, Pauli, and Olaf Thuestad. 2014. "Government-Enforced Self-Regulation." In *Risk Governance of Offshore Oil and Gas Operations*, edited by Preben Hempel Lindøe, Michael Baram, and Ortwin Renn, 243–73. New York: Cambridge University Press.

Basedau, Matthias, and Jann Lay. 2009. "Resource Curse or Rentier Peace? The Ambiguous Effects of Oil Wealth and Oil Dependence on Violent Conflict." *Journal of Peace Research* 46(6): 757–76.

Baunsgaard, Thomas. 2004. "Revenue Collection." In World Bank's *Petroleum Revenue Management Workshop*. Energy Sector Management Assistance Programme (ESMAP); Technical Paper Series No. 051, 83–93. Washington, DC: World Bank. Accessed April 14, 2015. http://documents.worldbank.org/curated/en/2004/03/5064714/petroleum-revenue-management-workshop.

Bergsli, Alf Tore. 2013. "Tariffavtaler: Fredsavtalene som aldri varer." *Arbeidslivet.no*, April 2. Accessed February 5, 2015. http://www.arbeidslivet.no/Lonn/Tariffavtaler/Tariffavtaler/.

Bhattacharyya, Sambit, and Roland Hodler. 2010. "Natural Resources, Democracy and Corruption." *European Economic Review* 54(4): 608–21.

BIS [Bank for International Settlements]. 2015. "BIS Effective Exchange Rate Indices." Narrow indices. Bank for International Settlements. Last modified April 16. Accessed May 6, 2015. http://www.bis.org/statistics/eer.htm.

Bjørnland, Hilde C. 1998. "The Economic Effects of North Sea Oil on the Manufacturing Sector." *Scottish Journal of Political Economy* 45(5): 553–85.

Bjørnland, Hilde C., and Leif Anders Thorsrud. 2013. "Boom or Gloom? Examining the Dutch Disease in a Two-Speed Economy." Centre for Applied Macro—and petroleum economics (CAMp) Working Paper Series No 6/2013.

Bloc, Fred. 2001. "Introduction." In *The Great Transformation,* by Karl Polanyi, xviii–xxxviii. Boston: Beacon Press.

Bloch, Harry, and David Sapsford. 2000. "Whither the Terms of Trade? An Elaboration of the Prebisch-Singer Hypothesis." *Cambridge Journal of Economics* 24(4): 461–81.

Bodin, Jean. 1962 [1576]. *The Six Books of a Commonwealth [Les six livres de la republique].* Translated by Robert Knolles; edited by K. D. McRae. Cambridge: Harvard University Press.

Booth, Michael. 2014. *The Almost Nearly Perfect People: Behind the Myth of the Scandinavian Utopia*. New York: Picador.

Boscheck, Ralf. 2007. "The Governance of Oil Supply: An Institutional Perspective on NOC Control and the Questions It Poses." *International Journal of Energy Sector Management* 1(4): 366–89.

BP. 2014. "BP Statistical Review of World Energy." BP. 63rd Edition. Last modified June. Accessed January 26, 2015. http://www.bp.com/en/global/corporate/energy-economics/statistical-review-of-world-energy.html.

Brunnschweiler, Christa N., and Erwin H. Bulte. 2008. "The Resource Curse Revisited and Revised: A Tale of Paradoxes and Red Herrings." *Journal of Environmental Economics and Management* 55(3): 248–64.

Brunstad, Rolf Jens, and Jan Morten Dyrstad. 1997. "Booming Sector and Wage Effects: An Empirical Analysis on Norwegian Data." *Oxford Economic Papers* 49(1): 89–103.

Burgis, Tom. 2015. *The Looting Machine: Warlords, Oligarchs, Corporations, Smugglers, and the Theft of Africa's Wealth*. New York: Public Affairs.

Bye, Torstein, Ådne Cappelen, Torbjørn Eika, Eystein Gjelsvik, and Øystein Olsen. 1994. "Noen konsekvenser av petroleumsvirksomheten for norsk økonomi." SSB Report 94/1. Statistics Norway, Oslo.

Cappelen, Ådne, Torbjørn Eika, and Inger Holm. 2000. "Resource Booms: Curse or Blessing?" Article presented at the Annual Meeting of American Economic Association. Oslo: Statistics Norway.

Cappelen, Ådne, Torbjørn Eika, and Jakim Blix Prestmo. 2014. "Virkninger på norsk økonomi av et kraftig fall i oljeprisen." *Økonomiske analyser* 3/2014: 31–41.

Cappelen, Ådne, and Lars Mjøset. 2013. "Can Norway be a Role Model for Natural Resource Abundant Countries?" In *Development Success: Historical Accounts from More Advanced Countries*, edited by Augustin K. Fosu, 44–73. Oxford: Oxford University Press.

Carrington, Damian. 2015. "Leave Fossil Fuels Buried to Prevent Climate Change, Study Urges." *The Guardian*. January 7. Accessed December 20, 2016. https://www.theguardian.com/environment/2015/jan/07/much-worlds-fossil-fuel-reserve-must-stay-buried-prevent-climate-change-study-says.

Carroll, Archie B. 1979. "A Three-Dimensional Conceptual Model of Corporate Social Performance." *Academy of Management Review* 4(4): 497–505.

Carroll, Archie B. 1991. "The Pyramid of Corporate Social Responsibility: Toward the Moral Management of Organizational Stakeholders." *Business Horizons* July–August: 39–48.

Cayford, Steven. 1996. "The Ogoni Uprising: Oil, Human Rights, and a Democratic Alternative in Nigeria." *Africa Today* 42(2): 183–97.

Chang, Ha-Joon. 2005. *Kicking Away the Ladder. Development Strategy in Historical Perspective*. London: Anthem Press.

Claes, Dag Harald. 2002. "Statoil—between Nationalisation, Globalisation and Europeanisation." ARENA Working Paper 02/34. ARENA, Oslo.

Climate Action Tracker. 2015. "Norway." Climate Action Tracker. Last modified July 8. Accessed July 16, 2015. http://climateactiontracker.org/countries/norway.html.

Collier, Paul. 2008. *The Bottom Billion: Why the Poorest Countries Are Failing and What Can Be Done About It*. New York: Oxford University Press.

Collins, Tom. 2003. "NOC Reform: Restructuring, Commercialization and Privatization." World Bank National Oil Companies Workshop, Current Roles and Future Prospects. May 27. Accessed April 1, 2015. http://siteresources.worldbank.org/INTOGMC/Resources/NOCReformCollins.pdf.

Corporate Knights. 2014. "2014 Global 100 Dataset." Sent on request, dated July 18, 2015. http://www.corporateknights.com/reports/2014-global-100/#requestReport.

Cotula, Lorenzo. 2008. "Reconciling Regulatory Stability and Evolution of Environmental Standards in Investment Contracts: Towards a Rethinking of Stabilization Clauses." *Journal of World Energy Law and Business* 1(2): 158–79.

Crouch, David. 2015. "Norwegian Businesses Warned of 'Wake-Up Call' on Corruption." *Financial Times*. January 2.

Daniel, Philip. 2004. "Overview of the Issues." Paper presented at the Petroleum Revenue Management Workshop, March. Arranged by Joint UNDP/World Bank Energy Sector Management Assistance Program.

Davis, Jeffrey, Rolando Ossowski, James A. Daniel, and Steven Barnett. 2001a. "Stabilization and Savings Funds for Nonrenewable Resources: Experience and Fiscal Policy Implications." IMF Occasional Paper No. 205. International Monetary Fund, Washington DC.

Davis, Jeffrey, Rolando Ossowski, James A. Daniel, and Steven Barnett. 2001b. "Oil Funds. Problems Posting as Solutions?" *Finance and Development* 28(4): 56–9.

Davis, Jeffrey, Rolando Ossowski, James A. Daniel, and Steven Barnett. 2003. "Stabilization and Savings Funds for Nonrenewable Resources: Experience and Fiscal Policy Implications." In *Fiscal Policy Formulation and Implementation in Oil-Producing Countries*, edited by J. M. Davis, R. Ossowski, and A. Fedelino, 273–315. Washington DC: International Monetary Fund.

Deloitte. 2014. "Extractive Industries Transparency Initiative: Cash Flows from the Petroleum Industry in Norway 2013." Report on the reconciliation of cash flows from petroleum activities as part of the implementation of Extractive Industries Transparency Initiative (EITI) in Norway. Translated from the original Norwegian version. Last modified December 2014. Accessed April 18, 2015. http://www.eiti.no/files/2015/02/2014_EITI_rapport_engelsk_for_2013.pdf.

DeMeritt, Jacqueline H. R., and Joseph K. Young. 2013. "A Political Economy of Human Rights: Oil, Natural Gas, and State Incentives to Repress." *Conflict Management and Peace Science* 30(2): 99–120.

Deyo, Fredric C. (ed.). 1987. *The Political Economy of the New Asian Industrialism*. Ithaca, NY: Cornell University Press.

Dunning, Thad. 2008. *Crude Democracy: Natural Resource Wealth and Political Regimes*. New York: Cambridge University Press.

Economist, The. 1977. "The Dutch Disease." November 26.

Economist, The. 2013. "The Lottery of Life." November 21.

Edigheji, Omano. 2005. "A Democratic Developmental State in Africa? A Concept Paper." Research Report No. 105. May. Centre for Policy Studies, Johannesburg. Accessed August 28, 2016. http://www.rrojasdatabank.info/devstate/edigheji.pdf.

Edigheji, Omano. 2007. "The State, State-Society Relations and Developing Countries' Economic Performance." PhD dissertation, Political Science, Norwegian University of Science and Technology (Trondheim, Norway).

Edigheji, Omano. 2010a. "Constructing a Democratic Developmental State in South Africa: Potential and Challenges." In *Constructing a Democratic Developmental State in South Africa: Potentials and Challenges*, edited by Omano Edigheji, 1–36. Cape Town: Human Sciences Research Council Press.

Edigheji, Omano (ed.). 2010b. *Constructing a Democratic Developmental State in South Africa: Potentials and Challenges*. Cape Town: Human Sciences Research Council Press.

Edigheji, Omano. 2013. "Toward a Conceptual Framework for a Developmental State in South Africa. A Discussion Document." April 15, draft. Public Service Commission, South Africa.

EIA. 2014. "Norway: Full Report." U.S. Energy Information Administration, Washington DC. Last modified April 28. Accessed January 29, 2015. http://www.eia.gov/countries/analysisbriefs/Norway/norway.pdf.

Eifert, Benn, Alan Gelb, and Nils Borje Tallroth. 2003. "Managing Oil Wealth." *Finance and Development* 40(1): 40–5.

EITI [The Extractive Industries Transparency Initiative]. 2016a. "EITI Standard 2016." EITI International Secretariat. February 23. Accessed December 11, 2016. https://eiti.org/sites/default/files/documents/english-eiti-standard_0.pdf.

EITI [The Extractive Industries Transparency Initiative]. 2016b. "Stakeholders: Companies." EITI International Secretariat. Accessed December 11, 2016. https://eiti.org/supporters/companies.

EITI [The Extractive Industries Transparency Initiative]. 2016c. "Stakeholders: Partner Organisations." EITI International Secretariat. Accessed December 11, 2016. https://eiti.org/supporters/partner-organisations.

EITI [The Extractive Industries Transparency Initiative]. n.d. "Promoting Public Awareness about How Countries Manage Their Oil, Gas and Mineral Resources." Accessed June 21, 2016. https://eiti.org/.

Esping-Andersen, Gösta. 1985. *Politics against Markets*. Ithaca, NY: Cornell University Press.

Evans, Peter. 1995. *Embedded Autonomy*. Princeton: Princeton University Press.

Evans, Peter. 2010a. *The Challenge of 21st Century Development: Building Capability-Enhancing States*. New York: United Nations Development Program.

Evans, Peter. 2010b. "Constructing the 21st Century Developmental State." In *Constructing a Democratic Developmental State in South Africa: Potentials and Challenges*, edited by Omano Edigheji, 37–58. Cape Town: Human Sciences Research Council Press.

Fagerberg, Jan, Ådne Cappelen, Lars Mjøset, and Rune Skarstein. 1990. "The Decline of Social-Democratic State Capitalism in Norway." *New Left Review* 1(181): 60–94.

Fagre, Nathan, and Louis T. Wells Jr. 1982. "Bargaining Power of Multinationals and Host Governments." *Journal of International Business Studies* 13(2): 9–23.

Forsythe, David P. 2012. *Human Rights in International Relations*. Third Edition. New York: Cambridge University Press.

Fox News. 2015. "Norway's Finance Minister Doubts Global-Warming Is Man-Made, Still Backs Climate." April 29. Accessed July 17, 2015. http://www.foxnews.com/world/2015/04/29/norway-finance-minister-doubts-global-warming-is-man-made-still-backs-climate/.

Frankel, Jeffrey A. 2012. "The Natural Resource Curse: A Survey of Diagnoses and Some Prescriptions." HKS Faculty Research Working Paper Series RWP12-014, John F. Kennedy School of Government, Harvard University.

Friedman, Milton. 1970. "A Friedman Doctrine—The Social Responsibility of Business is to Increase Its Profits." *New York Times Magazine*. September 13.

Frynas, Jędrzej George. 2009a. "Corporate Social Responsibility in the Oil and Gas Sector." *Journal of World Energy Law and Business* 2(3): 178–95.

Frynas, Jędrzej George. 2009b. *Beyond Corporate Social Responsibility: Oil Multinationals and Social Challenges*. New York: Cambridge University Press.

Garcia-Alonso, M., and P. Levine. 2004. "Strategic Procurement, Openness and Market Structure." Paper presented at the ECAAR Panel Session, annual meeting of the American Economic Association/ASSA, Atlanta, GA, September 26.

George, Henry. 1920 [1879]. *Progress and Poverty*. Garden City, NY: Doubleday, Page & Co.

Ghai, Dharam (ed.). 2006. *Decent Work: Objectives and Strategies*. Geneva, International Institute for Labour Studies and International Labour Office.

Gordon, Kathryn, and Joachim Pohl. 2011. "Environmental Concerns in International Investment Agreements: A Survey." OECD Working Papers on International Investment, No. 2011/01, OECD Investment Division. Paris: OECD.

Gran, Thorvald. 1994. *The State in the Modernization Process: The Case of Norway 1850–1970*. Oslo: Ad Notam Gylendal.

Grayson, Leslie E. 1981. *National Oil Companies*. Chichester: John Wiley and Sons.

Grosse, Robert. 2006. "The Bargaining Relationship between Foreign MNEs and Host Governments in Latin America." *International Trade Journal* 10(4): 467–99.

Gundersen, Ina. 2007. "Oil Veterans Look Back." Norwegian Petroleum Directorate. March 2. Accessed March 19, 2015. http://www.npd.no/en/news/News/2007/Oil-veterans-look-back-/.

Gylfason, Thorvaldur. 2001. "Natural Resources, Education, and Economic Development." *European Economic Review* 45(4–6): 847–59.

Haber, Stephen, and Victor Menaldo. 2011. "Do Natural Resources Fuel Authoritarianism? A Reappraisal of the Resource Curse." *American Political Science Review* 105(1): 1–26.

Hale, Andrew R., and J. Hovden. 1998. "Management and Culture: The Third Age of Safety. A Review of Approaches to Organizational Aspects of Safety, Health and Environment." In *Occupational Injury: Risk, Prevention and Intervention*, edited by A.-M. Feyer and A. Williamson, 129–66. London: Taylor & Francis Ltd.

Hamilton, Alexander. 1791. "Report on Manufactures." Communicated to the House of Representatives, December 5, 1791. Accessed January 31, 2015. www.constitution.org/ah/rpt_manufactures.pdf.

Hanisch, Tore Jørgen. 1992. "Overgang fra en avventende til en aktiv oljepolitikk, 1965–72." In *Norsk Oljehistorie. Fra vantro til overmot?*, edited by T. J. Hanisch and G. Nerheim. Oslo: Norsk Petroleumsforening.

Hanisch, Tore Jørgen. 1999. *Norsk Økonomisk Politikk i det 20. århundre*. Kristiansand: Høyskoleforlaget.

Hanisch, Tore Jørgen, and Gunnar Nerheim. 1992. *Fra vantro til overmot?* Volume 1 in *Norsk oljehistorie*, edited by T.J. Hanisch and Gunnar Nerheim. Oslo: Norsk Petroleumsforening, Leseselskapet.

Harvey, David I., Neil M. Kellard, Jakob B. Madsen, and Mark E. Wohar. 2010. "Erratum to 'The Prebisch-Singer Hypothesis: Four Centuries of Evidence.'" *Review of Economics and Statistics* 92(2): 367–77.

Hayman, Gavin. 2009. "Corruption and Bribery in the Extractive Industries." Transparency International Global Corruption Report 2009: Corruption and the Private Sector, Transparency International.

Hays. 2013. "Oil and Gas Global Salary Guide." Accessed May 25, 2015. http://www.hays.com/oil-and-gas/SalaryGuide/hays-oil-gas-global-salary-guide-2013-1269348.

Hendryx, Franck. 1959. "A Sovereign Nation's Legal Ability to Make and Abide by a Petroleum Concession Contract," PFAPC, OIC.

Herb, Michael. 2005. "No Taxation without Representation? Rents, Development, and Democracy." *Comparative Politics* 37(3): 297–317.

HGI [Henderson Global Investors]. 2005. *Carbon 100—Quantifying the Carbon Emissions, Intensities and Exposures of the FTSE 100*. London: Henderson Global Investors.

Hilgerdt, Folke. 1945. *Industrialization and Foreign Trade*. Geneva: League of Nations.

Høgsnes, Geir. 1999. *Krone for krone*. Oslo: Ad Notam Gyldendal.

Hotelling, Harald. 1931. "The Economics of Exhaustible Resources." *Journal of Political Economy* 39(2): 137–75.

Høyre. 1981. "Oljevirksomheten og Norges fremtid." Motmelding til *St. meld. nr. 53 (1979–80)*. Avgitt av Høyres oljepolitisk komite 18. mai 1981. Accessed April 1, 2015. http://www.rammcom.com/open/OPK81.pdf.

HR Watch [Human Rights Watch]. 1999. "The Price of Oil: Corporate Responsibility and Human Rights Violations in Nigeria's Oil Producing Communities." HR Watch Report. HRW Index No. 1-56432-225-4. Last modified January 1999. Accessed July 13, 2015. http://www.hrw.org/reports/1999/nigeria/nigeria0199.pdf.

Hudson, Michael. 2000. "Oil in Norway's Balance of Payments and the Prospective Impact of Privatization on the Krone." Speech to the Norwegian Shipowners' Association, August 17. Accessed April 14, 2015. http://michael-hudson.com/2000/08/oil-in-norways-balance-of-payments-and-the-prospective-impact-of-privatization-on-the-krone/.

Hudson, P. 2007. "Implementing Safety Culture in a Major Multi-National." *Safety Science* 45(6): 697–722.

Hutchinson, Michael M. 1994. "Manufacturing Sector Resiliency to Energy Booms: Empirical Evidence from Norway, the Netherlands, and the United Kingdom." *Oxford Economic Papers* 46(2): 311–29.

ICG [International Crisis Group]. 2008. "Nigeria: Ogoni Land after Shell." Africa Briefing No. 54. September 18. Accessed December 11, 2016. https://www.crisisgroup.org/file/1851/download?token=mrPbOrhi.

ILO. 1989. "Convention on Indigenous and Tribal Peoples in Independent Countries." ILO Convention No 169; administered by the International Labour Organization. Accessed December 11, 2016. http://www.ilo.org/dyn/normlex/en/f?p=NORMLEXPUB:12100:0::NO::P12100_ILO_CODE:C169.

ILO. 2001. "ILO-OSH-2001. Guidance on Occupational Safety and Health Management Systems." Second Edition. Geneva: ILO. Accessed June 22, 2015. http://www.ilo.org/wcmsp5/groups/public/@ed_protect/@protrav/@safework/documents/normative instrument/wcms_107727.pdf.

ILO. 2011. "OSH Management System: A Tool for Continual Improvement." Geneva: ILO. Last modified April 28. Accessed June 22, 2015. www.ilo.org/wcmsp5/groups/public/—ed_protect/—protrav/—safework/documents/publication/wcms_153930.pdf.

ILO. 2012. "Monitoring and Assessing Progress on Decent Work (MAP)—Project Brief." Last Modified January 1. Accessed June 21, 2015. http://www.ilo.org/integration/themes/mdw/map/WCMS_175756/lang–en/index.htm.

ILO. 2015a. "Decent Work." Accessed June 21, 2015. http://www.ilo.org/global/topics/decent-work/lang–en/index.htm.

ILO. 2015b. "Conventions and Recommendations." Accessed June 22, 2015. http://www.ilo.org/global/standards/introduction-to-international-labour-standards/conventions-and-recommendations/lang–en/index.htm.

ILO. n.d. "Ratifications of Fundamental Conventions and Protocols by Country." Accessed June 22, 2015. http://www.ilo.org/dyn/normlex/en/f?p=NORMLEXPUB:10011:0::NO::P10011_DISPLAY_BY,P10011_CONVENTION_TYPE_CODE:1,F.

IMF. 2007. "Code of Good Practices on Fiscal Transparency." Accessed June 21, 2016. https://www.imf.org/external/np/fad/trans/code.htm.

IMO [International Maritime Organization]. 1972. "Convention on the Prevention of Marine Pollution by Dumping of Wastes and Other Matter." Also known as the

London Convention. Accessed June 22, 2015. http://www.imo.org/en/About/Conventions/ListOfConventions/Pages/Convention-on-the-Prevention-of-Marine-Pollution-by-Dumping-of-Wastes-and-Other-Matter.aspx.

Industridepartementet. 1971. "Innstilling om organisasjon for statlige kontinentalsokkelsaker: fra et utvalg oppnevnt ved kgl. res. av 18. september 1970: innstillingen avgitt 24. mars 1971."

Innst. S. nr. 87 (2002–2003). Innstilling fra energi- og miljøkomiteen om olje- og gassvirksomheten. Oslo: Storting.

IOM[3] [Institute of Materials, Minerals and Mining]. 2015. "Norwegian Oil and Gas Association Guidelines." Oil & Gas Division. Last modified January 6. Accessed June 29, 2015. http://www.iom3.org/oil-gas-division/news/2015/jan/06/norwegian-oil-and-gas-association-guidelines.

IPIECA [The Global Oil and Gas Industry Association for Environmental and Social Issues]. 2007. "The Oil and Gas Industry and Climate Change." Last modified June. Accessed June 22, 2015. http://www.ipieca.org/sites/default/files/publications/siaf_climatechange.pdf.

IPIECA [The Global Oil and Gas Industry Association for Environmental and Social Issues]. 2008. "Human Rights and Ethics in the Oil and Gas Industry." London: IPIECA.

IPIECA [The Global Oil and Gas Industry Association for Environmental and Social Issues]. 2011. "Local Content Strategy." Accessed June 10, 2015. http://www.ipieca.org/sites/default/files/publications/Local_Content.pdf.

Isham, Jonathan, Michael Woolcock, Lant Pritchett, and Gwen Busby. 2005. "The Varieties of Resource Experience: Natural Resource Export Structures and the Political Economy of Economic Growth." *The World Bank Economic Review* 19(2): 141–74.

ITOPF [International Tanker Owners Pollution Federation]. 2014. "Oil Tanker Spill Statistics 2014." Accessed June 22, 2015. http://www.itopf.com/knowledge-resources/data-statistics/statistics/.

Jaén, Artur Colom. 2010. "Lessons from the Failure of Chad's Oil Revenue Management Model." ARI Working Paper 12/2010. Real Instituto Elcano. Accessed July 13, 2015. http://ARI12-2010_Colom_Chad_Oil_Revenue_Management_Model.pdf?MOD=AJPERES&CACHEID=8473080041b87f3a9de5ffe151fccd56.

Jakobsen, Jo. 2012. *Political Risk and the Multinational Company*. Trondheim: Tapir.

Jansen, Jan B., and Joachim M. Bjerke. n.d. "Norwegian Petroleum Taxation. An Introduction." BAHR Oslo. Accessed March 21, 2015. http://www.bahr.no/no/aktuelt/saker/_attachment/2869?_download=true&_ts=135436a4dc8.

Jensen, Nathan, and Leonard Wantchekon. 2004. "Resource Wealth and Political Regimes in Africa." *Comparative Political Studies* 37(7): 816–41.

Jensen, Stein, Bjørn Vedeler, and Egil F. Wulff. 1979. "Risk Assessment. A Study of Risk Levels within Norwegian Offshore Petroleum Activities." Report No. 26–27/2. "Safety Offshore." Norges teknisk-naturvitenskapelige forskningsråd (NTNF).

Johannesen, Finn Erhard, Asle Rønning, and Pål Thonstad Sandvik. 2005. *Nasjonal kontroll og industrielle fornyelse: Hydro 1945–1977. Bind to av Hydros historie 1905–2005*. Oslo: Pax.

Johnsen, Arve. 1988. *Utfordringen. Statoil år*. Oslo: Gyldendal.

Johnson, Chalmers. 1982. *MITI and the Japanese Miracle*. Stanford: Stanford University Press.

Johnson, Harry G. 1976. "The New International Economic Order." Selected Papers No. 49, Graduate School of Business, University of Chicago.

Johnston, Daniel. 2008 "Changing Fiscal Landscape." *Journal of World Energy Law & Business* 1(1): 31–54.

Johnston, Daniel. 2009. "Modern Reserve Disclosure." Discourses on Modern Reserve Disclosure article. Adapted from his 2005 column in *Petroleum Accounting and Financial Management Journal*, 24(2) Summer 2005, pp. 23–30. Accessed April 9, 2015. http://www.danieljohnston.com/pdf/modern_reserve_disclosure.pdf.

Johnston, David. 2007. "How to Evaluate the Fiscal Terms of Oil Contracts." In *Escaping the Resource Curse*, edited by Macartan Humphreys, Jeffrey D. Sachs, and Joseph E. Stiglitz, 53–88. New York: Columbia University Press.

Jones Luong Pauline, and Erika Weinthal. 2010. *Oil Is Not a Curse: Ownership Structure and Institutions in Soviet Successor States*. Cambridge: Cambridge University Press.

Kaasen, Knut. 2014. "Safety Regulations on the Norwegian Continental Shelf." In *Risk Governance of Offshore Oil and Gas Operations*, edited by Preben Hempel Lindøe, Michael Baram, and Ortwin Renn, 103–31. New York: Cambridge University Press.

Karl, Terry Lynn. 1997. *The Paradox of Plenty: Oil Booms and Petro-states*. Berkeley and Los Angles: University of California Press.

Karlsen, Jan Erik. 1982. *Arbeidervern på sokkelen*. Oslo: Universitetsforlaget.

Karlsen, Jan Erik, and Preben Hempel Lindøe. 2006. "The Nordic OSH Model at a Turning Point?" *Policy and Practice in Health and Safety* 4(1): 17–30.

Katzenstein, Peter. 1985. *Small States in World Markets*. Ithaca, NY: Cornell University Press.

Kemp, Alexander G. 1987. *Petroleum Rent Collection Around the World*. Halifax: Institute for Research on Public Policy.

Kindleberger, Charles, P. 1943. "Planning for Foreign Investment." *The American Economic Review: Papers and Proceedings* 33(2): 347–54.

Kindleberger, Charles, P. 1989. "Engel's Law" in his *Economic Laws and Economic History*. Raffaele Mattioli Lectures, 3–6, Cambridge: Cambridge University Press.

Knutsen, Sverre. 2007. "Staten og kapitalen i det 20. Århundre." Dr dissertation. Acta Humaniora No. 287. Oslo: Unipub.

Kobrin, Stephen J. 1987. "Testing the Bargaining Hypothesis in the Manufacturing Sector in Developing Countries." *International Organization* 41(4): 609–38.

Kolstad, Ivar, and Tina Søreide. 2009. "Corruption in Natural Resource Management: Implications for Policy Makers." *Resources Policy* 34(4): 214–26.

Kronenberg, Tobias. 2006. "Should We Worry about the Failure of the Hotelling Rule?" Working Paper at ETH Zürich University. Accessed May 1, 2015. https://www.ethz.ch/content/dam/ethz/special-interest/mtec/cer-eth/resource-econ-dam/documents/research/sured/sured-2006/sured_077_kronenberg.pdf.

Kuwait Fund. (n.d.) "Kuwait Fund Timeline, 1961–2014." Accessed July 18, 2015. https://www.kuwait-fund.org/timeLine/?lang=en.

Kvendseth, Stig S. 1988. *Funn! Historien om Ekofisks første 20 år*. Tananger: Phillips.

Lahn, Glada, Valérie Marcel, John Mitchell, Keith Meyers, and Paul Stevens. 2007. "Report on Good Governance of the National Petroleum Sector." Chatham House. April. London: Royal Institute of International Affairs.

Leftwich, Adrian. 2000. *States of Development: On the Primacy of Politics in Development.* Cambridge: Polity Press.

Leis, Jorge, John McCreery, and Juan Carlos Gay. 2012. "National Oil Companies Reshape the Playing Field." *Bain Brief.* Accessed April 5, 2015. http://www.bain.com/Images/BAIN_BRIEF_National_oil_companies_reshape_the_playing_field.pdf.

Leiserson, Mark Whittlesey. 1959. *Wages and Economic Control in Norway, 1945–1957.* London: Oxford University Press.

Leite, Carlos, and Jens Weidmann. 2002. "Does Mother Nature Corrupt? Natural Resources, Corruption, and Economic Growth." In *Governance, Corruption, and Economic Performance,* edited by George T. Abed and Sanjeev Gupta, 159–96. Washington, DC. IMF.

Letnes, Bjørn. 2002. "Foreign Direct Investment and Human Rights: An Ambiguous Relationship." *Forum for Development Studies* 29(1): 33–61.

Letnes, Bjørn. 2008. "Transnational Corporations, Economic Development and Human Rights: On the Importance of FDI Composition and Host Country Assets." PhD Thesis in Political Science. Trondheim: Norwegian University of Science and Technology (NTNU).

Lie, Einar. 2010. *Over Evne: Finansdepartementet 1965–1992.* Oslo: Pax.

Lindøe, Preben Hempel, Michael Baram, and Ortwin Renn (eds). 2014. *Risk Governance of Offshore Oil and Gas Operations.* New York: Cambridge University Press.

Lindøe, Preben Hempel, Jan Erik Karlsen, and T. Lie. 2001. *Et nordisk grep på arbeidsmiljøregulering?* Copenhagen: Nordic Council of Ministers.

List, Friedrich. 1885. *The National System of Political Economy.* Translated from the original (1841) edition by Sampson Lloyd. London: Longmans, Green and Company.

Mahdavi, Paasha. 2014. "Why Do Leaders Nationalize Their Oil Industry? The Politics of Resource Expropriation." *Energy Policy* 75: 228–43.

Mahdavy, Hussein. 1970. "The Patterns and Problems of Economic Development in Rentier States: The Case of Iran." In *Studies in Economic History of the Middle East from the Rise of Islam to the Present Day,* edited by M.A. Cook, 428–67. London: Oxford University Press.

Manley, David, and Thomas Lassourd. 2014. "Tanzania and Statoil: What Does the Leaked Agreement Mean for Citizens?" Natural Resource Governance Institute (NRGI) August Draft.

Martin, S., K. Hartley, and A. Cox. 1997. "Public Purchasing in the European Union: Some Evidence from Contracts Awards." *International Journal of Public Sector Management* 10(4): 279–93.

Matten, Dirk, and Andrew Crane. 2005. "Corporate Citizenship: Toward an Extended Theoretical Conceptualization." *The Academy of Management Review* 30(1): 166–79.

McGlade, Christophe, and Paul Ekins. 2015. "The Geographical Distribution of Fossil Fuels Unused when Limiting Global Warming to 2° C." *Nature* 517 (January 8): 187–90.

MDG [*Miljøpartiet De Grønne*]. 2013. "En ny Kyoto-avtale forplikter." December 12. Accessed July 16, 2015. https://www.mdg.no/nyheter/norge-har-brutt-kyoto-avtalen-ny-avtale-debattert-pa-stortinget-i-dag/.

Mehlum, Halvor, Karl Ove Moene, and Ragnar Torvik. 2006. "Institutions and Resource Curse." *The Economic Journal* 116(508): 1–20.

Melberg, Eldbjørg Vaage. 2010. "Determined to Learn from History." NPD. August 13. Accessed March 22, 2015. http://www.npd.no/en/Publications/Norwegian-Continental-Shelf/No1-2010/Determined-to-learn-from-history/.

Meld. St. 28 (2010–2011). "An Industry for the Future—Norway's Petroleum Activities." Oslo: Ministry of Petroleum and Energy.

Meld. St. 3 (2013–2014). "Statsrekneskapen 2013." Melding til Stortinget. Oslo: Det konglege finansdepartmentet.

Meld. St. 2 (2014–2015). "Revidert nasjonalbudsjett 2015." Oslo: Norwegian Government.

Meld. St. 20 (2014–2015). "Update of the Integrated Management Plan for the Barents Sea–Lofoten Area Including an Update of the Delimitation of the Marginal Ice Zone." Oslo: Ministry of Climate and Environment.

Midttun, Øyvind. 2003. "Konsesjonspolitikk på norsk sokkel: fire tiår med gulrot og pisk." *Norsk oljemuseum årbok 2003*: 11–22.

Mikkelsen, Solveig. 2014. "Etisk blikk på oljeforskning." *Universitetsavisa.no.* January 27. Accessed July 16, 2015. http://www.universitetsavisa.no/forskning/article40101.ece.

Milne, Richard. 2013. "Norway's Risky Good Life." Nordic and Baltic Correspondent for the *FT*, on-line video. September 9. Accessed May 2, 2015. http://video.ft.com/2659928351001/Norways-risky-good-life/World.

Ministry of Climate and Environment. 1981. "Pollution Control Act of 13 March 1981 No. 6 Concerning Protection Against Pollution and Concerning Waste Law." Translation based on the Norwegian text with latest amendments of 20 June 2003. Accessed July 3, 2015. Online at https://www.regjeringen.no/en/dokumenter/pollution-control-act/id171893/.

Ministry of Climate and Environment. 2014. "Norway's Sixth National Communication under the Framework Convention on Climate Change." Status report as of January 2014. Ministry of Climate and Environment, Oslo.

Ministry of Finance. 2007. "Norway's Position in the Debate on Sovereign Wealth Funds." Last modified on October 31. Accessed June 25, 2016. https://www.regjeringen.no/en/topics/the-economy/the-government-pension-fund/internt-bruk/Norways-position-in-the-debate-on-Sovere/id487466/.

Ministry of Finance. 2013. "Reduced Uplift in the Petroleum Tax System." Ministry of Finance, Press Release No. 19/2013. May 5. Accessed April 17, 2015. https://www.regjeringen.no/en/aktuelt/reduced-uplift-in-the-petroleum-tax-syst/id725999/.

Minor, Dylan, and John Morgan. 2011. "CSR as Reputation Insurance: Primum non nocere." *California Management Review* 53(3): 40–59.

Mitchell, Timothy. 2011. *Carbon Democracy: Political Power in the Age of Oil.* London: Verso.

Mjøset, Lars, and Ådne Cappelen. 2011. "The Integration of the Norwegian Oil Economy into the World Economy." *Comparative Social Research* 28: 167–264.

Moe, Thorvald G., Jon A. Solheim, and Bent Vale (eds). 2004. "The Norwegian Banking Crisis." Norges Bank Occasional Papers No. 33. Norges Bank, Oslo.

Moen, A., H. C. Blakstad, R. Rosness, and U. Forseth. 2009. "Disintegration and Revival of Tripartite Collaboration on HSE in the Norwegian Petroleum Industry." In *Reliability,*

Risk, and Safety: Theory and Applications, edited by C. Guedes Soares, Radim Bri, and Sebastián Martorell, Three Volume Set, 2191–8, Chapter 301. DOI: 10.1201/9780203859759. ch301.CRC Press.

Moran, Theodore H. 1974. *Multinational Corporations and the Politics of Dependence: Copper in Chile*. Princeton: Princeton University Press.

Moses, Jonathon W. 2000. *Open States in the Global Economy: The Political Economy of Small-State Macroeconomic Management*. Houndmills, Basingstoke, Hampshire: Macmillan Press.

Moses, Jonathon W. 2005. *Norwegian Catch-Up: Development and Globalization before World War II*. Aldershot: Ashgate.

Moses, Jonathon W. 2015. "Global Markets." *The SAGE Handbook of European Foreign Policy*, edited by Knud Erik Jorgensen, Aasne Kalland Aarstad, Edith Drieskens, Katie Laatikainen, and Ben Tonra, 998–1012, Los Angeles: Sage.

Moss, Laura. 2010. "The 13 Largest Oil Spills in History." Mother Nature Network. Last modified July 16. Accessed June 14, 2015. http://www.mnn.com/earth-matters/wilderness-resources/stories/the-13-largest-oil-spills-in-history.

MPE [Ministry for Petroleum and Energy]. 2004. "State Organization of Petroleum Operations." 2004 Facts Sheet on Norwegian Petroleum Activities, Ministry for Petroleum and Energy, Oslo.

MPE [Ministry for Petroleum and Energy]. 2013. "Ansvarsområder og oppgaver i Olje- og energidepartementet." Organization chart published by Ministry for Petroleum and Energy. Last modified November 1. Accessed March 22, 2015. https://www.regjeringen.no/globalassets/upload/oed/pdf_filer_2/oed_orgkart_nov13.jpg.

NBIM [Norges Bank Investment Management]. 2011. "Principles for Organization and Management of Norges Banks Investment Management." Laid down by Norges Banks Executive Board, June 8, 2011. Accessed May 15, 2015. http://www.nbim.no/globalassets/documents/governance/policies/principles-for-oganisation-and-management-of-nbim.pdf.

NBIM [Norges Bank Investment Management]. 2016a. "Management Mandate." Norges Bank Investment Management. Last modified on February 1. Accessed June 26, 2016. https://www.nbim.no/en/the-fund/governance-model/management-mandate/.

NBIM [Norges Bank Investment Management]. 2016b. "Market Value Development by Equity, Fixed Income and Real Estate." Norges Bank Investment Management. Accessed June 26, 2016. https://www.nbim.no/en/the-fund/market-value/.

NBIM [Norges Bank Investment Management]. 2016c. "The FUND." Norges Bank Investment Management. Accessed June 27, 2016. https://www.nbim.no/en/the-fund/.

Nelsen, Brent F. 1992. "Explaining Petroleum Policy in Britain and Norway, 1962–90." *Scandinavian Political Studies* 15(4): 307–28.

NENT [Den nasjonale forskningsetiske komite for naturvitenskap og teknologi]. 2014. "Vedrørende forskningsetisk vurdering av petroleumsforskning." Case 2014/3. Letter dated June 18.

Nergaard, Kristine. 2013. Personal e-mail correspondence about 1999–2011 data. April 29.

Nerheim, Gunnar. 1994. "Framveksten av det oljeindustrielle kompleks i Norge." TMV arbeidsnotat 80. Oslo: TMV.

Nerheim, Gunnar. 1996. *En gassnasjon blir til.* Volume 2 of *Norsk Oljehistorie.* Oslo: Leseselskapet.

NOAA. n.d. "Ekofisk-Bravo oil field." *NOAA Incident News.* Accessed March 22, 2015. http://incidentnews.noaa.gov/incident/6237.

Nooryani, Bunny. 2006. "Statoil Admits Iran Bribe; U.S. Sets $21 Million Fine (Update 2)." Bloomberg. October 13.

Nøra, Stig. 2011. "Statoil snart større ute enn hjemme." *Dagens persepektiv.* February 6.

Norad. 2013. "Facing the Resource Curse: Norway's Oil for Development Program." Report No. 6/2012. Norad. Accessed April 1, 2016. http://www.norad.no/en/toolspublications/publications/2013/facing-the-resource-curse-norways-oil-for-development-program/.

Nore, Petter. 2009. "Oil for Development–OfD." Presentation to the World Bank. Last modified March 3. Accessed April 1, 2015. http://siteresources.worldbank.org/EXTOGMC/Resources/336929-1237387264558/5930373-1237387327980/nore_eiti++.pdf.

Noreng, Øystein. 1980. *The Oil Industry and Government Strategy in the North Sea.* London and Boulder, CO: Croom Helm and ICEED.

Noreng, Øystein. 2004. "Norway: Economic Diversification and the Petroleum Industry." *Middle East Economic Survey* XLVII (45). November 8.

Noreng, Øystein. 2008. "Statoil som internasjonal aktør." *Samtiden* (2): 31–8.

Norges Bank. 2012. "Money, Credit and Banking." November 11. Accessed September 15, 2016. http://www.norges-bank.no/WebDAV/hms/data/Money_Credit.xlsx.

Norsk Olje&Gas. 2010. "The Licensing System." December 13. Accessed April 16, 2015. https://www.norskoljeoggass.no/en/Activities/Industry-policy/Licensing-policy/The-licensing-system/.

Norsk Petroleum. 2016. "Fakta/Felt." Accessed June 24, 2016. http://www.norskpetroleum.no/fakta/felt/.

North, Douglass C., and Robert P. Thomas. 1973. *The Rise of the Western World: A New Economic History.* New York: Cambridge University Press.

Norwegian Government. 2014a. "Konsesjonsverk." Oil and Energy Department's list of standard agreements in the petroleum sector. Last modified on January 15, 2014. Accessed April 18, 2015. https://www.regjeringen.no/nb/dokument/dep/oed/lover_regler/reglement/konsesjonsverk/id748087/.

Norwegian Government. 2014b. "Petroleumsprisrådet og normpriser." Oil and Energy Department's introduction to norm price system. Last modified on November 21, 2014. Accessed April 17, 2015. https://www.regjeringen.no/nb/tema/energi/olje-og-gass/petroleumsprisradet–og-fastsetting-av-n/id661459/.

Norwegian Government. 2014c. "Government Pension Fund Global (GPFG): Market Value." Ministry of Finance report on the GPFG. Last modified October 13, 2014. Accessed May 16, 2015. https://www.regjeringen.no/en/topics/the-economy/the-government-pension-fund/government-pension-fund-global-gpfg/market-value-and-capital-inflow/id696852/.

NOS [Norges Offentlig Statistikk]. 1972. *Nasjonalregnskap, 1954–1970,* NOS A474. Oslo: SSB.

NOU. 1972/15. *Ilandføring av petroleum*. Oslo: Norges Offentlige Utredninger.

NOU. 1974/40. *Rørledninger på dypt vann*. Oslo: Norges Offentlige Utredninger.

NOU. 1983/16. *Organisering av statens deltagelse i petroleumsvirksomhet*. Oslo: Norges Offentlige Utredninger.

NOU. 1983/27. *Petroleumsvirksomhetens framtid: det framtidige omfanget av petroleumsvirksomheten på norsk sokkel*. Oslo: Norges Offentlige Utredninger.

NOU. 2012/2. *Utenfor or innenfor—Norges avtaler med EU*. Oslo: Norges Offentlige Utredninger.

NPD [Norwegian Petroleum Directorate]. 1996. "Act 29 November 1996 No. 72 Relating to Petroleum Activities." Last amended by Act 24 June 2011 No 38. Last translated May 4, 2015. Accessed July 3, 2015. http://www.npd.no/en/Regulations/Acts/Petroleum-activities-act/.

NPD [Norwegian Petroleum Directorate]. 2010. "Guidelines for Plan for Development and Operation of a Petroleum Deposit (PDO) and Plan for Installation and Operation of Facilities for Transport and Utilisation of Petroleum (PIO)." Last modified February 4. Accessed June 28, 2016. http://www.npd.no/Global/Engelsk/5-Rules-and-regulations/Guidelines/PDO-PIO-guidelines_2010.pdf.

NPD [Norwegian Petroleum Directorate]. 2011. "The Norwegian Petroleum Directorate." Last Modified July 7. Accessed March 20, 2015. http://www.npd.no/en/About-us/.

NPD [Norwegian Petroleum Directorate]. 2012. "The Petroleum Sector–Norway's Largest Industry." *Facts 2012*. Oslo and Stavanger: Ministry of Petroleum and Energy and the Norwegian Petroleum Directorate.

NPD [Norwegian Petroleum Directorate]. 2013. *Fakta 2013*. Stavanger: Norsk Petroleumsverksemd.

NPD [Norwegian Petroleum Directorate]. 2014. *Facts 2014. Norwegian Petroleum Sector*, edited by Yngvild Tormodsgard. Oslo and Stavanger: Ministry of Petroleum and Energy and the Norwegian Petroleum Directorate.

NPD [Norwegian Petroleum Directorate]. 2015. "Sharing Data through Diskos for 20 Years." September 3. Accessed October 1, 2015. http://www.npd.no/en/news/News/2015/Sharing-petroleum-data-through-Diskos-for-20-years/.

NPD [Norwegian Petroleum Directorate]. 2016a. "Fact Pages." Last modified April 28. Accessed June 21, 2016. http://factpages.npd.no/factpages/.

NPD [Norwegian Petroleum Directorate]. 2016b. "The net Government Cash Flow from Petroleum Activities, 1971–2015." Last modified May 11. Accessed June 27, 2016. http://www.norskpetroleum.no/en/economy/governments-revenues/.

NRC [Natural Resource Charter]. (2009). "Natural Resource Charter: Draft for Consultation." November 2009 draft.

NRGI [National Resource Governance Institute]. 2014. "Natural Resource Charter." Second Edition. Accessed July 3, 2016. www.naturalresourcecharter.org.

NRGI [National Resource Governance Institute]. 2015. "Statoil Disclosures Undermine US Oil Lobby Arguments." March 19. Accessed 3 July 2016. http://www.resourcegovernance.org/news/statoil-disclosures-undermine-us-oil-lobby-arguments.

NRK. 2009. *Olje! Historien om det Norske Oljeeventyret*. Five-part documentary film. Oslo: NRK.

OECD. 2012. *OECD Economic Surveys. Norway*. February. Paris: OECD.

OECD. 2014. "Safety Performance Indicators, 2013 data." Report 2013s. July. OGP Data series. Accessed June 14, 2015. http://www.iogp.org/pubs/2013s.pdf.

OECD. 2015a. "Level of GDP Per Capita and Productivity." OECD.Stat. Accessed May 13, 2015. http://stats.oecd.org/Index.aspx?DataSetCode=PDB_LV.

OECD. 2015b. "Unit Labour Costs. Annual Indicators. Total Economy." OECD.Stat. Accessed May 13, 2015. http://stats.oecd.org/Index.aspx?DataSetCode=PDB_LV.

OECD. 2016a. "Unemployment Rate" OECD Data. Accessed June 18, 2016. https://data.oecd.org/unemp/unemployment-rate.htm.

OECD. 2016b. "Level of GDP Per Capita and Productivity." OECD.Stat. Accessed June 20, 2016. http://stats.oecd.org/index.aspx?DatasetCode=PDB_LV.

OGP [International Association of Oil and Gas Producers]. 2010. "Regulators' Use of Standards." Report No. 426. March. Accessed June 17, 2015. www.ogp.org.uk/pubs/426.pdf.

OGP [International Association of Oil and Gas Producers]. 2014. "Safety Performance Indicators, 2013 data." Report 2013s. July. OGP Data series. Accessed June 14, 2015. http://www.iogp.org/pubs/2013s.pdf.

Oljedirektoratet. 1998. *Oljedirektoratets Årsberetning 1998*. Stavanger: Oljedirektoratet.

Olsen, Johan P. 1983. *Organized Democracy. Political Institutions in a Welfare State—The Case of Norway*. Oslo: Universitetsforlaget.

Olsen, Øystein. 2013. "Tale av sentralbanksjef Øystein Olsen til Norges Banks representantskap og inviterte gjester, torsdag 14. februar 2013." *Økonomiske Perspektiver*. May 17, 2015. http://www.norges-bank.no/pages/93164/aarstalen_2013.pdf.

Open Oil. 2015. "Imagine an Open Oil Industry . . . " Open Oil. Accessed June 21, 2016. http://openoil.net.

Open Society Foundation. 2016. "Covering Oil: A Reporter's Guide to Energy and Development." Open Society Foundation. Accessed June 21, 2016. http://www.opensocietyfoundations.org/publications/covering-oil-reporters-guide-energy-and-development.

OSHA. n.d. "Safety and Health Topics. Oil & Gas Extraction." U.S: Occupational Safety and Health Administration. Accessed June 22, 2015. https://www.osha.gov/SLTC/oilgaswelldrilling/.

Osmundsen, Terje. 1981. *Gjøkungen: Skal Statoil styre Norge?* Oslo: Dreyer.

Petoro. 2015. "Key Duties." Petoro. Accessed March 20, 2015. https://www.petoro.no/about-petoro/key-duties.

Petoro. 2016. "SDØE-fakta." Petoro. Accessed June 21, 2016. https://www.petoro.no/om-petoro/sd%C3%B8e-fakta/liste-over-lisenser.

Plato. 1987 [*c*.360 BCE]. *The Republic*. Translated by Desmond Lee, Second edition. Harmondsworth: Penguin.

Polanyi, Karl. 2001 [1944]. *The Great Transformation*. Boston: Beacon Books.

Polity. 2014. "Polity IV Project dataset. Political Regime Characteristics and Transitions, 1800–2013." Principal authors: Monty G. Marshall, Ted Robert Gurr, and Keith Jaggers. Accessed December 11, 2016. http://www.systemicpeace.org/inscrdata.html.

Porter, Michael E., and Claas van der Linde. 1995. "Green and Competitive Ending the Stalemate." *Harvard Business Review* 73(5): 120–34.

Prebisch, Raúl. 1950. "The Economic Development of Latin America and its Principle Problems." Economic Commission for Latin America. Lake Success, NY: UN Dept. of Economic Affairs. Accessed February 11, 2015. http://archivo.cepal.org/pdfs/cdPrebisch/002.pdf.

Prebisch, Raúl. 1964. "Towards a New Trade Policy for Development." Document prepared for the 1964 Geneva Convention on World Trade and Development. Accessed on February 14, 2015. http://www.washingtontradereport.com/Prebisch_UNCTAD_1964.pdf.

PSA(N). 2016a. "About the HSE Regulations." Accessed June 27, 2016. http://www.psa.no/about-the-hse-regulation/category929.html.

PSA(N). 2016b. "Personal Injuries on the Norwegian Continental Shelf 2004–2015." Accessed June 27, 2016. http://www.psa.no/facts-and-statistics/personal-injuries-on-the-norwegian-continental-shelf-2004-2015-article12078-921.html.

PWYP [Publish What You Pay]. 2016. "Who Publishes Extractive Contracts?" Accessed June 21, 2016. http://www.publishwhatyoupay.org/.

Qvigstad, Jan F. 2011. "Om å forvalte rikdom." Visesentralbanksjef Jan F. Qvigstads tale i Det Norske Videnskaps-Akademi. Accessed June 24, 2016. http://www.norges-bank.no/Publisert/Foredrag-og-taler/2011/8112011-JFQ-Videnskapsakademiet/.

Rabe, Stephen G. 1982. *The Road to OPEC. United States Relations with Venezuela, 1919–1976*. Austin: University of Texas Press.

Radcliffe Committee. 1959. "Committee on the Working of the Monetary System: Report." London: Her Majesty's Stationary Office.

Radon, Jenik. 2005. "The ABCs of Petroleum Contracts: License-Concession Agreements, Joint Ventures, and Production-Sharing Agreements." In *Governing Oil*, edited by Svetlana Tsalik and Anya Schiffrin, 61–80. New York: Open Society Institute.

Reich, Robert B. 2008. "The Case Against Corporate Social Responsibility." Goldman School Working Paper Series, NO. GSPP08-003. August 1.

Reinert, Erik S. 2007. *How Rich Countries Got Rich . . . and Why Poor Countries Stay Poor*. London: Constable.

Reinertsen, Maria. 2009. "Oljefondets utspring." *Morgenbladet*, May 22.

Remøe, Svend Otto. 1986. "Arbeidskonflikter i norsk oljevirksomhet: En statistisk oversikt og analyse." Rogalandsforskning Report No. 56.1123/2.

Ricardo, David. 1971 [1817]. *Principles of Political Economy and Taxation*. Harmondsworth: Penguin.

Riksrevisjonen. 2010. "Riksrevisjonens undersøkelse av forvaltningspraksis ved tildeling av utvinningstillatelser i petroleumssektoren." Riksrevisjonens administrative report 1 2010, Stortinget, Oslo.

Rivkin, David W., Sophie J. Lamb, and Nicola K. Leslie. 2015. "The Future of Investor-State Dispute Settlement in the Energy Sector: Engaging with Climate Change, Human Rights and the Rule of Law." *Journal of World Energy Law and Business* 8(2): 130–53.

RNNP [Risikonivå i norsk petroleumsvirksomhet]. 2001. "Utvikling i risikonivå - norsk sokkel." Pilotprosjektrapport for 2000. Last modified April 24. Accessed June 28, 2015. http://www.ptil.no/getfile.php/PDF/Pilotprosjekt%20-%20hovedrapport2000.pdf.

Røed Larsen, Erling. 2005. "Are Rich Countries Immune to the Resource Curse? Evidence from Norway's Management of Its Oil Riches." *Resources Policy* 30(2): 75–86.

Rommetvedt, Hilmar. 1998. "Norwegian Parliamentary Committees: Performance, Structural Change and External Relations." *Journal of Legislative Studies* 4: 60–84.

Rosness, Ragnar, and Ulla Forseth. 2014. "Boxing and Dancing: Tripartite Collaboration as an Integral Part of a Regulatory Regime." In *Risk Governance of Offshore Oil and*

Gas Operations, edited by Preben Hempel Lindøe, Michael Baram, and Ortwin Renn, 309–39. New York: Cambridge University Press.

Ross, Michael L. 1999. "The Political Economy of the Resource Curse." *World Politics* 51(2): 297–322.

Ross, Michael L. 2001. "Does Oil Hinder Democracy?" *World Politics* 53(3): 321–65.

Ross, Michael L. 2012. *The Oil Curse. How Petroleum Wealth Shapes the Development of Nations*. Princeton: Princeton University Press.

Routley, Laura. 2014. "Developmental States in Africa? A Review of Ongoing Debates and Buzzwords." *Development Policy Review* 32(2): 159–77.

Ruggie, John Gerard. 1982. "International Regimes, Transactions, and Change: Embedded Liberalism in the Postwar Economic Order." *International Organization* 36(2): 379–415.

Ruggie, John Gerard. 2013. *Just Business. Multinational Corporations and Human Rights*. New York: Norton and Company.

Ryggvik, Helge. 1992. "Det første møtet mellom norske myndigheter og multinasjonale oljeselskap (1963–65)." Masters thesis, University of Oslo.

Ryggvik, Helge. 2010. "The Norwegian Oil Experience: A Toolbox for Managing Resources?" Report number 2. Oslo: Senter for Teknologi, Innovasjon og Kultur, TIK.

Ryggvik, Helge. 2014. "Norsk Oljehistorie." *Store norske leksikon*. Last modified October 16. Accessed June 9, 2015. https://snl.no/Norsk_oljehistorie.

Ryggvik, Helge, and Marie Smith-Solbakken. 1997. *Blod, svette og olje*. Volume 2 of *Norsk oljehistorie*, edited by in Tore Jørgen Hanisch. Oslo: Ad notam Gyldendal.

Rystad Energy. 2012. "Internasjonal omsetning fra norske oljeserviceselskaper." Report to the Olje og Energidepartementet. Last modified August 21. Accessed June 9, 2015. https://www.regjeringen.no/globalassets/upload/oed/rapporter/rapport_rystad-energy_internasjonal-omsetning-fra-norske-oljeserviceselskapert.pdf.

Rystad Energy. 2014. "Internasjonal omsetning fra norske oljeserviceselskaper." Report to the Olje-og energidepartementet. Official version: Last modified October 23. Accessed June 9, 2015. https://www.regjeringen.no/globalassets/upload/oed/pdf_filer_2/rystad_energy_internasjonal_omsetning_fra_norske_oljeserviceselskaper_rapport_2014.pdf.

Sabahi, Guly. 2015. "Local Content—Smart Legislation and Global Capital for a Vibrant African E&P Industry." Presentation at the AIPN Spring Meeting. April 21, Austin Texas.

Sachs, Jeffrey D., and Andrew M. Warner. 1995. "Natural Resource Abundance and Economic Growth." NBER Working Paper 5398.

Sachs, Jeffrey D., and Andrew M. Warner. 2001. "Natural Resources and Economic Development. The Course of Natural Resources." *European Economic Review* 45: 827–38.

Saksvik, Per Øystein, and Michael Quinlan. 2003. "Regulating Systematic Occupational Health and Safety Management: Comparing the Norwegian and Australian Experience." *Relations industrielles/Industrial Relations* 58(1): 33–59.

Sala-i-Martin, Xavier, and A. Subramanian. 2003. "Addressing the Natural Resource Curse: An Illustration from Nigeria." NBER Working Paper, No. 9804.

Sampson, Anthony. 1975. *The Seven Sisters. The Great Oil Companies and the World They Shaped*. New York: Viking Press.

Sassen, Saskia. 1996. *Losing Control? Sovereignty in an Age of Globalization*. New York: Columbia University Press.

Sæter, Kjetil. 2014. "Statoil vet ikke hvem partneren er." *Dagens Næringsliv*. September 21.

Schwartz, John. 2015. "Norway Will Divest from Coal in Push Against Climate Change." *New York Times*. June 5.

Seip, Jens Arup. 1963. *Fra embedsmannsstat til ettpartistat og andre essays*. Oslo: Universitetsforlaget.

Senghaas, Dieter. 1985. *The European Experience*. Berg: Leamington.

Singer, Hans W. 1950. "The Distribution of Gains between Investing and Borrowing Countries." *American Economic Review* 40(2): 473–85.

Sjøberg, Jeanette. 2015. "Studentene vender oljebransjen ryggen." *Aftenposten*. May 21.

Skancke, Martin. 2003. "Fiscal Policy and Petroleum Fund Management in Norway." In *Fiscal Policy Formulation and Implementation in Oil-Producing Countries*, edited by J. M. Davis, R. Ossowski, and A. Fedelino, 316–38. Washington DC: International Monetary Fund.

Skjærseth, Jon Birger, Kristian Tangen, Philip Swanson, Atle Christer Christiansen, Arild Moe, and Leiv Lunde. 2004. "Limits to Corporate Social Responsibility: A comparative study of four major oil companies." Lysaker, Norway: Fridtjof Nansen Institute.

Skule, Sveinung, and Toril Grytli. 1997. "Teknologisk utvikling og samfunnsendring. Eksempler fra oljehistorien og bankhistorien." FAFO Report 217. Oslo: FAFO.

Slyngstad, Yngve. 2016. "Innledning til høring om forvaltningen av Statens pensjonsfond utland 2015 i Stortingets finanskomité." Last modified April 29. Accessed June 26, 2016. http://www.norges-bank.no/Publisert/Foredrag-og-taler/2016/2016-04-29-Slyngstad/.

Smith, Adam. 1776. *Wealth of Nations*. Accessed June 20, 2016. http://www.econlib.org/library/Smith/smWNCover.html.

Smith, Benjamin. 2004. "Oil Wealth and Regime Survival in the Developing World: 1960–1999." *American Journal of Political Science* 48(2): 232–46.

Smith-Solbakken, Marie. 1997. "Oljearbeiderkulturen: historien om cowboyer og rebeller." PhD dissertation, Department of History, Norwegian University of Science and Technology (NTNU). Trondheim Norway.

Søbye, Espen. 2001. "En opera i uka." *Samfunnsspeilet* 1: 70–9.

Solow, Robert M. 1974. "The Economics of Resources or the Resources of Economics." *The American Economic Review* 64(2): 1–14.

Sonangol E.P. 2004. "Republic of Angola LAW No. 10/04 of 12 November 2004." Petroleum Activities Law. Translation provided by Miranda, Correia, Amenoeira and Associates. Accessed December 11, 2016. http://www.eisourcebook.org/cms/files/attachments/other/Angola%20Petroleum%20Activities%20Law,%202004.pdf.

Spar, Debora. 1999. "Foreign Investment and Human Rights." *Challenge* 42(1): 55–80.

Spense, David R. 2011. "Corporate Social Responsibility in the Oil and Gas Industry: The importance of reputational risk." *Chicago-Kent Law Review* 86(1): 59–85.

SSB. n.d. Diverse issues of "Arbeidsmarkedsstatistikk." Listed under 06 Arbeidsliv, yrkesdeltaking. Accessed June 30, 2015. https://www.ssb.no/a/histstat/publikasjoner/ereg77–96.html#06.

SSB. 2015a. "Tabell: 06308: Innvandringer, etter innvandringsgrunn og innflyttingsår." Accessed May 17, 2015. *Statistikkbanken* (SSB 2016).

SSB. 2015b. "Table 08435: Employed Persons, by Resident Status, Industry (SIC2007) and Country Background." 4th quarter (2008K4–2013K4). Accessed May 17, 2015. *Statistikkbanken* (SSB 2016).

SSB. 2015c. "Tabell: 05869: Flyttinger fra og til utlandet." Accessed May 17, 2015. *Statistikkbanken* (SSB 2016).

SSB. 2016. *Statistikkbanken*. https://www.ssb.no/statistikkbanken.

SSB. 2016 [07952]. "Tabell: 07952: Arbeidstakarer i arbeidskonflikt og tapte arbeidsdagar, etter næringshovudområde." (SN2007) Olje- og gassutvinning og bergverksdrift. *Statistikkbanken* (SSB 2016).

SSB. 2016 [9189–1]. "Bruttonasjonalprodukt, markedsverdi." In Table: 09189: Makroøkonomiske hovedstørrelser. *Statistikkbanken* (SSB 2016).

SSB. 2016 [09189]. "Gross Domestic Product, Market Values & Gross Domestic Product, Mainland Norway, Market Values." In Table: 09189: Final expenditure and gross domestic product. *Statistikkbanken* (SSB 2016).

SSB. 2016 [10909]. "Total Expenditures." In Table 10909: General Government Historical Data. Revenue and Expenditure by Sector and Type (NOK Millions). *Statistikkbanken* (SSB 2016).

Ssennoga, Francis. 2006. "Examining Discriminatory Procurement Practices in Developing Countries." *Journal of Public Procurement* 6(3): 218–49.

St. meld. nr. 11 (1968–9). "Undersøkelse om boring etter petroleumsforekomster på den norske kontinentalsokkel i Nordsjøen." Oslo: Industridepartement.

St. meld. nr. 76 (1970–1971). "Om undersøkelse etter og utvinning av undersjøiske naturforekomster på den norske kontinentalsokkel m.m." Særskilt vedlegg: Oversikt over oljepolitiske spørsmål (Jens Evensen).

St. meld. nr. 25 (1973–74). "Petroleumsvirksomhetens plass i det norske samfunnet." Oslo: Finansdepartementet.

St. meld. nr. 53 (1979–80). "Om virksomheten på den norske kontinentalsokkelen." Oslo: Olje- og energidepartementet.

St. meld. nr. 73 (1983–84). "Om organiseringen av statens deltagelse i petroleumsvirksomheten." Oslo: Olje- og energidepartementet.

St. meld. nr. 58 (1996–1997). "Miljøvernpolitikk for en bærekraftig utvikling. Dugnad for framtida." Oslo: Miljøverndepartementet.

St. meld. nr. 39 (1999–2000). "Olje- og gassvirksomheten." Oslo: Olje- og energidepartementet.

St. meld. nr. 29 (2000–2001). "Retningslinjer for den økonomiske politikken." Oslo: Finansdepartementet.

St. meld. nr. 7 (2001–2002). "Om helse, miljø og sikkerhet i petroleumsvirksomheten." Oslo: Arbeids- og sosialdepartementet.

St. meld. nr. 17 (2002–2003). "Om statlige tilsyn." Oslo: Arbeids- og administrasjonsdepartementet.

St. meld. nr. 26 (2006–2007). "Regjeringens miljøpolitikk og rikets miljøtilstand." Oslo: Miljøverndepartementet.

St. meld. nr. 10 (2008–2009). "Corporate Social Responsibility in a Global Economy." Oslo: Ministry of Foreign Affairs.

St. meld. nr. 28 (2010–2011). "En næring for framtida–om petroleumsvirksomheten." Oslo: Olje- og energidepartementet.

St. meld. nr. 3 (2013–2014). "Revidert nasjonalbudsjett 2013." Oslo: Finansdepartementet.

STAR [Stolen Asset Recovery Initiative]. 2015. "Medhi Hashemi Rafsanjani/Statoil." Case Control Number 132. World Bank. Accessed July 18, 2015. http://star.worldbank.org/corruption-cases/node/18720.

Statoil. 2012. "Anti-Corruption Compliance Program." Statoil. December version. Accessed July 18, 2015. http://production.presstogo.com/fileroot/gallery/StatoilASA/files/preview/7fcb84c0e1ac8d6de040007f01001c34/7fcb84c0e1aa8d6de040007f0100 1c34.pdf.

Statoil. 2013b. "The In Amenas Attack. Report of the Investigation into the Terrorist Attack on In Amenas." Statoil. Accessed July 19, 2015. http://www.statoil.com/en/NewsAndMedia/News/2013/Downloads/In%20Amenas%20report.pdf.

Statoil. 2014a. "Statoil Worldwide." Statoil. Accessed July 18, 2015. http://www.statoil.com/en/about/worldwide/pages/default.aspx.

Statoil. 2014b. "Ethics Code of Conduct." Version 2.9. Statoil. Accessed July 18, 2015. http://www.statoil.com/en/About/EthicsValues/Downloads/Ethics%20Code%20of%20Conduct%20(Aug2014).pdf.

Statoil. 2014c. "Ethics and Anti-corruption e-Learning Programme." Statoil. Last modified September 25. Accessed July 8, 2016. http://www3.statoil.com/statoilaetikk/menu.htm.

Statoil. 2014d. "Statoil's Ethics Hotline." Statoil. Last modified October 6. Accessed July 18, 2015. http://www.statoil.com/en/about/ethicsvalues/pages/etikkhjelpelinjen.aspx.

Statoil. 2015a. "Payments to Governments 2014." Statoil. Accessed July 17, 2015. http://www.statoil.com/no/InvestorCentre/AnnualReport/AnnualReport2014/Documents/DownloadCentreFiles/01_KeyDownloads/2014%20Payments%20to%20governments.pdf.

Statoil. 2015b. "Can an Oil and Gas Company Be Part of a Sustainable Energy Future?" Statoil. Accessed July 19, 2015. http://www.statoil.com/en/environmentsociety/pages/default.aspx.

Statoil. 2015c. "Statoil in Brief." Statoil. Last modified May 29. Accessed July 18, 2015. http://www.statoil.com/en/about/inbrief/pages/default.aspx.

Statoil. 2015d. "Safety, Sustainability and Ethics Committee." Statoil. Last modified July 1. Accessed July 18, 2015. http://www.statoil.com/en/About/CorporateGovernance/GoverningBodies/SafetySustainabilityAndEthicsCommittee/Pages/default.aspx.

Statoil. 2016. *The Statoil Book*. Statoil. Version 3.3. Accessed July 8, 2016. http://www.statoil.com/en/About/TheStatoilBook/Downloads/The%20Statoil%20Book%20(January%202016).pdf.

Stevens, Harley C. 1959. "Some Reflections on the First Arab Petroleum Congress." *Middle East Journal* 13(3): 273–80.

Stevens, Paul. 2003. "Resource Impact: Curse or Blessing—A Literature Survey." *The Journal of Energy Literature* 9(1), June: 3–42.

Stiglitz, Joseph. 2003. "Do as the US Says, Not as It Does." *The Guardian*. October 28.

Stokke, Torgeir Aarvaag. 2000. "Organisasjonsgrader i norsk arbeidsliv, 1945–1998." FAFO Notat 2000:10. Accessed February 7, 2015. http://fafo.no/media/com_netsukii/643.pdf.

Stopford, John, and Susan Strange. 1991. *Rival States, Rival Firms: Competition for World Market Shares*. Cambridge: Cambridge University Press.

Svensen, Thomas K., Sandra Simonsen, and Kristian M. Lind. 2014. "Oil and Gas Regulation in Norway: Overview." Practical Law, *Multi-Jurisdictional Guide 2014*. Accessed December 11, 2016. http://global.practicallaw.com/6-529-5206.

SWFI [Sovereign Wealth Fund Institute]. 2015. "Sovereign Wealth Fund Ranking. Largest Sovereign Wealth Funds by Assets under Management." Sovereign Wealth Fund Institute. Last modified May 2. Accessed May 2, 2015. http://www.swfinstitute.org/fund-rankings/.

Taverne, Bernard G. 2008. *Petroleum, Industry and Governments: A Study of the Involvement of Industry and Governments in the Production and Use of Petroleum*, second edition. Austin, TX: Wolters Kluwer.

Taylor, Ben. 2014. "Tanzania: Leaked Agreement Shows Tanzania May Not Get a Good Deal for Gas." *AllAfrica.com*. July 4.

Thurber, Mark C., David R. Hults, and Patrick R.P. Heller. 2011. "Exporting the 'Norwegian Model': The Effect of Administrative Design on Oil Sector Performance." *Energy Policy* 39(9): 5366–78.

Thurber, Mark C., and Benedicte Tangen Istad. 2010. "Norway's Evolving Champion: Statoil and the Politics of State Enterprise." Program on Energy and Sustainable Development Working Paper, No. 92, May.

Tordo, Silvana, Michael Warner, Osmel E. Manzano, and Yahua Anouti. 2013. *Local Content Policies in the Oil and Gas Sector.* Washington DC: World Bank.

Toye, John, and Richard Toye. 2004. *United Nations Intellectual History Project: UN and Global Political Economy: Trade, Finance and Development.* Bloomington, IN: Indiana University Press.

Transparency International. 2008. *Promoting Revenue Transparency—2008 Report on Revenue Transparency of Oil and Gas Companies*. Berlin: Transparency International.

Transparency International. 2014a. "Corruption Perceptions Index 2014." Accessed January 27, 2015. http://www.transparency.org/cpi2014/infographic/global.

Transparency International. 2014b. "Visualising the Corruption Perception Index 2014." Accessed July 19, 2015. http://www.transparency.org/cpi2014/infographic/global.

Tsui, Kevin K. 2011. "More Oil, Less Democracy: Evidence from Worldwide Crude Oil Discoveries." *The Economic Journal* 121(551): 89–115.

UD Prop. 1S (2010–2011). Proposisjon til Stortinget for budsjettåret 2011. Oslo: Norwegian Government.

Ulfelder, Jay. 2007. "Natural-Resource Wealth and the Survival of Autocracy." *Comparative Political Studies* 40(8): 995–1018.

UN. 1958. "1958 Geneva Conventions on the Law of the Sea. Geneva, 29 April 1958." Accessed April 14, 2015. http://legal.un.org/avl/ha/gclos/gclos.html.

UN. 1974a. "Declaration for the Establishment of a New International Economic Order." United Nations General Assembly A/RES/S-6/3201 of May 1, 1974. Accessed February 14, 2015. http://www.un-documents.net/s6r3201.htm.

UN. 1974b. "Charter of Economic Rights and Duties of States." United Nations General Assembly A/RES/29/3281 of December 12, 1974. Accessed February 14, 2015. http://www.un-documents.net/a29r3281.htm.

UN. n.d. "Global Compact." Accessed July 12, 2015. https://www.unglobalcompact.org/participation.

UN General Assembly. 1966. "Permanent Sovereignty over Natural Resources." Resolution adopted by the 21st session of the UN General Assembly. Resolution number 2158. November 25. Accessed April 7, 2015. http://daccess-ods.un.org/access.nsf/Get?OpenAgent&DS=A/RES/2158(XXI)&Lang=E&Area=RESOLUTION.

UNCBD. 1992. "1992 UN Convention on Biological Diversity." Accessed June 22, 2015. https://www.cbd.int/convention/text/default.shtml.

UNCLOS. 1982. "United Nations Convention on the Law of the Sea of 10 December 1982 Overview and full text." Accessed June 22, 2015. http://www.un.org/depts/los/convention_agreements/convention_overview_convention.htm.

UNCTAD. 2014. "Recent Developments in Investor-State Dispute Settlement (ISDS)." IIA issue Note, No.1 (April). Accessed February 13, 2015. http://unctad.org/en/PublicationsLibrary/webdiaepcb2014d3_en.pdf.

UNFCCC. 1992. 1992 UN Framework Convention on Climate Change (UNFCCC). Accessed June 22, 2015. http://unfccc.int/resource/docs/convkp/conveng.pdf.

US Congress. 1975. Subcommittee on Multinational Corporations of the Committee on Foreign Relations. *Multinational Corporations in Brazil and Mexico: Structural Sources of Economic and Noneconomic power*. 94th Congr., 1st Sess. Washington: Government Printing Office.

US Senate. 1974. *Multinational Corporations and US Foreign Policy. Hearings before the Subcommittee on Multinational Corporations of the Committee on Foreign Relations* (Senator Frank P. Church, Chairman) 93rd Congress, 2d session, 11 volumes. "Multinational Hearings," Part 5. Senator Church's Multinational Subcommittee, published as parts 4–9 of the series on Multinational Corporations and US Foreign Policy. Washington: US Senate.

van der Linde, Coby. 2000. *The State and the International Oil Market: Competition and the Changing Ownership of Crude Oil Assets*. Boston, MA: Kluwer Academic.

Velculescu, Delia. 2008. "Norway's Oil Fund Shows the Way for Wealth Funds." *IMF Survey Magazine: Policy*. July 9.

Vernon, Raymond. 1971. *Sovereignty at Bay: The Multinational Spread of US Enterprises*. New York: Basic Books.

Victor, David G., David R. Hults, and Mark C. Thurber. 2012. "Major Conclusions and Implications for the Future of the Oil Industry." In *Oil and Governance: State-Owned Enterprises and the World Energy Supply*, edited by Victor, David G., David R. Hults, and Mark C. Thurber, 887–928. New York: Cambridge University Press.

Vogel, Laurent. 1998. *Prevention at the Workplace*. Brussels: European Trade Union Technical Bureau for Health and Safety.

Wade, Robert. 1990. *Governing the Market: Economic Theory and the Role of the Government in East Asian Industrialization*. Princeton: Princeton University Press.

Wagner, Jay, and Kit Armstrong. 2010. "Managing Environmental and Social Risks in International Oil and Gas Projects: Perspectives on Compliance." *Journal of World Energy Law and Business* 3(2): 140–65.

Warner, Michael. 2011. *Local Content in Procurement*. Sheffield: Greenleaf.

Watts, Michael J. 2005. "Righteous Oil? Human Rights, the Oil Complex, and Corporate Social Responsibility." *Annual Review of Environment and Resources* 30 (November): 373–407.

Wicken, Olav. 2007. "The Layers of National Innovation Systems: The Historical Evolution of a National Innovation System in Norway." TIK Working Paper on Innovation Studies No. 20070601. Last modified June 25.

Wiig, Vibeke and Svein Olav Nås. 1992. "Teknologiavtalene som insentiv i norsk forskningspolitikk." Framtek-notat 14. Oslo: NFR.

Willoch, Kåre. 1981. "Erklæring ved møte med oljeselskapene." Oslo, December 11. Accessed June 25, 2015. https://www.regjeringen.no/globalassets/upload/smk/vedlegg/taler-og-artikler-av-tidligere-statsministre/kare-willoch/1981/erklaering_ved_moete_med_oljeselskapene.pdf.

World Bank. 2003. "Striking a Better Balance—The World Bank Group and Extractive Industries: The Final Report of the Extractive Industries Review." Open Knowledge Depository. Last modified December. Accessed May 4, 2015. https://openknowledge.worldbank.org/handle/10986/17705.

World Bank. 2010a. "World Bank Group Approves Support for Chad-Cameroon Petroleum Development and Pipeline Project—June 6, 2000." News and Events. Accessed July 14, 2015. http://web.worldbank.org/archive/website01210/WEB/0__CON-2.HTM.

World Bank. 2010b. "Project Description." Accessed July 14, 2015. http://web.worldbank.org/archive/website01210/WEB/0__CO-15.HTM.

World Bank. 2010c. "Chad-Cameroon Pipeline Project." Accessed July 14, 2015. http://web.worldbank.org/archive/website01210/WEB/0__MENUP.HTM.

World Bank. 2011. *World Bank Development Report. Conflict, Security and Development*. Washington DC: World Bank.

World Bank. 2015. "GNI (Current US$)." Data tables. Accessed July 18, 2015. http://data.worldbank.org/indicator/NY.GNP.MKTP.CD.

World Bank. 2016. "Doing Business" databank. Accessed June 24, 2016. http://databank.worldbank.org/data/reports.aspx?source=doing-business.

World Commission on Environment and Development. 1987. *Our Common Future*. Oxford. Oxford University Press.

World Today, The. 1959. "The First Arab Petroleum Congress." *The World Today* 15(6): 246–53.

Wright, Joseph, Erica Frantz, and Barbara Geddes. 2015. "Oil and Autocratic Regime Survival." *British Journal of Political Science* 45(2): 287–306.

WTO. 2013. "Special and Differential Treatment Provisions in WTO Agreements and Decisions." Committee on Trade and Development. Last modified June 14. WT/COMTD/W/196.

WTO. 2015a. "Regional Trade Agreements: Facts and Figures." World Trade Organization information sheet on RTAs. Accessed February 13, 2015. http://www.wto.org/english/tratop_e/region_e/regfac_e.htm.

WTO. 2015b. "Agreement on Government Procurement. Parties, Observers and Accessions." World Trade Organization's information sheet on the Agreement on Government Procurement. Accessed February 14, 2015. http://www.wto.org/english/tratop_e/gproc_e/memobs_e.htm.

WTO. 2016a. "Members and Observers." World Trade Organization's information sheet on current members. Accessed July 2, 2016. http://www.wto.org/english/thewto_e/whatis_e/tif_e/org6_e.htm.

WTO. 2016b. "Trade and Investment: Technical Information. Agreement on Trade Related Investment Measures." World Trade Organization's information sheet on TRIMs. Accessed June 24, 2016. https://www.wto.org/english/tratop_e/invest_e/invest_info_e.htm.

WTO. 2016c. "Agreement on Subsidies and Countervailing Measures ('SCM Agreement')." World Trade Organization's information sheet on the SCM Agreement. Accessed June 24, 2016. https://www.wto.org/english/tratop_e/scm_e/subs_e.htm.

WTO. 2016d. "Canada-Autos (DS139, 142)." In *2006 WTO Dispute Settlement: One-Page Case Summaries. 1995–September 2006*. Geneva: WTO.

WTO. 2016e. "Special and Differential Treatment Provisions." World Trade Organization Trade and Development Committee's information sheet on SDTPs. Accessed June 24, 2016. https://www.wto.org/english/tratop_e/devel_e/dev_special_differential_provisions_e.htm.

WTO Secretariat. 2014. "All RTAs_Cumulative." Accessed February 13, 2015. http://www.wto.org/english/tratop_e/region_e/facts_and_figure_chart_2014.xls.

Yamada, Hiroshi, and Gawon Yoon. 2014. "When Grilli and Yang Meet Prebisch and Singer: Piecewise Linear Trends in Primary Commodity Prices." *Journal of International Money and Finance* 42: 193–207.

Yates, Douglas Andrew. 2012. *The Scramble for African Oil: Oppression, Corruption and War for Control of Africa's Natural Resources*. London: Pluto Press.

Yergin, Daniel. 2009. *The Prize: The Epic Quest for Oil, Money and Power*. New York: Free Press.

Zwetsloot, Gerard. 2013. "What Are Occupational Safety and Health Management Systems and Why Do Companies Implement Them?" European Agency for Health and Safety at Work. OSHwiki. Last Modified April 6. Accessed December 11, 2016. https://oshwiki.eu/wiki/What_are_occupational_safety_and_health_management_systems_and_why_do_companies_implement_them%3F.

Index

The manufacturer's authorised representative in the EU for product safety is Oxford University Press España S.A. of El Parque Empresarial San Fernando de Henares, Avenida de Castilla, 2 - 28830 Madrid (www.oup.es/en or product.safety@oup.com). OUP España S.A. also acts as importer into Spain of products made by the manufacturer.
Printed and bound by CPI Group (UK) Ltd, Croydon, CR0 4YY
14/09/2026
02216306-0010